Manual of Water Supply Practices

M62

Membrane Applications for Water Reuse

First Edition

American Water Works Association

Manual of Water Supply Practices—M62

Membrane Applications for Water Reuse

Disclaimer
The authors, contributors, editors, and publisher do not assume responsibility for the validity of the content or any consequences of its use. In no event will AWWA be liable for direct, indirect, special, incidental, or consequential damages arising out of the use of information presented in this book. In particular, AWWA will not be responsible for any costs, including, but not limited to, those incurred as a result of lost revenue. In no event shall AWWA's liability exceed the amount paid for the purchase of this book.

If you find errors in this manual, please email books@awwa.org. Possible errata will be posted at www.awwa.org/resources-tools/resource.development.groups/manuals-program.aspx.

Senior Managing Editor/Project Manager: Melissa Valentine
Cover art: Melanie Yamamoto
Production: Janice Benight
Manuals Specialist: Sue Bach

Library of Congress Cataloging-in-Publication Data

Names: Wong, Joseph, active 2018, author. | Alspach, Brent, author. | Chalmers, Bruce, author. | American Water Works Association, issuing body.
Title: M62 - Membrane applications for water reuse / by Joseph Wong, Brent Alspach, Bruce Chalmers.
Other titles: Membrane applications for water reuse
Description: First edition. | Denver, CO : American Water Works Association, [2018]
Identifiers: LCCN 2018008413 | ISBN 9781625762627
Subjects: LCSH: Water reuse. | Membranes (Technology)
Classification: LCC TD429 .W66 2018 | DDC 628.1/64--dc23
LC record available at https://lccn.loc.gov/2018008413

Printed in the United States of America

ISBN 978-1-62576-262-7 eISBN-13 978-1-61300-448-7

American Water Works Association

American Water Works Association
6666 West Quincy Avenue
Denver, CO 80235-3098
awwa.org

Contents

This page intentionally blank.

Figures

This page intentionally blank.

Tables

Acknowledgments

The AWWA Technical and Educational Council, the Water Quality and Technology Division, and the Membrane Processes committee gratefully acknowledge the contributions of the volunteers who drafted, edited, and provided the significant and critical commentary essential to the development of AWWA M62. The Technical Review Board members dedicated many hours in the final stages of preparation of this edition to ensure the overall technical quality, consistency, and accuracy of the manual.

Technical Review Board Members

Joseph Wong, Chair, Brown and Caldwell, Walnut Creek, Calif.
Brent Alspach, ARCADIS, Carlsbad, Calif.
Bruce Chalmers, CDM Smith, Irvine, Calif.
Dawn Flancher, AWWA Staff, Denver, Colo.
Judith Herschell, Herschell Environment, Pittsburgh, Penn.
Kenneth Mercer, AWWA Staff, Denver, Colo.

Authors of the First edition

*A special thank you to these volunteers who served as chapter coordinators.

*Joseph Wong,** Chair, Brown and Caldwell, Walnut Creek, Calif.
*Jorge Aguinaldo,** RWL Water, Tampa, Fla.
Rick Bond, Black & Veatch, Kansas City, Mo.
Thomas Broderick, Loudoun Water, Ashburn, Vir.
Charlie Cruz, Separation Processes Inc., Carlsbad, Calif.
Emily Davis, Separation Processes Inc., Carlsbad, Calif.
James (Jay) DeCarolis, Black & Veatch, San Marcos, Calif.
Andrew Findlay, MWH Australia, Manly West, Qld Australia
Val Frenkel, EKI, Erler & Kalinowski Inc., Burlingame, Calif.
*Silvana M. Ghiu,** Hazen and Sawyer, San Diego, Calif.
Wesley Harijanto, California State Polytechnic University, Pomona, Calif.
Turaj Hosseini, Separation Processes Inc., Carlsbad, Calif.
Gary L. Hunter, Black & Veatch, Kansas City, Mo.
Curtis Kiefer, CDM Smith, Fort Lauderdale, Fla.
Tom Knoell, Orange County Water District, Fountain Valley, Calif.
Edmund A. Kobylinski, Black & Veatch, Cary, N.C.
Natalie W. La, California State Polytechnic University, Pomona, Calif.
*Jim Lozier,** ch2m, Tempe, Ariz.
Robert McCandless, Brown and Caldwell, Phoenix, Ariz.
*Chandra Mysore,** JACOBS™, Atlanta, Ga.
Jeff Neemann, Black & Veatch Corp., Kansas City, Mo.

Key Wee Ong, Singapore Public Utilities Board, Singapore
Eric Owens, West Basin Municipal Water District, Carson, Calif.
Mehul Patel, Orange County Water District, Fountain Valley, Calif.
Alan E. Rimer, Black & Veatch, Cary, N.C.
Cheryl Ross, West Basin Municipal Water District, Long Beach, Calif.
Larry Schimmoller, ch2m, Englewood, Colo.
Rich Stratton, HDR, Shingle Springs, Calif.
Arun Subramani, MWH, Arcadia, Calif.
Jennifer Thompson, CDM Smith, Carlsbad, Calif.
Ralph Valencia, United Water, Los Angeles, Calif.
*Don Vandertulip,** Arcadis, Dallas, Texas
*Vasu Veerapaneni,** Black & Veatch, Kansas City, Mo.
Alex Wesner, Separation Processes Inc., Carlsbad, Calif.
*Greg Wetterau,** CDM Smith, Rancho Cucamonga, Calif.
Wyatt Won, West Basin Municipal Water District, Carson, Calif.
Joe Zhao, URS Corporation, Santa Ana, Calif.
Christine Zheng, California State Polytechnic University, Pomona, Calif.

Reviewers to the First Edition

Robert Bergman, ch2m, Gainesville, Fla.
James Crook, Environmental Consultant, Norwell, Mass.

Chapter Authors

Chapter 1: Development of Water Reuse Practices

Don Vandertulip and Val Frenkel

Chapter 2: Planning for Reuse Applications

Don Vandertulip and Lisa Prieto

Chapter 3: Water Reuse Guidelines and Regulations

Joseph Wong and Cheryl Ross

Chapter 4: Source and Treated Water Quality

Chandra Mysore and Joseph Zhao

Chapter 5: Membrane Process Treatment Facility Design

Jorge T. Aguinaldo, James De Carolis, Srinivas Veerapaneni, Scott Freeman, Alex Wesner, Robert Bergman, Jeff Neemann, and James Lozier

Chapter 6: Operations

Silvana Ghiu, Eric Owens, Emily Davis, Turaj Hosseini, Ralph Valencia, Tom Knoell, and Wyatt Won

Chapter 7: Residuals Management

Chandra Mysore, Rick Bond, and Edmund Kobylinski

Chapter 8: Cost of Treatment

Greg Wetterau and Curt Kiefer

Chapter 9: Case Studies

Authors are listed in each case study

Chapter 10: Future Technology Trends and Contaminants of Emerging Concerns

Joseph Wong, Arun Subramani, and Robert McCandless

This page intentionally blank.

Preface

Reverse osmosis (RO) membranes have been used in water reuse since the 1960s. Cellulose acetate membranes for treating conventionally clarified municipal effluent was initially applied to small industrial applications and as irrigation water for golf courses. In the 1970s, Orange County Water District in southern California used cellulose acetate membranes to produce 5 mgd of RO permeate that was blended with imported water for injection into seawater intrusion barrier wells. The use of membranes in full-scale reuse applications has changed dramatically based on research performed in the 1980s and 1990s. Those efforts demonstrated that microfiltration–ultrafiltration (MF–UF) membranes can offer superior pretreatment compared to RO when treating municipal effluents. Those efforts also incorporated polyamide membrane, which has all but replaced cellulose acetate in this application.

In the 1990s, many municipal agencies began operating full-scale MF–UF and polyamide RO membrane systems to treat secondary and tertiary municipal effluents. At that time, early adopters of large-scale membrane treatment processes for water reuse were rare. These membrane users transitioned the industry from theoretical and pilot-scale investigations into full-scale operations, bringing about a new facet of water reuse. In the years since, the industry has learned much about membrane performance and sustainability over long-term operation, including handling unanticipated operational challenges brought on by organic-laden, variable feed sources that can change not only year to year but sometimes day to day.

Operational considerations for low- and high-pressure membrane technologies in water reuse applications are similar to their potable system analogs. However, there are subtle differences that can pose additional problems or issues to the water reuse operator if these are not considered or anticipated. Membrane system operators in reuse applications need to understand that "industry guidance" has historically been based on potable water treatment applications. Irreversible fouling and flux loss that lead to increased cleaning intervals and reduced membrane life are constant challenges in this environment. Finding the balance between cleaning frequency and chemical and energy costs is often the goal for membrane facilities in water reuse applications. Fiber breakage and loss of RO membrane rejection are significant problems that can be accelerated by this source water. This manual presents a comprehensive description of the issues related to applying membrane technologies in water reuse projects.

This page intentionally blank.

Abbreviations

ACH	aluminum chlorohydrate
ANL	Argonne National Laboratory
ANSI	American National Standards Institute
AOC	assimilable organic carbon
AOP	advanced oxidation process
ASR	aquifer storage and recovery
AWT	advanced water treatment
AWWA	American Water Works Association
BAC	biologically active carbon
BIRM	biological immune response modulator
BNR	biological nutrient removal
BOD	biological oxygen demand
CA	cellulose acetate
$Ca(OCl)_2$	calcium hypochlorite
CEC	chemicals of emerging concern
CIP	clean in place
COC	cycles of concentration
COD	chemical oxygen demand
CTA	cellulose triacetate
DAF	dissolved air flotation
DALY	disability-adjusted life year
DOE	Department of Energy
DPR	direct potable reuse
EC	electrical conductivity
ED	electrodialysis
ED–EDR	electrodialysis–electrodialysis reversal
EDC	endocrine-disrupting chemical
EDR	electrodialysis reversal
efOM	effluent organic matter
ESA	Endangered Species Act
FAC	free available chlorine
FeCl3	ferric chloride
GRRP	groundwater recharge reuse projects
GWI	Global Water Intelligence
GWRRDR	Groundwater Replenishment Reuse Draft Regulation
HPM	high-pressure membrane
IMS	integrated membrane system

IPR	indirect potable reuse
LPM	low-pressure membrane
LRV	log-reduction value
MBR	membrane bioreactor
MCL	maximum contaminant limit
MF	microfiltration
MFI	modified fouling index
MGD	million gallons per day
MIC	microbiologically induced corrosion
MMFI	mini-plugging factor index
NaOCl	sodium hypochlorite
NDMA	*N*-nitrosodimethylamine
NF	nanofiltration
NOM	natural organic matter
NPDES	National Pollutant Discharge Elimination System
NTU	nephelometric turbidity unit
NWRI	National Water Research Institute
O&M	operations and maintenance
PA	polyamide
PAC	powdered activated carbon
PACl	polyaluminum chloride
PES	polyether sulfone
PP	polypropylene
PPCP	pharmaceuticals and personal care products
ppm	parts per million
PSU	polysulfone
PVDF	polyvinyl fluoride
RIB	rapid infiltration basins
RO	reverse osmosis
RWC	recycled water contribution
RWQCB	regional water quality control board
SAR	soil adsorption ratio
SAT	soil aquifer treatment
SBS	sodium bisulfite
SDI	silt density index
SMP	soluble microbial products
TCEQ	Texas Commission on Environmental Quality
TDC	total direct cell
TDS	total dissolved solids

TKN	total Kjeldahl nitrogen
TMDL	total maximum daily load
TOC	total organic carbon
TSS	total suspended solids
UF	ultrafiltration
USEPA	US Environmental Protection Agency
VOC	volatile organic compound
WHO	World Health Organization
WRA	WateReuse Association
WTP	water treatment plant
WWTP	wastewater treatment plant

This page intentionally blank.

Chapter 1

Development of Water Reuse Practices

HISTORY OF WATER REUSE

Reuse of water has occurred for centuries beginning with reuse of liquid waste in agricultural practice. Sewage farming was practiced in the United States in the 1800s and peaked in California in 1923, with 70 communities applying their municipal wastewater directly on food crops (NRC 2012). Agricultural water reuse was prominent in Texas south of San Antonio starting in the 1880s, with a formal contract between the City of San Antonio and the San Antonio Irrigation Company in 1901. Other Texas cities followed—Amarillo in 1920, Lubbock in 1930, Odessa in 1940, and Abilene in 1960—providing reclaimed water to farmers and ranchers. As cities grew, centralized wastewater treatment was more widely used and improved water quality, allowing for the first small urban water reuse system for irrigation of Golden Gate Park in San Francisco in 1912 (WRA 2012).

Industrial reuse of water reclaimed from treated municipal wastewater was documented in the 1940s, with the city of Big Spring, Texas, supplying the Cosden Oil and Chemical refinery in 1944. The cities of Odessa and Amarillo followed in the 1950s and the 1960s; this was the beginning of reclaimed water use for power plant cooling in several Texas cities (Texas Water Development Board 2011). In the early 1970s, Bethlehem Steel in Baltimore, Md., began using 100 mgd of reclaimed water for industrial purposes (USEPA 2004). By 2010, 57 power plants in 16 states were using reclaimed water for power plant cooling (DOE–ANL 2007).

TERMINOLOGY

ANSI/AWWA G481 (2014), *Reclaimed Water Program for Operation and Management*, describes the critical requirements for effective operation and management of a reclaimed water program and defines reclaimed water as water recovered following treatment of domestic wastewater to the level appropriate for the end use. This AWWA Standards Committee for Reclaimed Water formally approved the following: "Throughout the industry the terms *reclaimed water, recycled water,* and *water reuse* are used, in some instances reflecting a geographic preference, in other cases due to regulatory definition and requirements, and still other cases due to end use. The terms are often used interchangeably, but the end user should be cautioned that there are specific instances when one term is more appropriate than the other."

UNPLANNED WATER REUSE

The July 2012 National Academy of Science report, "Water Reuse: Potential for Expanding the Nation's Water Supply Through Reuse of Municipal Wastewater" (NRC 2012), identified the unplanned reuse of return flows from upstream dischargers as de facto reuse, which has been occurring since wastewater systems began discharging upstream of other communities. In 1980, of water utilities serving 76 million residents, 20 million were served from source waters with 1 percent or more from a wastewater-derived source and another 20 million were served from sources that contained more than 10 percent wastewater during low flow periods (NRC 2012). The US Environmental Protection Agency (USEPA) has published guidelines for water reuse (2012) that provide extensive discussion of both unplanned and planned reuse projects. The Guidelines define de facto reuse as "a situation where reuse of treated wastewater is, in fact, practiced but is not officially recognized (for example, a drinking water supply intake located downstream from a wastewater treatment plant [WWTP] discharge point)."

PLANNED WATER REUSE

Planned reuse refers to intentional nonpotable reuse or indirect or direct potable reuse. The USEPA Guidelines provide the following definitions:

Indirect potable reuse (IPR): Augmentation of a drinking water source (surface or groundwater) with reclaimed water followed by an environmental buffer that precedes drinking water treatment.

Direct potable reuse (DPR): The introduction of reclaimed water (with or without retention in an engineered storage buffer) directly into a drinking water treatment plant, either collocated or remote from the advanced wastewater treatment system. A key distinction is the absence of an environmental buffer, although an engineered buffer may be included. If the advanced water treatment processes are collocated at the reclamation plant, this approach is often referred to as "pipe to pipe" or "flange to flange."

There are IPR projects in many countries, with more than 20 in the United States as of August 2014. Examples include the Orange County Water District (Calif.) Groundwater Replenishment Project that augments the groundwater aquifer and provides water to several seawater barrier injection wells; the Upper Occoquan Service Authority (Va.) that provides supplemental water to the Occoquan Reservoir for Fairfax, Va., water supply; and the Singapore Public Utility Board NEWater supply that augments surface water storage. While there are only three DPR plants worldwide to date, several are in the planning stage. The three current DPR facilities (as of August 2014) are located in Windhoek, Namibia (Africa; 1969); Big Spring, Texas (April 2013); and an emergency operating permit for Wichita Falls, Texas (July 2014). A facility in Cloudcroft, N.M., is currently under construction, and a project in Brownwood, Texas, has approval for construction with funding pending. El Paso Water Utilities (EPWU) has successfully completed a direct potable pilot project and is moving to the next phases of the project. EPWU has operated an IPR facility since 1985. Table 1-1, reprinted from Table 3-9 of the USEPA Guidelines, provides a partial list of

IPR and DPR reuse projects worldwide. The Case Study column references the case studies that are provided in the appendix of the 2012 USEPA Guidelines, with the exception of the Namibia Case Study, which is reported in the 2012 NAS report. Some project capacities may have increased over time, notably the Orange County Calif., project, which operates at 70 mgd and is currently (February 2015) under construction for expansion to 100 mgd.

THE ROLE OF MEMBRANES IN WATER REUSE

Membranes play the following roles in water reuse:

- High-pressure membrane (HPM) reverse osmosis (RO) and nanofiltration (NF) are used to desalinate reclaimed water or to remove specific dissolved contaminants (organic or inorganic).
- Low-pressure membrane (LPM) microfiltration (MF) and ultrafiltration (UF), collectively referred to as membrane filtration, are used to control suspended and colloidal particular matter in reclaimed water or applied as a pretreatment to HPM.
- Membrane bioreactors (MBRs), typically using membranes, consolidate secondary and tertiary treatment into one unit process to provide effluent water quality that is comparable to the tertiary effluent produced by LPM.

Table 1-1 Selected worldwide indirect and direct potable reuse facilities

Country	City	Project Capacity, mgd	Description of Advanced System for Potable Reuse	Case Study
Belgium	Wulpen	1.9	Reclaimed water is returned to the aquifer before being reused as a potable water source	Belgium-Recharge
India	Bangalore (planned)	36	Reclaimed water will be blended in the reservoir, which is a major drinking water source	India-Bangalore
Namibia	Windhoek	5.5	Reclaimed water is blended with conventionally treated surface water for potable reuse	NAS, 2012
United States	Big Spring, Texas	3	Reclaimed water is blended with raw surface water for potable reuse	US-TX-Big Spring
United States	Upper Occoquan, Va.	54	Reclaimed water is blended in the reservoir, which is a major drinking water source	US-VA-Occoquan
United States	Orange County, Calif.	40	Reclaimed water is returned to the aquifer before being reused as a potable water source	US-CA-Orange County
United Kingdom	Langford	10.5	Reclaimed water is returned upstream to a river, which is the potable water source	United Kingdom-Langford
Singapore	Singapore	122	Reclaimed water is blended in the reservoir, which is a major drinking water source	Singapore-NEWater
South Africa	Malahleni	4.2	Reclaimed water from a mine is supplied as drinking water to the municipality	South Africa-Malahleni Mine

Source: Adapted from Von Sperling and Chernicharo (2002).

Membrane desalination (RO or NF) became the leading industrial technology for reducing reclaimed water salinity and specific contaminants of concern due to the high salt and minerals rejection rates, efficiency, and economics of the process. Membrane desalination has become an integrated part of IPR and DPR processes where concentrate disposal allows its application.

Membrane filtration (MF or UF) may not be an economical solution when compared to conventional filtration processes. Membrane filtration typically provides superior effluent water quality consistently, which made membrane filtration a preferred technology as a pretreatment to RO or NF applications. Membrane filtration processes are often selected for pathogen removal and log-removal credits at water treatment plants. When combined using LPM followed by HPM, it is referred to as an integrated membrane system (IMS), which is a key concept for IPR and DPR. This concept was successfully implemented at the Orange County, Calif., new water reuse recharge plant and the Big Spring, Texas, DPR plant, and it is expected to be followed by many other water reuse treatment plants.

MBRs produce effluent water quality characteristics that are similar those produced by MF. MBRs can be followed by membrane desalination when salinity or specific contaminants are a concern. This type of IMS is gaining increased interest, particularly for industrial applications.

HOW TO USE THIS MANUAL

This manual introduces the topic of membrane application for water reuse. While not an exhaustive reference, the manual can be used to identify typical membrane applications that will meet desired water quality objectives, as in the use of MF or UF membranes to pretreat water prior to treatment through NF or RO membrane units.

The chapter summaries provided here introduce the reader to the contents included in this manual and facilitate selection of chapters of immediate interest to the reader.

Chapter 2, Planning for Reuse Applications, introduces the reader to approaches to planning agricultural, nonpotable, commercial/industrial, environmental, and potable reuse projects, with reference to case studies that support the multiple facets of effective project planning. This chapter describes various ways to identify the quality of reclaimed water that is appropriate for the intended use. This is followed by selection of appropriate treatment technologies to achieve the required water quality. The roles of various membrane processes, including MBRs, in achieving the intended water quality or providing pretreatment for subsequent advanced treatment processes are described. Chapter 2 concludes by introducing regulatory, permitting, and environmental trade-offs when selecting membrane treatment options and provides cautions in system design based on variable flow delivered to typical wastewater treatment facilities.

Chapter, 3 Water Reuse Guidelines and Regulations, summarizes guidelines and regulations developed in countries around the world, including the United States, to address water reuse in general and reuse specifically for potable water augmentation. Currently, there are no international regulations that cover the multiple potential water reuse applications. Many resource-constrained countries adopt the World Health Organization guidelines as an initial step in protecting public health and reducing waterborne illness. Guidelines established in Australia, Japan, the United States, the European Union, Singapore, and Namibia are provided. Finally, the approaches used in California for IPR and DPR are highlighted.

Chapter 4, Source and Treated Water Quality, provides an overview of source water qualities typically considered for reuse, with a focus on municipal and industrial applications. Wastewater effluent quality varies based on regional regulations and stream quality standards. The concept of "fit-for-purpose" treatment is used to identify the level of treatment needed to meet the water quality goals of the intended end use. For membrane applications, variations in source water quality require the designer to select the type and size of pretreatment and membrane components to achieve the desired end use water quality. Specific water quality parameters for typical reclaimed water applications are provided. Multiple barrier treatment approaches including membrane treatment options for potable reuse are discussed.

Chapter 5, Membrane Process Treatment Facility Design, discusses membrane pretreatment and operational considerations in process and membrane selection to meet water quality goals for the intended reuse application. Pretreatment strategies for membrane processes include removing material that may damage the membrane, enhancing membrane performance, minimizing membrane fouling, and changing operations to prolong membrane life. Membrane selection for reuse applications is described, with a focus on chemical requirements and wastewater disinfection in advance of the membrane process.

Chapter 6, Operations, provides information on the operational considerations for LPM and HPM systems and recommendations for standard operating procedures. Strategies to minimize the incidence of membrane fouling and for membrane cleaning to restore capacity and specific flux intended in the design are included. Integrity testing of membrane systems, both pressure decay and vacuum decay test methods, and continuous water quality testing are described as effective in controlling quality and monitoring the treatment integrity of the membrane system. Membrane storage, replacement, and procurement frequency, along with maintenance of membranes, are identified as critical components in facility operations.

Chapter 7, Residuals Management, presents information regarding the management of concentrated waste streams. The volume of concentrated reject water may be 10 to 30 percent of the feedwater, have up to 10 times the concentration of salts and other contaminants, and represent much greater capital and operating costs than for the primary membrane treatment system. Concentrate management is described in terms of health, safety, regulations, energy demands, and cost. Final discharge options to surface water, a sanitary sewer, underground injection, and land application and evaporation ponds are evaluated. Current research activities intended to expand new approaches to concentrate management and reduce the cost of concentrate management are also discussed.

Chapter 8, Cost of Treatment, notes that costs for membrane systems vary significantly depending on source water quality and whether the final product water is to be used for nonpotable or potable applications. Because actual costs for facilities need to account for site-specific differences that cannot be generalized, this chapter provides a framework for developing preliminary costs. Costs fall into the following two major categories: cost of potable reuse facilities and cost of membrane-based nonpotable reuse facilities. Costs are presented only for treatment facilities and do not incorporate costs of transmission, distribution, or injection into aquifers or reservoirs, which are beyond the scope of this manual. Costs presented include capital cost, equipment, operation and maintenance, indirect, and contingency allowances. Most costs are for IPR facilities since there are only three DPR facilities in operation. The DPR and IPR facilities often incorporate the same treatment unit processes, with the IPR facilities including an environmental buffer not present in a DPR facility.

Chapter 9, Case Studies, presents nine large-scale membrane projects in the United States, Singapore, Australia, and Taiwan. These case studies compare approaches, project scale, and cost data to assist planners with development of cost-effective supply solutions for the variety of physical conditions that may be encountered.

Chapter 10, Future Technology Trends and Contaminants of Emerging Concern, describes current trends in the materials used to manufacture membranes for reclaimed water processes. As fouling is a significant operational issue, new cleaning processes to maintain membrane flux are identified. Advantages and disadvantages are discussed to support process selection. Two novel membrane processes, forward osmosis and membrane distillation, are described. The chapter concludes with a discussion of removing contaminants of concern and the capability of membranes to remove these contaminants in the wastewater source waters.

REFERENCES

ANSI/AWWA. 2014. Standard G481, *Reclaimed Water Program for Operation and Management.* Denver, CO: AWWA.

DOE–ANL (Department of Energy–Argonne National Laboratory). 2007. *Use of Reclaimed Water for Power Plant Cooling.* ANL/EVS/R-07/3. Argonne, IL: Argonne National Laboratory.

NRC (National Research Council). 2012. *Water Reuse: Potential for Expanding the Nation's Water Supply Through Reuse of Municipal Wastewater.* Washington, DC: National Academies Press.

Texas Water Development Board. 2011. *History of Water Reuse in Texas.* Austin, TX: Texas Water Development Board.

USEPA (US Environmental Protection Agency). 2004. *2004 EPA Guidelines for Water Reuse.* EPA/625/R-04/108. Cincinnati, OH: USEPA.

———. 2012. *2012 EPA Guidelines for Water Reuse.* EPA/600/R-12/618. Cincinnati, OH: USEPA.

Chapter 2

Planning for Reuse Applications

Planning for municipal use of an alternative water source or expansion of an existing water source is important, especially for systems that are reaching full allocation of their current water resources. The planning steps for industrial projects are likely to be different from those for municipal projects due to differences in project drivers, economics, and schedules. While each planning component is appropriate for each type of reclaimed water use, some planning steps require considerably more effort for applications in specific geographic areas.

REUSE APPLICATIONS

It has been reported that as much as 12 bgd (45 MCM/d) of the 32 bgd (121 MCM/d) produced in the United States could be beneficially reclaimed and reused (NRC 2012). Estimates by the WateReuse Association and Global Water Intelligence indicate that approximately 7 to 8 percent of treated wastewater in the United States is reused (Miller 2006; GWI 2009). Based on these estimates, there is tremendous potential for expanding the use of reclaimed water. Challenges to increased reuse include cost-effective treatment, delivery of reclaimed water to new users, and retrofitting facilities to use reclaimed water. Figure 2-1 provides a summary of uses of reclaimed water in the United States.

NONPOTABLE REUSE

Based on a 2011 survey, almost all water reuse in the United States is for nonpotable applications. Nonpotable reuse includes irrigation (municipal or agricultural) and commercial applications. A small percentage (less than 1 percent) is used for planned potable reuse applications—indirect potable reuse (IPR) or direct potable reuse (DPR). The following sections describe each of these in more detail.

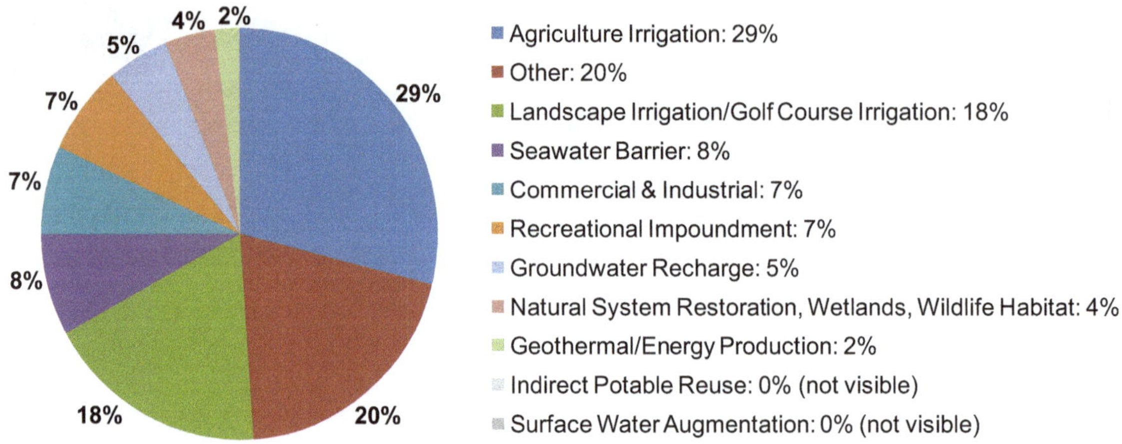

Reprinted from *2012 EPA Guidelines for Water Reuse*
Source: USEPA (2012).

Figure 2-1 Summary of water reuse applications in the United States

Agricultural Reuse

As shown in Figure 2-1, 29 percent of reclaimed water is used for agricultural purposes. Individual states adopt their own regulations for specific uses and appropriate water quality. As of 2012, 27 states had adopted regulations or guidelines for food crop applications, while 43 states had regulations or guidelines for processed food crops and nonfood crops. In the United States, nearly 37 percent (128,000 mgd) of freshwater use is for agricultural irrigation. Significant discussion on individual state regulations and water quality requirements for agricultural use, including recent research results, are presented elsewhere (USEPA 2012). Parameters important in agricultural reuse projects include the level of required disinfection, the levels of total dissolved solids (TDS), and the sodium adsorption ratio (SAR), which is the ratio of the concentration of sodium ions to the concentration of calcium plus magnesium ions.

Removal of TDS and reduction in SAR both require desalination. Pretreatment to protect the nanofiltration (NF) or reverse osmosis (RO) membranes often includes use of microfiltration (MF) or ultrafiltration (UF) membranes to minimize the size and number of particles.

Commercial and Industrial Applications

As shown in Figure 2-1, approximately 7 percent of reclaimed water use in the United States, including large-scale power cooling, is for commercial and industrial applications. Reclaimed water can be used for the following:

- Dust control,
- Odor and contaminant control,
- Scrubber applications,
- Product mixing paints and carpet/fabric dyes,
- Boiler applications,
- Refineries,
- Cooling towers, and
- Manufacturing.

POTABLE REUSE

Indirect potable reuse (IPR) refers to the planned decision to supplement a utility's existing water supplies with highly treated reclaimed water. Four states (California, Washington, Florida, and Hawaii) have regulations or guidelines specific to IPR. Texas and Virginia handle IPR projects on a case-by-case basis. Georgia and other states are developing guidelines to expand options for reclaimed water use to help meet water supply demands under long-term drought conditions.

When considering augmentation of groundwater with reclaimed water, common requirements include the following: public hearings; an advanced wastewater pretreatment program; and groundwater monitoring, where the injected water quality meets or exceeds current potable water quality parameters. California's draft regulations have stringent regulatory requirements for groundwater augmentation, and similar draft regulations are anticipated for surface water augmentation based on state approval of an IPR pilot study for San Diego (USEPA 2012). Treatment process options are discussed in Chapter 4 and elsewhere (USEPA 2012). Most, though not all, treatment trains use membrane processes as one of the multiple treatment barriers, followed by an advanced oxidation process (AOP) to destroy specific contaminants. The AOP typically includes medium-pressure ultraviolet (UV) radiation followed by hydrogen peroxide or ozone to create scavenging hydroxyl free radicals. In addition to organic compound removal, both processes also add additional disinfection. RO–AOP is sometimes referred to as full advanced treatment (FAT).

Groundwater Recharge

Groundwater recharge consists of direct injection or surface recharge infiltration via ponds or basins. Direct injection entails injection at the aquifer or radose zone via wells. The Orange County (Calif.) Groundwater Replenishment Project uses both direct injection wells for seawater barrier protection and infiltration through spreading basins to support the groundwater supply with 70 mgd of highly treated water. This project has been expanded to 100 mgd with build-out to 130 mgd. Surface recharge can include land spreading that takes advantage of soil aquifer treatment processes or rapid infiltration basins. Infiltration in the Los Angeles County Montebello Forebay has been practiced for 52 years and currently supplies 40 percent of the Los Angeles County groundwater supply.

Groundwater augmentation has also been practiced in El Paso, Texas, since 1985 where advanced wastewater treatment is followed by ozone and carbon filtration (biologically active carbon, (BAC) with ozone) and final chlorine disinfection. Treated water is transported to injection wells along the Texas–New Mexico border. The extraction wells are located such that the time for the injected water to migrate to the entrance well is one year.

Surface Water Augmentation

Surface water augmentation is the intentional addition of reclaimed water to increase the firm yield of a surface water supply. Examples of this practice in the United States include the Occoquan Reservoir in northern Virginia, Lake Lanier in Georgia, and a new project near Dallas, Texas. The Dallas project increases the firm yield of Lavon Lake from 93 mgd to 184 mgd by "polishing" 91 mgd of mixed Trinity River–Dallas South wastewater treatment plant (WWTP) effluent through a 1,840-acre constructed wetland system. Internationally, the Singapore Public Utility Board augments surface water supplies with highly treated water known as NEWater and supplies reclaimed water for industrial processes.

Direct Potable Reuse

Direct potable reuse (DPR) refers to the introduction of reclaimed water directly into a municipal water supply system or to a water treatment plant without an environmental buffer. Several publications provide additional guidance on implementation of and regulations for DPR projects (Tchobanoglous et al. 2011; NRC 2012; Crook 2010).

Examples of DPR include a plant in Windhoek, Namibia, that has been in operation since 1968; during the fifth plant expansion project in 1997, the process train added UF for increased production and reliability. A small DPR facility is currently under construction in Cloudcroft, N.M., using RO–AOP followed by wastewater treatment via membrane bioreactors (MBRs). AOP product water will be blended with natural raw water and then treated using UF membranes and carbon filtration. Big Spring, Texas, has a 2-mgd DPR plant that started full-scale production in April 2013, discharging reclaimed water following MF–RO–AOP into a lake water transmission main that delivers the blended water to six surface water treatment plants. Wichita Falls, Texas, began a 7-mgd DPR project under an emergency permit issued by the Texas Commission on Environmental Quality in July 2014 (Lawler 2014). Effluent from the WWTP travels 12 miles in a temporary *high-density polyethylene* main to an RO membrane treatment system at the city's water treatment plant. Reclaimed water is blended 50:50 with raw lake water and then treated through the conventional surface water treatment units at the same water treatment plant (Moody 2014).

TREATMENT OPTIONS FOR WATER REUSE

Domestic and industrial wastewater treatment processes are required to meet individual discharge permit requirements and can include pretreatment and primary and secondary treatment processes. Nitrification–denitrification and phosphorus removal for nutrient reduction may be required for permit compliance. For landscape and agricultural irrigation, there is typically no need for removal of nutrients as they provide value to the irrigators. Disinfection is required for most discharges, but disinfection requirements for reclaimed water typically require higher levels of disinfection for pathogen inactivation. For IPR and DPR applications, disinfection to meet potable water quality standards is typically required. In these cases, the choice of disinfectant is critical as disinfectant by-product concentrations may become elevated above regulated levels.

Advanced treatment processes may be required to meet discharge permits, but they are typically required to meet potable applications. Filtration is typically required, even for nonpotable applications, to consistently meet turbidity limits of 2 to 3 ntu. In order to meet drinking water quality criteria, more robust filtration processes may be required and often include membrane processes. Coagulation and sedimentation may be used in reclamation processes, especially if lime softening is applied to increase the number of cycles in cooling tower applications. Other advanced processes such as ion exchange, granular activated carbon, biologically activated carbon, and AOP may also be necessary for IPR or DPR applications to enhance water quality and final product stabilization.

MBR treatment offers many advantages, producing high-quality permeate water that can be used directly for nonpotable water reuse applications. MBR units may be used for satellite treatment units to provide point-of-use application of the reclaimed water, either as an onsite treatment and reuse application or for mining raw wastewater from a sewer outfall. Midland, Texas, has a satellite MBR facility under construction; it will use a 0.2-mgd MBR treatment unit sited in a city park within a building that includes odor control. Reclaimed water will be supplied to the adjacent Midland College for landscape irrigation and cooling tower makeup water. A satellite plant was selected because the alternative would have required construction on a significant amount of developed city property. MBR units typically include MF or UF membranes for nonpotable quality applications.

Membrane desalination using NF or RO membranes is typically required for TDS reduction and greater removal of microorganisms and organics. NF and RO are common in IPR and DPR applications, with RO membranes able to reduce levels of most, but not all, organic contaminants. Most NF and RO unit processes require pretreatment, which is typically provided by MF or UF systems. An AOP unit process is typically added for disinfection and to destroy recalcitrant organics that cannot be removed through the RO process. RO is also widely applied in industrial treatment units to provide ultrapure water for the electronics industry and for boiler feedwater. Reuse of water used in food and beverage manufacturing following treatment with a combination of membranes with or without AOP can minimize the water footprint of those industries and reduce demands on municipal water supplies.

EFFECTIVENESS OF TREATMENT

A comparison of membrane and conventional treatment processes is recommended during process design for a water reclamation plant. Comparison of physical treatment process effectiveness, including membrane processes, is described more fully in Chapter 4.

An important focus of treatment evaluation is bacteriological quality. The designer must ensure that the final water quality is appropriate for the specific intended use. This also includes multiple disinfectants to follow conventional or membrane unit processes.

In meeting regulatory log-removal criteria, treatment with membranes results in greater log removals of bacteria, protozoa, and viruses, which may allow for use of less disinfectant than for a conventional treatment system. Table 2-1 provides indicative log reductions for various stages of treatment and unit processes. While secondary treatment can reduce *Escherichia coli* by 1 to 3 logs, membrane filtration can be expected to reduce *E. coli* by 4 to 6 logs. For *Giardia,* secondary treatment can be expected to provide 0.5- to 1.5-log reduction, while membrane filtration can be expected to provide greater than 6-log reduction. Therefore, if the specific end use requires a 6-log reduction, minimal disinfection would be required following membrane filtration, while high disinfectant doses would be required to achieve an additional 4.5- to 5.5-log inactivation of *Giardia* for secondary treatment.

The effectiveness of filtration depends on the pore sizes of the membrane filters and can generally produce better effluent quality than either surface or depth filters. However, MF and UF systems have higher capital and operation and maintenance costs compared to depth and granular media filtration systems. NF and RO life-cycle costs can be substantially higher due to high energy use and specialized equipment. See Chapter 8 for an expanded discussion of membrane process costs.

RESIDUALS, DISPOSAL, AND REGULATORY REQUIREMENTS

For a reclamation plant that treats WWTP effluent, there is a stream of water to be treated and recycled or discharged to a unit process in the WWTP. The WWTP may also have backwash water treatment and recycling from conventional filtration systems.

Conventional media filtration systems generate only one waste stream, a waste filter backwash that varies between 2 and 6 percent of the total plant intake source water volume. Granular media filters require feedwater to be preconditioned with coagulant prior to filtration, typically adding an additional 60 to 80 percent to the solids load, which typically results in higher solids handling costs. In addition to the solids that were originally in the source water, the conventional filter backwash waste stream likely contains coagulant (typically iron or aluminum salt) and polymer. See AWWA Manual M46 (2007) for more details on residuals and concentrate management.

Table 2-1 Indicative log removals for various stages of treatment and unit processes

	Indicator Microorganisms				Pathogenic Microorganisms			
Type of Microorganism	*Escherichia coli* (indicator bacteria)	*Clostridium perfringens*	Phage (indicator virus)	Enteric Bacteria (e.g., *Campylobacter*)	Enteric Viruses	*Giardia lamblia*	*Cryptosporidium parvum*	Helminths
Bacteria	x	x		x				
Protozoa and helminths						x	x	x
Viruses			x		x			
Indicative Log Reductions in Various Stages of Wastewater Treatment*								
Secondary treatment	1–3	0.5–1	0.5–2.5	1–3	0.5–2	0.5–1.5	0.5–1	0–2
Dual-media filtration†	0–1	0–1	1–4	0–1	0.5–3	1–3	1.5–2.5	2–3
Membrane filtration (ultrafiltration, nanofiltration, and reverse osmosis)‡	4–>6	>6	2–>6	>6	2–>6	>6	4–>6	>6
Reservoir storage	1–5	N/A	1–4	1–5	1–4	3–4	1–3.5	1.5–>3
Ozonation	2–6	0–0.5	2–6	2–6	3–6	2–4	1–2	N/A
Ultraviolet disinfection	2–>6	N/A	3–>6	2–>6	1–>6	3–>6	3–>6	N/A
Advanced oxidation	>6	N/A	>6	>6	>6	>6	>6	N/A
Chlorination	2–>6	1–2	0–2.5	2–>6	1–3	0.5–1.5	0–0.5	0–1

Sources: Bitton (1999); EPHC (2008); Mara & Horan (2003); NRC (1998, 2012); Rose et al. (1996, 2001); USEPA (1999, 2003, 2004); WHO (1989).

* Reduction rates depend on specific operating conditions such as retention times, contact times, concentrations of chemicals used, pore size, filter depths, pretreatment, and other factors. Ranges given should not be used as design or regulatory bases—they are meant to show relative comparisons only.

† Including coagulation.

‡ Removal rates vary dramatically depending on the installation and maintenance of the membranes.

Abbreviation: N/A, not available.

Membrane residual streams include the minerals and silt particles retained on the membranes, inorganic and biological foulants that accumulated on the membrane surface, and cleaning solutions used to restore production capacity. The quantity and quality of membrane residual waters depend on the efficiency of pretreatment. In wastewater and reclamation systems, pretreatment for membranes seldom includes coagulation. The volume of the MF–UF membrane wash-water stream is typically 5 to 10 percent of the plant intake volume, or approximately twice the waste filter backwash volume generated by conventional pretreatment systems (AWWA 2007).

For reclaimed water systems, biological fouling is more pronounced than in freshwater water treatment applications because the organic content in the wastewater effluent provides a higher concentration of trace organics and a continuous food source. A membrane's large surface area provides an ideal environment for growth of microorganisms, which secrete a polysaccharide that forms a biofilm that covers the membrane, causing deterioration in flux and salt rejection. NF and RO membranes may be damaged by the oxidants used to control microorganisms in MF–UF pretreatment systems, which may lead to higher biological fouling rates.

Table 2-2 Filter pore size comparison*

Filter Type	Filtration Driving Force	Nominal Pore Size, μm	Contaminants Targeted for Removal
		Depth	
Noncompressible media	Gravity or pressure differential	60–300	TSS, turbidity, some protozoan oocysts and cysts
Compressible media			
		Surface Filtration	
Surface filtration	Gravity	5–20	TSS, turbidity, some protozoan oocysts and cysts
		Membrane†	
Microfiltration	Pressure differential	0.05	TSS, turbidity, some protozoan oocysts and cysts, some bacteria and viruses
Ultrafiltration	Pressure differential	0.002–0.050	Macromolecules, colloids, most bacteria, some viruses, proteins
Nanofiltration	Pressure differential	<0.002	Small molecules, some hardness, viruses
Reverse osmosis	Pressure differential	<0.002	Very small molecules, color, hardness, sulfates, nitrate, sodium, other ions

Reprinted from USEPA (2012).
*Information taken from the California Department of Public Health (2011) and Metcalf & Eddy (2003).
†Information taken from *Water Treatment Membrane Processes* (AWWA 1996).
Abbreviation: TSS, total suspended solids.

Both MF–UF and NF–RO systems use a chemical clean-in-place (CIP) process to clear inorganic and organic contaminants from the membrane surface. A typical CIP system consists of one or more CIP tanks. These tanks typically include mixers and/or a solution heater and, for specific process temperature control, a cooling system, pump(s), cartridge filter, feed and recirculation piping, instrumentation, power supply, and control equipment. Nonchlorinated permeate is typically used to supply the CIP tank for cleaning solution makeup. Powdered or liquid chemicals may be fed directly into the CIP tank, a separate smaller solution tank, or a CIP feed–recycle line used for chemical mixing (AWWA 2007).

RO concentrate creates significant operations, management, and regulatory challenges that are typically more significant for inland facilities. Coastal reclamation plants often have access to a seawater outfall where the concentrate can be directed. While this increases the TDS concentration in the wastewater effluent, the concentrate constituents were in the wastewater effluent prior to RO treatment and are returned to the ocean outfall in the same quantity as would be discharged without RO removal. The increased TDS may also buffer the TDS gradient between the freshwater, wastewater effluent and seawater. Most discharge permits require that TDS in the discharge plume closely match the ambient seawater TDS concentration. Inland facilities are often faced with more costly options including deep well injection into nonpotable groundwater, typically a brackish source with more than 10,000 mg/L TDS. These sources and the well casing must isolate the concentrate from potential potable water supplies. Concentrate may also be stored on site in evaporation ponds if climatic conditions support net evaporative loss and the land is available and affordable. An example of this is found at the Palo Verde Nuclear

Generation Station in Arizona, where reclaimed water from a regional WWTP in Phoenix supplies up to 85 mgd for power plant cooling. The reclamation plant provides additional onsite softening and recirculates the cooling water to obtain 25 cycles of operation. The blowdown (that is, concentrate) TDS is basically seawater (approximately 32,000 mg/L) and is stored in evaporation ponds that cover 650 acres on the site (Carpenter, Cavazos, & Cook 2012). Other more costly zero liquid discharge alternatives may require the use of crystallizers, which have high capital and energy costs.

After feasibility and cost, regulations typically dictate the choice of concentrate discharge or disposal option. Discharge to surface waters is typically controlled by state wastewater discharge regulations. Many states limit high TDS discharges to maintain receiving stream water quality. TDS or specific constituents can be limited in concentration or mass in the discharge. Many cities and utilities have industrial pretreatment regulations that limit the TDS level of discharges to the municipal treatment system. This is done to reduce operating costs at the wastewater treatment facility and potentially improve reclaimed water quality. Deep well injection requires a permit for injection from the primary regulatory agency; these may require audit option public hearings similar to those required for wastewater discharge permits. Deep well injection is becoming an increasingly popular alternative, but it can be expensive.

FLOW EQUALIZATION AND STORAGE REQUIREMENTS FOR NONPOTABLE REUSE

Managing variable reclamation influent flows to optimize the supply for reuse requires effective design of wastewater effluent storage, membrane production capacity, permeate product water storage, and distribution system pumping and storage capacity. Each system must be sized to provide cost effectiveness. Because of the high capital costs, including cost of the membrane systems, it is not economical to provide excess peak production capacity to meet peak-period user demands. Effluent supply to a membrane reclamation plant may also be limited during low flow periods. Onsite storage to equalize the reclaimed water supply is beneficial because it can preclude reduction in membrane treatment throughput. Sizing the membrane units for continuous operation also prevents shutdown of individual trains when their capacity is not required, which protects the membrane systems. Equalization storage for reclaimed water in advance of the high-service distribution system pumping should be sized to adequately meet user demands.

As with potable water systems, nonpotable distribution system storage can be used to balance both system pressure and localized demands. With reclaimed water systems, it is also common to include a portion of the system storage in proximity to high-demand users. This may reduce the required size of the distribution system pumps and piping while allowing constant delivery to high-demand users at average daily rates. This provides reliability for the high-volume users as they have onsite storage and can better control their reclaimed water supply.

ENVIRONMENTAL IMPACTS

Selection of membrane treatment process units over conventional physical or chemical process units for water reclamation does not eliminate environmental impacts; however, membrane units may have lower environmental impacts compared to conventional alternatives. Typically, membrane unit (MF–UF and NF–RO) housing has a smaller footprint, which reduces site disturbance and grading. Membrane filtration can result in increased log removal of bacteria, protozoa, and viruses compared to granular media filtration, reducing the use of disinfecting chemicals. If chlorine gas is the disinfectant of choice, reduction of the stored volume of chlorine gas cylinders reduces health risks to workers and neighbors near the treatment plant. MF–RO–AOP treatment processes were selected for several southern California reclamation plants including the West Basin Water District, Orange County Water District, San Diego demonstration pilot plant (USEPA 2012), to meet treatment process requirements and reduce net energy use. Selection was based on lower energy demands for reclamation compared to pumping raw water over long distances, which was an alternative in these specific cases.

LEGAL AND INSTITUTIONAL ISSUES

Regulations and design criteria for IPR and DPR facilities are evolving, and no state has developed DPR regulations. Four states have guidelines, regulations, or draft regulations for IPR facilities, and several states handle IPR projects on a case-by-case basis. California issued revised (draft) groundwater replenishment regulations in 2011 after adopting a new recycled water policy in 2009 and issuance of reports from an expert panel specific to monitoring for contaminants of emerging concern (CDPH 2011). Part of these current draft regulations define specific treatment unit processes to be included to meet minimum water quality standards and provide redundancy in the treatment process. To date (September 2014), the FAT process, which requires RO with appropriate pretreatment, must be used or pilot studies with documentation that supports an alternate process must be reviewed and approved by the Division of Drinking Water and Environmental Management (formerly CDPH) as providing equivalent public health protection and equal water quality. Other state regulations are not as stringent as the draft from California, which is the only state to require a specific treatment process such as FAT. New Mexico and Texas each have one authorized, constructed DPR facility in initial operation. El Paso (Texas) Water Utilities began full-scale operation of its IPR facility using ozone and biologically active carbon filtration in 1985, so there is precedence in the United States for non-RO treatment processes.

REGULATORY REQUIREMENTS AND PERMITTING, INCLUDING FUTURE REGULATORY IMPACTS

With the exception of IPR and DPR facilities, permitting of a reclamation plant would be similar using conventional secondary treatment processes or using membrane processes if both processes meet the same water quality criteria for the end use. The distinction between conventional and membrane processes becomes more evident when RO is needed to remove TDS, creating concentrate or a brine stream that must be managed. In a project under construction in California, sanitary sewer capacity to transport increased volumes of concentrate was not available for the MF–RO–AOP plant expansion from 3 to 8 mgd. The Vander Lans Reclamation Plant in Orange County (Calif.), has added a third-stage RO unit to increase process recovery and limit concentrate discharge to the same volume discharged from the 3-mgd facility. In this example, permit limitations to the sanitary sewer required process innovation to facilitate the expansion.

PUBLIC EDUCATION AND OUTREACH PROGRAMS

Stakeholder identification and inclusion in project development are critical components of any new reclamation project. As the public and water users become more aware of advances in treatment, membrane treatment processes may be perceived as a requirement by many stakeholders. While a membrane unit process may be a good selection, stakeholders and the public must be educated on available treatment processes, and the relative costs for construction and operations must be communicated. Pilot studies with comparable conventional and membrane units can identify relative construction and operations and maintenance costs for the quality of water produced by both systems. Inclusion of the stakeholders in the selection of treatment options and identification of the impact on individual water rates may help influence process selection based on the science at a specific facility. Beyond the unit process selection and cost impacts, public education and outreach for a membrane treatment process should be comparable to any other project effort in the community.

REFERENCES

ANSI/AWWA. 2014. Standard G481, *Reclaimed Water Program for Operation and Management.* Denver, CO: AWWA.

AWWA. 2007. M46, *Reverse Osmosis and Nanofiltration.* Denver, CO: AWWA.

Carpenter, G.; Cavazos, D.; & Crook, J. 2012. *Sale of the Century: Determining the Price of Reclaimed Water to be leased to the Largest Nuclear Power Plant in the U.S.* WateReuse Association Symposium, September. Hollywood, FL.

CDPH (California Department of Public Health). 2011. *Groundwater Replenishment Reuse Draft Regulation.* Title 22, California Code of Regulations, Division 4. Environmental Health Chapter 3. Recycling Criteria. Sacramento, CA: CDPH.

Crook, J. 2010. *Regulatory Aspects of Direct Potable Reuse in California.* Fountain Valley, CA: National Water Research Institute.

GWI (Global Water Intelligence). 2009. *Municipal Water Reuse Markets 2010.* Oxford, UK: Media Analytics Ltd.

Lawler, D. 2014. "Toilet to Tap" Water Can Be Better. *Fort Worth Star-Telegram*, July 25.

Miller, W.G. 2006. *Integrated Concepts in Water Reuse: Managing Global Water Needs, Desalination* (187: 65-75).

Moody, M. 2014. No Rain? No Problem: Emergency Water Reuse Pipeline Helps Mitigate Drought-Related Diasters in Wichita Falls. *WaterWorld.com*, April 1.

NRC (National Research Council). 2012. *Water Reuse: Potential for Expanding the Nation's Water Supply Through Reuse of Municipal Wastewater.* Washington, DC: National Academies Press.

Tchobanoglous, G.; Leverenz, H.L.; Nellor, M.H.; & Crook, J. 2011. *Direct Potable Reuse: The Path Forward.* Washington, DC: WateReuse Research Foundation and Water Reuse California.

USEPA (US Environmental Protection Agency). 2012. *2012 EPA Guidelines for Water Reuse.* EPA/600/R-12/618. Cincinnati, OH: USEPA.

WRA (WateReuse Association). 2009. *Manual of Practice: How to Develop a Water Reuse Program.* Alexandria, VA: WRA.

Chapter 3

Water Reuse Guidelines and Regulations

Water recycling and reuse have been practiced for more than 100 years, initially on an informal, unregulated basis. As the practice gained acceptance around the world as a way to stretch diminishing freshwater supplies, the need for guidance and/or regulations became a necessity from human health and environmental perspectives. In recent decades, controlled uses for reclaimed water for nonpotable and industrial applications have become more common in the United States and other countries. Guidelines and regulations for nonpotable and indirect potable uses of reclaimed water have evolved and vary widely from state to state and country to country. This chapter summarizes guidelines and regulations for reclaimed water use in representative states and countries and is not meant to be exhaustive. As the emphasis of this manual is on membrane applications, discussions on water reuse categories that do not involve membrane usage may not be included.

WORLD HEALTH ORGANIZATION GUIDELINES

No international regulations exist for the myriad of reclaimed water uses. Several countries with water reuse programs have referenced or adopted elements of the World Health Organization's (WHO's) *Guidelines for the Safe Use of Wastewater, Excreta and Greywater*, volumes 1 to 4 (WHO 2006a, 2006b, 2006c, 2006d) and *Water Quality—Guidelines, Standards and Health: Assessment of Risk and Risk Management for Water-related Infectious Disease* (WHO 2006). These documents provide guidance to economically constrained countries for establishing reuse programs that use domestic wastewater and waste. These programs can help in the conservation of existing freshwater supplies; promote the use of waste streams as a resource; and reduce poverty, hunger, and disease in the absence of a treatment infrastructure.

Hazards and Health Risks

The primary human health risks related to water reuse are excreta-borne and vector-borne pathogens that often survive long enough to be a threat to humans. To a lesser extent, chemical-borne disease can also pose a threat through uptake by plants and presence in soil. The WHO developed a risk management outline for its guidelines and standardized health risk by using disability-adjusted life-years (DALYs) as the unit to compare a population's health with its disease burden or other risk factors. WHO set a tolerable risk value of 1×10^{-6} DALY/person/year for pathogenic microorganisms and chemical constituents. The DALY value is the basis for log reduction of microbials in the WHO guidelines for agriculture and aquaculture. Microbial log-reduction goals are based on crop type (leaf or root crops, high-growing or low-growing crops), irrigation type (restricted, unrestricted, localized), and any wastewater/waste treatment prior to application.

Risk Management Targets

Table 3-1 shows the corresponding log reductions for pathogens and helminth ova to achieve the WHO DALY goal.

Soil and food crops can uptake and absorb chemicals and contribute to health risks. Table 3-2 shows maximum soil concentrations for trace metals and organics based on WHO health risk assessment.

Table 3-1 Health-based targets and helminth reduction targets for wastewater use in agriculture

Exposure Scenario	Health-based Target, DALY/person/year	Log_{10} Pathogen Reduction Needed	Number of Helminth Eggs per Liter
Unrestricted irrigation			
Leaf crops	$\leq 10^{-6,*}$	6	$\leq 1^{\dagger,\ddagger}$
Root crops	$\leq 10^{-6,*}$	7	$\leq 1^{\dagger,\ddagger}$
Restricted irrigation			
Highly mechanized	$\leq 10^{-6,*}$	3	$\leq 1^{\dagger,\ddagger}$
Labor intensive	$\leq 10^{-6,*}$	4	$\leq 1^{\dagger,\ddagger}$
Localized (drip) irrigation			
High-growing crops	$\leq 10^{-6,*}$	2	No recommendation§
Low-growing crops	$\leq 10^{-6,*}$	4	$\leq 1^{\dagger,\ddagger}$

Source: WHO (2006a).

* Rotavirus reduction. The health-based target can be achieved for unrestricted and localized irrigation by a 6–7 log unit pathogen reduction (obtained by a combination of wastewater treatment and other health protection measures); for restricted irrigation, it is achieved by a 2–3 log unit pathogen reduction.

† When children aged <15 years are exposed, additional health protection measures should be used (for example, treatment to ≤0.1 egg/L, protective equipment such as gloves or shoes/ boots, or chemotherapy).

‡ An arithmetic mean should be determined throughout the irrigation season. The mean value of ≤1 egg/L should be obtained for at least 90 percent of samples in order to allow for the occasional high-value sample (that is, with >10 eggs/L). With some wastewater treatment processes (for example, waste stabilization ponds), the hydraulic retention time can be used as a surrogate to ensure compliance with ≤1 egg/L.

§ No crops to be picked up from the soil.

Table 3-2 Maximum tolerable soil concentrations of various toxic chemicals based on human health protection

Chemical	Soil Concentration, mg/kg
Trace metals	
Antimony	36
Arsenic	8
Barium*	302
Beryllium*	0.2
Boron*	1.7
Cadmium	4
Fluorine	635
Lead	84
Mercury	7
Molybdenum*	0.6
Nickel	107
Selenium	6
Silver	3
Thallium*	0.3
Vanadium*	47
Organics	
Aldrin	0.48
Benzene	0.14
Chlordane	3
Chlorobenzene	211
Chloroform	0.47
2,4-D	0.25
DDT	1.54
Dichlorobenzene	15
Dieldrin	0.17
Dioxins	0.00012
Heptachlor	0.18
Hexachlorobenzene	1.4
Lindane	12
Chemical	
Methoxychlor	4.27
Polyaromatic hydrocarbons (as benzoapyrene)	16
Polychlorinated biphenyls	0.89
Pentachlorophenol	14
Phthalate	13,733
Pyrene	41
Styrene	0.68
2,4,5-T	3.82
Tetrachloroethane	1.25
Tetrachloroethylene	0.54
Toluene	12
Toxaphene	0.0013
Trichloroethane	0.68

Source: WHO (2006).

Implementing Mitigation Measures

The WHO guidelines also describe basic, universal protection measures to reduce risks from water reuse, such as promoting health and hygiene, handling and preparing food properly, reducing vector contact and improving vector control, washing and cooking produce, accessing safe drinking water, treating wastewater/excreta, immunizing as appropriate, timing of waste application, and using personal protective equipment (for example, footwear, gloves).

For example, the highest pathogen reduction of 7 log is suitable for root and nonroot crops that are irrigated with wastewater and may be eaten uncooked. Pathogen reduction of up to 7 log can be achieved with the following options:

- Allowing pathogen die-off between the last irrigation and human consumption (2-log reduction);
- Washing crops with clean water prior to consumption (1-log reduction);
- Peeling produce (2-log reduction);
- Cooking produce (6- to 7-log reduction);
- Using localized irrigation such as subsurface or drip irrigation (2- to 4-log reduction); and
- Using conventional wastewater treatment, if available (4- to 7-log reduction).

System Assessment and Monitoring

Implementation of a risk management plan includes confirming that any new system can operate within acceptable limits, ongoing operational monitoring to ensure continued performance within design specifications, and verifying that end-use products meet acceptable health limits. Table 3-3 provides recommended minimum monitoring frequencies and target limits for agriculture and aquaculture.

NATIONAL GUIDELINES

The two primary factors that drive countries to develop water reuse guidance are increasing demands on freshwater supplies and limits of freshwater supplies. Population demographics; geography; resources; economics; infrastructure; social, cultural, and institutional constraints; and weather trends also factor into the nature, content, and scope of any national guidance.

Economically constrained countries that do not have institutions or infrastructure to protect human health may choose to adopt the WHO guidelines using "fit-for-purpose" principles to suit their unique needs. Conversely, economically developed countries can more easily develop, implement, and support guidelines through existing institutions.

Representative water reuse guidelines are presented here for Australia, the United States, and Japan.

Australia Guidelines

According to the National Program for Sustainable Irrigation (2006), Australia's per capita annual water use of 343,000 gal is the third highest in the world. More than 65 percent of that use is for agriculture (primarily crops and livestock grazing). The burden of using potable supplies for agriculture is offset by increased use of reclaimed water.

No national water reuse regulations exist in Australia. In lieu of national regulations, national guidelines were developed in two phases using management and policy documents for overall water resources and drinking water as the basis. States and territories establish reuse regulations based on the guidelines.

Table 3-3 Recommended minimum verification monitoring for wastewater and excreta use in agriculture and aquaculture

Activity or Exposure	Water Quality Monitoring Parameters*	
Agriculture	*Escherichia coli*/100 mL† (arithmetic mean)	Helminth Eggs/Liter† (arithmetic mean)
Unrestricted irrigation		
Root cops	≤10^3	≤1
Leaf crops	≤10^4	
Drip irrigation, high-growing crops	≤10^5	
Restricted irrigation		
Labor-intensive, high-contact agriculture	≤10^4	≤1
Highly mechanized agriculture	≤10^5	
Septic tank	≤10^6	
Aquaculture	*Escherichia coli*/100 mL† (arithmetic mean)	Viable Trematode Eggs/Liter†, b
Produce consumers		
Pond	≤10^4	Not detected
Wastewater	≤10^5	Not detected
Excreta	≤10^6	Not detected
Workers, local communities		
Pond	≤10^3	No viable eggs
Wastewater	≤10^4	No viable eggs
Excreta	≤10^5	No viable eggs

Source: WHO (2006).

*Monitoring should be conducted at the point of use or the point of effluent discharge. Frequency of monitoring is as follows:

- Urban areas: one sample every two weeks for *Escherichia coli* and monthly for helminth eggs.
- Rural areas: monthly sample for *E. coli* and one sample very one to two weeks for helminth eggs.

Five-liter composite samples are required for helminth eggs prepared from grab samples taken six times per day. Monitoring for trematode eggs is difficult due to a lack of standardized procedures. The inactivation of trematode eggs should be evaluated as part of system validation.

†For excreta, weights may be used instead of volumes, depending on the type of excreta: 100 mL of wastewater is equivalent to 1 to 4 g of total solids; 1 L = 10 to 40 g of total solids. The required *E. coli* or helminth numbers would be the same per unit of weight.

Phase 1, published in 2006 (NRMMC/EPHC/AHMC 2006), covers guidance and management practices for recycled water quality and end uses for treated sewage and graywater, as follows:

- Residential uses (landscape irrigation, toilet flushing, and laundry);
- Irrigation for agriculture and horticulture, open spaces, and urban recreation;
- Firefighting;
- Industrial uses (cooling towers, process water, wash-down water, dust control); and
- Onsite business and residential graywater treatment for nonpotable use.

The phase 1 guidance is founded on a risk management framework of 12 elements grouped into the following four categories: commitment to the reuse program and water quality, analysis of the reuse system and its management, support of the system and management, and review and improvement of the system. The framework is modeled after a similar risk management framework for Australia's 2004 Drinking Water Guidelines (NHRMC 2004).

Phase 2, published in 2008–2009, consists of the following three modules and expands the material covered in phase 1 to include recharge and drinking water supply augmentation and stormwater harvesting:

- Module 1: covers the planned use of reclaimed water, either from treated sewage or stormwater, to augment drinking water supplies (NRMMC/EPHC/NHMRC 2008);
- Module 2: provides guidelines for stormwater nonpotable reuse (NRMMC/EPHC/NHMRC 2009); and
- Module 3: provides guidelines for groundwater recharge with all water types including recycled water (NRMMC/EPHC/NHMRC 2008).

Regional Australian regulations and guidance. The individual states and territories of Victoria, New South Wales, Queensland, South Australia, Western Australia, Tasmania, Capital Territory, and Northern Territory have statutes and controls that govern phase 1 and phase 2 uses of recycled water for urban, agricultural, industrial, augmentation, and recharge uses. State regulatory criteria may be more stringent than the national guidelines. National and local codes from animal husbandry, plumbing, water supply, and sewerage agencies may also apply. The end user conducts additional treatment for water quality–specific end uses such as MF and reverse osmosis (RO) for boiler feedwater or nitrification for cooling tower makeup water. Examples of phase 1 reuse regulations for two representative states, Victoria and New South Wales, are presented below.

Victoria regulations and guidance. Table 3-4 provides a summary of Victoria's phase 1 water reuse categories, water quality objectives, and range of nonpotable end uses.

Table 3-4 Victoria, Australia, reclaimed water classes, treatment, and pathogen reduction

Class	Water Quality Objective (annual median unless specified)	Treatment Process	Range of Uses
A	*Escherichia coli*: <10 cfu/100 mL Turbidity: <10 ntu BOD_5: <10 mg/L TSS: <5 mg/L pH: 6–9 Chlorine residual: 1 mg/L or equivalent at point of use	Tertiary treatment; sufficient log reduction to achieve: <10 *E. coli*/100 mL <1 helminth/L <1 protozoa/50 L <1 virus/50 L	Urban nonpotable: with uncontrolled public access Agricultural: all crops including food crops eaten raw Industrial: open systems with worker exposure potential
B	*E. coli*: <100 cfu/100 mL BOD_5: <20 mg/L TSS: <30 mg/L pH: 6–9	Secondary treatment; pathogen reduction (including helminth reduction)	Agricultural: for example, dairy cattle grazing Industrial: for example, wash-down water
C	*E. coli*: <1,000 cfu/100 mL BOD_5: <20 mg/L TSS: <30 mg/L pH: 6–9	Secondary treatment; pathogen reduction (including helminth reduction)	Urban nonpotable: with controlled public access Agricultural: for example, food crops eaten cooked or processed; grazing/livestock fodder Industrial: systems with no worker exposure potential
D	*E. coli*: <10,000 cfu/100 mL BOD_5: <20 mg/L TSS: <30 mg/L pH: 6–9	Secondary treatment	Agricultural: nonfood crops, including turf, wood lots, flowers

Source: EPA Victoria (2003).

Abbreviations: BOD_5, 5-day biochemical oxygen demand; TSS, total suspended solids.

New South Wales regulations and guidance. Table 3-5 provides a summary of New South Wales' phase 1 water reuse guidelines and controls for nonpotable end uses.

Table 3-5 New South Wales, Australia, guidelines for reclaimed water use from municipal sewage treatment plants

Reuse Type—Agricultural	Treatment	Water Quality Criteria	Effluent Monitoring	Control
Food crops: crops to be eaten raw and in direct contact with irrigation (for example, salad greens)	Tertiary pathogen reduction (via chlorine or detention)	pH: 6.5–8.5 Predisinfection turbidity: ≤2 ntu Chlorine residual: 1 mg/L or equivalent Thermotolerant coliforms: <10 cfu/100 mL Intestinal nematode egg or larva: <1/L	pH weekly Turbidity continuous Disinfection system daily (chlorine, UV, other) Coliforms weekly	Limit application rates to protect groundwater quality Salinity a factor in irrigation Control helminths with minimum of 25 days ponding or equivalent treatment (for example, sand filtration)
Food crops: crops to be eaten cooked or processed and not in direct contact with irrigation	Secondary pathogen reduction (via chlorine or detention)	pH: 6.5–8.5 Thermotolerant coliforms: <1,000 cfu/100 mL	pH weekly BOD weekly TSS weekly Coliforms weekly	Limit application rates to protect groundwater quality Salinity a factor in irrigation Do not harvest dropped crops Cook crops at >70°C for 2 minutes commercially
Food production: pasture and fodder for grazing animals (exc. pigs and dairy animals-cattle, sheep, and goats)	Secondary pathogen reduction (via chlorine or detention)	pH: 6.5–8.5 Thermotolerant coliforms: <1,000 cfu/100 mL	pH weekly TSS weekly Disinfection system daily (chlorine, UV, other) Coliforms weekly	Limit application rates to protect groundwater quality Withhold access 4 hours or until irrigated space is dry Helminth controls
Food production: pasture and fodder for dairy animals (with withholding period)	Secondary pathogen reduction (via chlorine or detention)	pH: 6.5–8.5 Thermotolerant coliforms: <1,000 cfu/100 mL	pH weekly TSS weekly Disinfection system daily (chlorine, UV, other)	Limit application rates to protect groundwater quality Withhold access 5 days

(Continued)

Table 3-5 New South Wales, Australia, guidelines for reclaimed water use from municipal sewage treatment plants *(Continued)*

Reuse Type—Agricultural	Treatment	Water Quality Criteria	Effluent Monitoring	Control
Food production: pasture and fodder for dairy animals (without withholding period); drinking water (all stock exc. pigs); wash-down water for dairies	Secondary pathogen reduction (via chlorine or detention)	pH: 6.5–8.5 Thermotolerant coliforms: <100 cfu/100 mL	pH weekly TSS weekly Disinfection system daily (chlorine, UV, other) Coliforms weekly	Limit application rates to protect groundwater quality No withholding period Helminth controls
Nonfood crops: for example, silviculture, turf, cotton	Secondary pathogen reduction (via chlorine or detention)	pH: 6.5–8.57 Thermotolerant coliforms: <10,000 cfu/100 mL	pH weekly BOD weekly TSS weekly Coliforms weekly	Limit application rates to protect groundwater quality Restrict public access Withhold access 4 hours or until irrigated space is dry

Source: NSW Department of Environment and Conservation (2003).

Japan Guidelines

Even with Japan's overall high rainfall (>60 in./year), the country faces varied precipitation patterns and periodic drought. Water shortages can occur in heavily populated urban areas. Wastewater has been deliberately reused in Japan for more than 50 years. With the exception of rural areas and small urban sewerage works, municipal wastewater is treated at large, centralized sewage treatment plants to secondary or tertiary quality and is suitable for reuse. Approved uses include toilet flushing, recreational impoundments, melting snow, landscape irrigation, and agriculture. Two additional urban uses that were added to the existing approved list to bolster demand were road dampening to mitigate heat island effect and stream restoration.

Table 3-6 summarizes Japan's water reuse guidelines.

Table 3-6 Japan's water reuse standards

Parameter	End Use Type			
	Toilet Flushing	Road Application to Reduce Heat Island Effect	Landscape Irrigation	Recreational Impoundments
Escherichia coli	ND/100 mL	ND/100 mL	1,000 cfu/qoo mL	ND/100 mL
Turbidity	Control target: 2 ntu	Control target: 2 ntu	Control target: 2 ntu	2 ntu
pH	5.8–8.6	5.8–8.6	5.8–8.6	5.8–8.6
Appearance	Shall not be unpleasant	Shall not be unpleasant	Shall not be unpleasant	Shall not be unpleasant
Chromaticity (color quality)	–	–	40 degrees	10 degrees
Odor	Shall not be unpleasant	Shall not be unpleasant	Shall not be unpleasant	Shall not be unpleasant
Chlorine residual (point of use)	Control targets: free, 0.1 mg/L; combined, 0.4 mg/L	Control targets: free, 0.1 mg/L; combined, 0.4 mg/L	No numerical target at point of use (no human contact assumed)	Control targets: free, 0.1 mg/L; combined, 0.4 mg/L
Tertiary treatment type	Sand filtration	Sand filtration	Sand filtration	Coagulation and sand filtration

Source: 5th Japan–US Government Conference on Drinking Water Quality Management and Wastewater Control, March 4, 2009.

US Guidelines

The US Environmental Protection Agency (USEPA) first published *Guidelines for Water Reuse* in 1980 as a research document. Subsequent revisions were issued in 1992, 2004, and 2012 in response to technological advancements and the increasing number of states that were adopting reuse regulations. The 2012 edition (USEPA 2012) summarizes current state regulations, integrates nontraditional water sources into water management scenarios, incorporates findings from recent research studies, and provides suggested guidelines for various reclaimed water applications. Table 3-7 provides a summary of these guidelines.

Table 3-7 Suggested guidelines for water reuse[1]

Reuse Category and Description	Treatment	Reclaimed Water Quality[2]	Reclaimed Water Monitoring	Setback Distances[3]	Comments
			Urban Reuse		
Unrestricted The use of reclaimed water in non-potable municipal settings where public access is not restricted.	• Secondary[4] • Filtration[5] • Disinfection[6]	• pH = 6.0–9.0 • ≤10 mg/L BOD[7] • ≤2 ntu[8] • No detectable fecal coliform/100 mL[9, 10] • 1 mg/L Cl_2 residual (min.)[11]	• pH – weekly • BOD – weekly • Turbidity – continuous • Fecal coliform – daily • Cl_2 residual – continuous	• 50 ft (15 m) to potable water supply wells; increased to 100 ft (30 m) when located in porous media[18]	• At controlled-access irrigation sites where design and operational measures significantly reduce the potential of public contact with reclaimed water, a lower level of treatment, for example, secondary treatment and disinfection to achieve <14 fecal coli/100 mL may be appropriate. • The reclaimed water should not contain measurable levels of pathogens.[12] • Reclaimed water should be clear and odorless. • Higher chlorine residual and/or a longer contact time may be necessary to ensure that viruses and parasites are inactivated or destroyed. • Chlorine residual >0.5 mg/L in the distribution system is recommended to reduce odors, slime, and bacteria. • See Section 3.4.3 in the 2004 USEPA Guidelines for recommended treatment reliability requirements.

(Continued)

Table 3-7 Suggested guidelines for water reuse[1] *(Continued)*

Reuse Category and Description	Treatment	Reclaimed Water Quality[2]	Reclaimed Water Monitoring	Setback Distances[3]	Comments
			Urban Reuse		
Restricted The use of reclaimed water in nonpotable applications in municipal settings where public access is controlled or restricted by physical or institutional barriers such as fencing, advisory signage, or temporal access restriction.	• Secondary[4] • Disinfection[6]	• pH = 6.0–9.0 • ≤30 mg/L BOD[7] • ≤30 mg/L TSS • ≤200 fecal coliform/100 mL[9, 13, 14] • 1 mg/L Cl_2 residual (min.)[11]	• pH – weekly • BOD – weekly • TSS – daily • Fecal coliform – daily • Cl_2 residual – continuous	• 300 ft (90 m) to potable water supply wells • 100 ft (30 m) to areas accessible to the public (if spray irrigation)	• If spray irrigation, TSS <30 mg/L may be necessary to avoid clogging of sprinkler heads. See Section 3.4.3 in the 2004 USEPA Guidelines for recommended treatment reliability requirements. • For use in construction activities including soil compaction, dust control, washing aggregate, and making concrete, worker contact with reclaimed water should be minimized and a higher level of disinfection (for example, <14 fecal coli/100 mL) should be provided when frequent worker contact with reclaimed water is likely.

(Continued)

Table 3-7 Suggested guidelines for water reuse[1] *(Continued)*

Reuse Category and Description	Treatment	Reclaimed Water Quality[2]	Reclaimed Water Monitoring	Setback Distances[3]	Comments
			Agricultural Reuse		
Food Crops[15] The use of reclaimed water for surface or spray irrigation of food crops that are intended for human consumption, consumed raw.	• Secondary[4] • Filtration[5] • Disinfection[6]	• pH = 6.0–9.0 • ≤10 mg/L BOD[7] • ≤2 ntu[8] • No detectable fecal coliform/100 mL[9, 10] • 1 mg/L Cl_2 residual (min.)[11]	• pH – weekly • BOD – weekly • Turbidity – continuous • Fecal coliform – daily • Cl_2 residual – continuous	• 50 ft (15 m) to potable water supply wells; increased to 100 ft (30 m) when located in porous media[18]	• See 2012 USEPA Guidelines, Table 3-5, for other recommended chemical constituent limits for irrigation. • Chemical (coagulant and/or polymer) addition prior to filtration may be necessary to meet water quality recommendations. • The reclaimed water should not contain measurable levels of pathogens.[12] • Higher chlorine residual and/or a longer contact time may be necessary to ensure that viruses and parasites are inactivated or destroyed. • High nutrient levels may adversely affect some crops during certain growth stages. • See Section 3.4.3 in the 2004 USEPA Guidelines for recommended treatment reliability requirements.

(Continued)

Table 3-7 Suggested guidelines for water reuse[1] *(Continued)*

Reuse Category and Description	Treatment	Reclaimed Water Quality[2]	Reclaimed Water Monitoring	Setback Distances[3]	Comments
			Agricultural Reuse		
Processed Food Crops[15] The use of reclaimed water for surface irrigation of food crops that are intended for human consumption, commercially processed.	• Secondary[4] • Disinfection[6]	• pH = 6.0–9.0 • ≤30 mg/L BOD[7] • ≤30 mg/L TSS • ≤200 fecal coliform/100 mL[9, 13, 14] • 1 mg/L Cl_2 residual (min.)[11]	• pH – weekly • BOD – weekly • TSS – daily • Fecal coliform – daily • Cl_2 residual – continuous	• 300 ft (90 m) to potable water supply wells • 100 ft (30 m) to areas accessible to the public (if spray irrigation)	• See 2012 USEPA Guidelines, Table 3-5, for other recommended chemical constituent limits for irrigation. • If spray irrigation, TSS <30 mg/L may be necessary to avoid clogging of sprinkler heads. • High nutrient levels may adversely affect some crops during certain growth stages.
Non-Food Crops The use of reclaimed water for irrigation of crops that are not consumed by humans, including fodder, fiber, and seed crops, or to irrigate pasture land, commercial nurseries, and sod farms.					• See Section 3.4.3 in the 2004 USEPA Guidelines for recommended treatment reliability requirements. • Milking animals should be prohibited from grazing for 15 days after irrigation ceases. A higher level of e.g., to achieve <14 fecal coli/100 mL, should be provided if this waiting period is not adhered to.

(Continued)

Table 3-7 Suggested guidelines for water reuse[1] *(Continued)*

Reuse Category and Description	Treatment	Reclaimed Water Quality[2]	Reclaimed Water Monitoring	Setback Distances[3]	Comments
			Impoundments		
Unrestricted The use of reclaimed water in an impoundment in which no limitations are imposed on body contact.	• Secondary[4] • Filtration[5] • Disinfection[6]	• pH = 6.0-9.0 • ≤10 mg/L BOD[7] • ≤2 ntu[8] • No detectable fecal coliform/100 mL[9, 10] • 1 mg/L Cl_2 residual (min.)[11]	• pH – weekly • BOD – weekly • Turbidity – continuous • Fecal coliform – daily • Cl_2 residual – continuous	• 500 ft (150 m) to potable water supply wells (min.) if bottom not sealed	• Dechlorination may be necessary to protect aquatic species of flora and fauna. • Reclaimed water should be non-irritating to skin and eyes. • Reclaimed water should be clear and odorless. • Nutrient removal may be necessary to avoid algae growth in impoundments. • Chemical (coagulant and/or polymer) addition prior to filtration may be necessary to meet water quality recommendations. • Reclaimed water should not contain measurable levels of pathogens.[12] • Higher chlorine residual and/or a longer contact time may be necessary to ensure that viruses are inactivated or destroyed. • Fish caught in impoundments can be consumed. • See Section 3.4.3 in the 2004 USEPA Guidelines for recommended treatment reliability requirements.
Restricted The use of reclaimed water in an impoundment where body contact is restricted.	• Secondary[4] • Disinfection[6]	• ≤30 mg/L BOD[7] • ≤30 mg/L TSS • ≤200 fecal coliform/100 mL[9, 13, 14] • 1 mg/L Cl_2 residual (min.)[11]	• pH – weekly • TSS – daily • Fecal coliform – daily • Cl_2 residual – continuous	• 500 ft (150 m) to potable water supply wells (min.) if bottom not sealed	• Nutrient removal may be necessary to avoid algae growth in impoundments. • Dechlorination may be necessary to protect aquatic species of flora and fauna. • See Section 3.4.3 in the 2004 USEPA Guidelines for recommended treatment reliability requirements.

(Continued)

Table 3-7 Suggested guidelines for water reuse[1] *(Continued)*

Reuse Category and Description	Treatment	Reclaimed Water Quality[2]	Reclaimed Water Monitoring	Setback Distances[3]	Comments
			Environmental Reuse		
Environmental Reuse The use of reclaimed water to create wetlands, enhance natural wetlands, or sustain stream flows.	• Variable • Secondary[4] and disinfection[6]	Variable, but not to exceed: • ≤30 mg/L BOD[7] • ≤30 mg/L TSS • ≤200 fecal coliform/100 mL[9, 13, 14] • 1 mg/L Cl_2 residual (min.)[11]	• BOD – weekly • SS – daily • Fecal coliform – daily • Cl_2 residual – continuous	N/A	• Dechlorination may be necessary to protect aquatic species of flora and fauna. • Possible effects on groundwater should be evaluated. • Receiving water quality requirements may necessitate additional treatment. • Temperature of the reclaimed water should not adversely affect ecosystem. • See Section 3.4.3 in the 2004 USEPA Guidelines for recommended treatment reliability requirements.
			Industrial Reuse		
Once-Through Cooling	• Secondary[4]	• pH = 6.0–9.0 • ≤30 mg/L BOD[7] • ≤30 mg/L TSS • ≤200 fecal coliform/100 mL[9, 13, 14] • 1 mg/L Cl_2 residual (min.)[11]	• pH – weekly • BOD – weekly • TSS – daily • Fecal coliform – daily • Cl_2 residual – continuous	• 300 ft (90 m) to areas accessible to the public	• Windblown spray should not reach areas accessible to workers or the public.
Recirculating Cooling Towers	• Secondary[4] • Disinfection[6] (chemical coagulation and filtration[5] may be needed)	Variable, depends on recirculation ratio: • pH = 6.0–9.0 • ≤30 mg/L BOD[7] • ≤30 mg/L TSS • ≤200 fecal coliform/100 ml[9, 13, 14] • 1 mg/L Cl_2 residual (min.)[11]		• 300 ft (90 m) to areas accessible to the public. May be reduced if high level of disinfection is provided	• Windblown spray should not reach areas accessible to workers or the public. • Additional treatment by user is usually provided to prevent scaling, corrosion, biological growths, fouling, and foaming. • See Section 3.4.3 in the 2004 USEPA Guidelines for recommended treatment reliability requirements.

(Continued)

Table 3-7 Suggested guidelines for water reuse[1] *(Continued)*

Reuse Category and Description	Treatment	Reclaimed Water Quality[2]	Reclaimed Water Monitoring	Setback Distances[3]	Comments
Groundwater Recharge: Nonpotable Reuse					
The use of reclaimed water to recharge aquifers that are not used as a potable drinking water source.	• Site specific and use dependent • Primary (min.) for spreading • Secondary[4] for injection	• Site specific and use dependent	• Depends on treatment and use	• Site specific	• Facility should be designed to ensure that no reclaimed water reaches potable water supply aquifers. • See Section 2.5 of the 2004 USEPA Guidelines for more information. • For injection projects, filtration and disinfection may be needed to prevent clogging. • For spreading projects, secondary treatment may be needed to prevent clogging. • See Section 3.4.3 in the 2004 USEPA Guidelines for recommended treatment reliability requirements.

(Continued)

Table 3-7 Suggested guidelines for water reuse[1] *(Continued)*

Reuse Category and Description	Treatment	Reclaimed Water Quality[2]	Reclaimed Water Monitoring	Setback Distances[3]	Comments
			Indirect Potable Reuse		
Groundwater recharge by spreading into potable aquifers	• Secondary[4] • Filtration[5] • Disinfection[5] • Soil aquifer treatment	Includes, but not limited to, the following: • No detectable total coliform/100 mL[9, 10] • 1 mg/L Cl_2 residual (min.)[11] • pH = 6.5–8.5 • ≤2 ntu[8] • 2 mg/L TOC of wastewater origin • Meet drinking water standards after percolation through vadose zone	Includes, but not limited to, the following: • pH – daily • Total coliform – daily • Cl_2 residual – continuous • Drinking water standards – quarterly • Other[17] – depends on constituent • TOC – weekly • Turbidity – continuous • Monitoring is not required for viruses and parasites; their removal rates are prescribed by treatment requirements	• Distance to nearest potable water extraction well that provides a minimum of 2 months retention time in the underground.	• Depth to groundwater (that is, thickness to the vadose zone) should be at least 6 ft (2 m) at the maximum groundwater mounding point. • The reclaimed water should be retained underground for at least 2 months prior to withdrawal. • Recommended treatment is site specific and depends on factors such as type of soil, percolation rate, thickness of vadose zone, native groundwater quality, and dilution. • Monitoring wells are necessary to detect the influence of the recharge operation on the groundwater. • Reclaimed water should not contain measurable levels of pathogens after percolation through the vadose zone.[12] • See Section 3.4.3 in the 2004 USEPA Guidelines for recommended treatment reliability requirements. • Recommended log reductions of viruses, *Giardia*, and *Cryptosporidium* can be based on challenge tests or the sum of log-removal credits allowed for individual treatment processes. Monitoring for these pathogens is not required. • Dilution of reclaimed water with waters of nonwastewater origin can be used to help meet the suggested TOC limit.

(Continued)

Table 3-7 Suggested guidelines for water reuse[1] *(Continued)*

Reuse Category and Description	Treatment	Reclaimed Water Quality[2]	Reclaimed Water Monitoring	Setback Distances[3]	Comments
			Indirect Potable Reuse		
Augmentation of surface water supply reservoirs	• Secondary[4] • Filtration[5] • Disinfection[6] • Advanced wastewater treatment[16]	Includes, but not limited to, the following: • No detectable total coliform/100 mL[9, 10] • 1 mg/L Cl_2 residual (min.)[11] • pH = 6.5–8.5 • ≤2 ntu[8] • 2 mg/L TOC of wastewater origin • Meet drinking water standards	Includes, but not limited to, the following: • pH – daily • Turbidity – continuous • Total coliform – daily • Cl_2 residual – continuous • TOC – weekly • Drinking water standards – quarterly • Other[17] – depends on constituent • Monitoring is not required for viruses and parasites; their removal rates are prescribed by treatment requirements	• Site specific based on providing 2 months retention time between introduction of reclaimed water into a raw water supply reservoir and the intake to a potable water treatment plant.	• The reclaimed water should not contain measurable levels of pathogens.[12] • Recommended level of treatment is site specific and depends on factor such as receiving water quality, time and distance to point of withdrawal, dilution, and subsequent treatment prior to distribution for potable uses. • Higher chlorine residual and/or a longer contact time may be necessary to ensure virus and protozoa inactivation. • See Section 3.4.3 in the 2004 USEPA Guidelines for recommended treatment reliability requirements. • Recommended log reductions of viruses, *Giardia*, and *Cryptosporidium* can be based on challenge tests or the sum of log-removal credits allowed for individual treatment processes. Monitoring for these pathogens is not required. • Dilution of reclaimed water with water of nonwastewater origin can be used to help meet the suggested TOC limit.

(Continued)

Table 3-7 Suggested guidelines for water reuse[1] *(Continued)*

Source: USEPA (2012).

Abbreviations: BOD, biological oxygen demand; TOC, total organic carbon; TSS, total suspended solids.

[1] Guidelines are based on water reclamation and reuse practices in the United States.

[2] Unless otherwise noted, recommended quality limits apply to the reclaimed water at the point of discharge from the treatment facility.

[3] Setback distances are recommended to protect potable water supply sources from contamination and to protect humans from unreasonable health risks due to exposure to reclaimed water.

[4] Secondary treatment processes include activated sludge processes, trickling filters, rotating biological contractors, and stabilization pond systems. Secondary treatment should produce effluent in which both the BOD and SS do not exceed 30 mg/L.

[5] Filtration means the passing of wastewater through natural undisturbed soils or filter media such as sand and/or anthracite or the passing of wastewater through microfilters or other membrane processes.

[6] Disinfection means the destruction, inactivation, or removal of pathogenic microorganisms by chemical, physical, or biological means. Disinfection may be accomplished by chlorination, ozonation, other chemical disinfectants, ultraviolet irradiation, membrane processes, or other processes.

[7] As determined from the 5-day BOD test.

[8] The recommended turbidity should be met prior to disinfection. The average turbidity should be based on a 24-hour time period. The turbidity should not exceed 5 ntu at any time. If SS is used in lieu of turbidity, the average SS should not exceed 5 mg/L. If membranes are used as the filtration process, the turbidity should not exceed 0.2 ntu and the average SS should not exceed 0.5 mg/L.

[9] Unless otherwise noted, recommended coliform limits are median values determined from the bacteriological results of the last 7 days for which analyses have been completed. Either the membrane filter or fermentation tube technique may be used.

[10] The number of total or fecal coliform organisms (whichever one is recommended for monitoring in the table) should not exceed 14/100 mL in any sample.

[11] This recommendation applies only when chlorine is used as the primary disinfectant. The total chlorine residual should be met after a minimum actual modal contact time of at least 90 minutes unless a lesser contact time has been demonstrated to provide indicator organism and pathogen reduction equivalent to those suggested in these guidelines. In no case should the actual contact time be less than 30 minutes.

[12] It is advisable to fully characterize the microbiological quality of the reclaimed water prior to implementation of a reuse program.

[13] The number of fecal coliform organisms should not exceed 800/100 mL in any sample.

[14] Some stabilization pond systems may be able to meet this coliform limit without disinfection.

[15] Commercially processed food crops are those that, prior to sale to the public or others, have undergone chemical or physical processing sufficient to destroy pathogens.

[16] Advanced wastewater treatment processes include chemical clarification, carbon adsorption, reverse osmosis and other membrane processes, advanced oxidation, air stripping, ultrafiltration, and ion exchange.

[17] Monitoring should include inorganic and organic compounds, or classes of compounds, that are known or suspected to be toxic, carcinogenic, teratogenic, or mutagenic and are not included in the drinking water standards.

[18] See USEPA Guidelines, Section 4.4.3.7, for additional precautions that can be taken when a setback distance of 100 ft (30 m) to potable water supply wells in porous media is not feasible.

State and local regulations and guidance. No specific national regulations exist for reuse of reclaimed water in the United States. Thirty states have adopted regulations, and 15 states have guidelines and/or design standards for water reuses. As with Australian state regulations, local codes from animal husbandry, plumbing, water supply, and sewerage agencies may pertain to or impact individual state reuse law. States with no regulations or guidance may permit projects on a case-by-case basis. Regulatory citations, water reuse types, and end uses are given below for the following states with long-term experience in water reuse programs: Arizona, California, Florida, Nevada, Texas, and Washington. Links to these state regulations are listed in the section Reuse Requirements, which appears later in this chapter.

Arizona regulations. Arizona Administrative Code, title 18, chapter 9, article 7, governs the uses of reclaimed water. Five classes of water are approved for reuse: A+, A, B+, B, and C. Table 3-8 shows the minimum class of reclaimed water required for a specific application. Special requirements may be mandated to protect groundwater supplies if total nitrogen exceeds 10 mg/L. Class A+ and B+ reuse water must not exceed 10 mg/L in total nitrogen under these circumstances.

Table 3-8 Arizona water reuse types and classes

Type of Reuse	Minimum Class of Reclaimed Water Required
Irrigation of food crops	A
Recreational impoundments	A
Residential landscape irrigation	A
School ground landscape irrigation	A
Open access landscape irrigation	A
Toilet and urinal flushing	A
Fire protection systems	A
Spray irrigation of an orchard or vineyard	A
Commercial closed-loop air conditioning systems	A
Vehicle and equipment washing (does not include self-service vehicle washes)	A
Snowmaking	A
Surface irrigation of an orchard or vineyard	B
Golf course irrigation	B
Restricted access landscape irrigation	B
Landscape impoundment	B
Dust control	B
Soil compaction and similar construction activities	B
Pasture for milking animals	B
Livestock watering (dairy animals)	B
Concrete and cement mixing	B
Materials washing and sieving	B
Street cleaning	B
Pasture for nondairy animals	C
Livestock watering (nondairy animals)	C
Irrigation of sod farms	C
Irrigation of fiber, seed, forage, and similar crops	C
Silviculture	C

Several groundwater recharge projects have been operating in Arizona, including in Scottsdale, Tucson, Gilbert, and Chandler. However, the state has not established any specific regulations for this category of reuse, and projects are approved on a case-by-case basis.

California Title 22 Code of Regulations. The California regulations for water reuse criteria are prescribed in California Title 22 Code of Regulations (Title 22). The water reuse criteria are established by the Division of Drinking Water and Environmental Management (DDWEM or DDW), formerly the California Department of Public Health, which also deals with water reuse and has been moved to be under the State Water Resources Control Board. The state agency that regulates water recycling projects is the California Regional Water Quality Control Board. Current regulations are compiled in *California Health Laws Related to Recycled Water* (referred to as "the Purple Book"), which was updated in June 2001 (CDPH 2001). The Purple Book consists of excerpts from the Health and Safety Code, Water Code, and Titles 22 and 17 of the California Code of Regulations.

Title 22 provides definitions and uses of four categories of recycled water quality according to treatment level and disinfection performance requirements: un-disinfected, disinfected secondary–2.2, disinfected secondary–23, and disinfected tertiary. Table 3-9 summarizes the reuse categories and end uses.

Table 3-9 California water reuse types and classes

Type of Reuse	Minimum Class of Reclaimed Water Required
Food crops including all edible root crops where the recycled water comes into contact with the edible portion of the crop	Disinfected tertiary
Parks, playgrounds, schoolyards, residential landscaping, unrestricted-access golf courses	Disinfected tertiary
Unrestricted surface impoundments (additional pathogen monitoring required)	Disinfected tertiary
Any other irrigation use not specified in this section and not prohibited by other sections of the California Code of Regulations	Disinfected tertiary
Cemeteries, freeway landscaping, restricted-access golf courses	Disinfected secondary–23
Ornamental nursery stock and sod farms where access by the general public is not restricted	Disinfected secondary–23
Pasture for animals that produce milk for human consumption	Disinfected secondary–23
Any nonedible vegetation where access is controlled	Disinfected secondary–23
Surface irrigation of food crops where the edible portion is produced above ground and not contacted by the recycled water	Disinfected secondary–2.2
Restricted surface impoundments	Disinfected secondary–2.2
Orchards and vineyards where the recycled water does not come into contact with the edible portion of the crop	Un-disinfected, oxidized
Non-food–bearing trees (Christmas tree farms are included in this category provided no irrigation with recycled water occurs for a period of 14 days prior to harvesting or allowing access by the general public)	Un-disinfected, oxidized
Fodder and fiber crops and pasture for animals that do not produce milk for human consumption	Un-disinfected, oxidized
Seed crops not eaten by humans	Un-disinfected, oxidized
Food crops that must undergo commercial pathogen-destroying processing before being consumed by humans	Un-disinfected, oxidized
Ornamental nursery stock and sod farms provided no irrigation with recycled water occurs for a period of 14 days prior to harvesting, retail sale, or allowing access by the general public	Un-disinfected, oxidized

Disinfected tertiary recycled water must include tertiary filtration and subsequent disinfection. Tertiary filtration can be accomplished through natural undisturbed soil, granular media bed or equivalent (for example, cloth media), or membrane filtration, including MF, ultrafiltration (UF), nanofiltration, and RO. For conventional filtration, the average effluent turbidity must not exceed 2.0 ntu in a 24-hour period, not exceed 5.0 ntu more than 5 percent of the time, and not exceed 10 ntu any time. For membrane filtration, the effluent turbidity must not exceed 0.2 ntu more than 5 percent of the time in 24 hours and not exceed 0.5 ntu any time. Due to cost reasons, membranes are not expected to be used for the less-stringent types of reuse, but certain users may choose to use membranes to achieve better-quality recycled water.

DDW has produced several draft versions of regulations regarding groundwater replenishment (indirect potable reuse [IPR]). The most recent version, Groundwater Replenishment Reuse Draft Regulation (GWRRDR; CDPH 2013), was released on Mar. 28, 2013. The 2013 GWRRDR is most relevant to the application of membranes for reuse. Based on a bill passed by the California Senate and approved by the governor in 2010 (California State Senate 2010), the California Water Code (SWRCB 2011) was amended in 2010 to require DDW to develop and adopt uniform water reuse criteria for IPR for groundwater recharge by Dec. 13, 2013, and to develop and adopt uniform water reuse criteria for IPR for surface water augmentation by Dec. 31, 2016, if a specified expert panel convened pursuant to the bill finds that the criteria would adequately protect public health. A brief history of the GWRRDR development and the key provisions are presented below. The final regulation was adopted by DDW on June 30, 2014. As of Dec. 31, 2016, the regulation for surface water augmentation had not been adopted.

GWRRDR's history of development is described here. The existing groundwater replenishment regulations in California, which became effective in 1978, consist of only three paragraphs and do not include detailed criteria. DDW formed the Groundwater Recharge Committee in 1986 to develop a regulation package, which was first proposed in 1988 and dealt with only groundwater recharge by spreading. The 1989 draft included groundwater injection projects. The 2001 and 2002 revisions of the draft regulations involved the type of organics treatment and total organic carbon (TOC) levels needed to deal with *N*-nitrosodimethylamine and 1,4-dioxane. Since 2002 to 2011, minor revisions have been added to the draft regulations to deal with chemicals of emerging concern such as endocrine disruptors and personal care products. According to a 1996 memorandum of agreement, DDW sets standards and requirements for wastewater reuse to protect public health, and the regional water quality control boards (RWQCBs) have the permitting and ongoing oversight authority of groundwater recharge reuse projects (GRRPs). DDW reviews recycled water proposals for compliance with Title 22 criteria and provides requirements and recommendations. DDW requirements and recommendations for permit approval are to be incorporated in RWQCB permits.

Groundwater replenishment projects, also called indirect potable reuse projects, produce a source of drinking water that is stored in an aquifer with some natural treatment and retention time to identify and respond to a treatment failure if it occurs. A GRRP is defined as a project that involves the planned use of recycled municipal wastewater and is operated for the purpose of replenishing a groundwater basin designated in the water quality control plan (as defined in Water Code, section 13050j) for use as a source of municipal and domestic water supply. The requirements for surface spreading and subsurface application (injection) projects are different in that the former is without full advanced treatment (FAT) but relies on soil aquifer treatment (SAT), while the latter relies on continuous advanced treatment of the entire flow. The FAT includes RO and an advanced oxidation process (AOP) as multiple barriers.

The key provisions of the GWRRDR are described here. There are common and different requirements for groundwater replenishment by surface application and by subsurface application in the 2013 GWRRDR. These are summarized in the following subsections.

Following are common requirements for surface and subsurface applications:

- The wastewater management agency must implement and maintain a source control program.
- The treatment system must achieve a 12-log enteric virus reduction, a 10-log *Giardia* cyst reduction, and a 10-log *Cryptosporidium* oocysts reduction using at least three treatment barriers.
- For each pathogen, a separate treatment process can only be credited up to a 6-log reduction and has to achieve at least 1-log reduction.
- There is a retention time credit for virus of 1-log per month.
- Recycled water must be retained underground for a period of time sufficient to identify treatment failure and implement actions, with a minimum time of two months.
- Total nitrogen must be <10 mg/L in recycled water or recharge water before or after application.
- All DDW drinking water maximum contaminant levels (MCLs; except nitrogen) and secondary MCLs (except color) must be met.
- Diluent water (required to provide dilution of recycled water prior to application) must also meet DDW MCLs, secondary MCLs, and notification levels (NLs).
- A maximum TOC concentration = 0.5 mg/L ÷ RWC (recycled water contribution) in recycled water must be met.
- Following are the monitoring well requirements:
  - At least one well with no less than two weeks but no more than six months of travel time in the saturated zone;
  - At least one well between the GRRP and the nearest down gradient well; and
  - At least one well at least 30 days up gradient of the nearest drinking water well.
- Quarterly monitoring of recycled water and groundwater from down-gradient monitoring wells for priority toxic pollutants, chemicals with NLs, and other chemicals specified by DDW is required.
- Annual monitoring of recycled water for indicator compounds that may represent characteristics of pharmaceuticals, endocrine-disrupting chemicals, and personal care products is required.

Following are the requirements specific for surface application:

- FAT is only needed for that portion of recycled water to meet the TOC/RWC requirements.
- Initial maximum RWC is 20 percent with an averaging period of 120 months.
- RWC incremental increase to 50 percent and then up to 75 percent maximum only with approval by DDWEM and RWQCB.
- Weekly monitoring of TOC or indicator compounds for SAT process.

Following are the requirements specific for subsurface application:

- FAT using RO and an AOP for treatment of oxidized wastewater is required.

- RO membrane system performance requirements include:
  - Sodium chloride average rejection ≥99.2 percent and
  - Permeate TOC ≤0.25 mg/L during the first month of full-scale operation.

- Perform an occurrence study to identify indicator compounds and select a total of at least nine indicator compounds, with at least one from each of the following functional groups:
  - Hydroxyl aromatic,
  - Amino/acylamino aromatic,
  - Nonaromatic with carbon double bonds,
  - Deprotonated amine,
  - Alkoxy polyaromatic,
  - Alkyl aromatic,
  - Alkoxy aromatic,
  - Saturated aliphatic, and
  - Nitro aromatic.

- The treatment removal efficiency of the AOP shall be ≥0.5 log (69 percent) for each indicator compound in categories 1–7 above and ≥0.3 log (50 percent) for each indicator compound in categories 8 and 9 above; or ≥0.5 log for 1,4-dioxane.

The future of direct potable reuse in California. As demand increases due to dwindling freshwater supplies, population pressure, and climate fluctuations, interest in DPR continues to grow among water utilities, the environmental community, and other interested parties in California. Direct potable reuse could be a feasible option to supplement conventional supplies, IPR, seawater desalination, and conservation. The National Water Research Institute published a white paper in April 2010 (Crook 2010) in which they identified the following 10 factors to be addressed for regulatory approval of DPR in the state:

- A clear definition of direct potable reuse (DPR);
- Safeguards for lack of an environmental buffer that provide detention time;
- Define the treatment processes for multiple barriers;
- Determine if dilution with potable water will be required as an additional safeguard;
- Define water quality monitoring requirements to demonstrate reduction in microbial pathogens and chemicals that impact human health;
- Describe the health risk assessment process and mitigation measures;
- Determine the role and requirements of DPR project review by expert advisory panel(s);
- Evaluate current regulatory framework's applicability to DPR projects;
- Clarify regulatory agency roles in overseeing DPR projects; and
- Develop rapid emergency communications system between water agencies and regulators to prevent serving undertreated/noncompliant water due to treatment process upset or failure.

The California Water Code required DDW to investigate the feasibility of developing uniform water reuse criteria for DPR and provide a final report to the legislature by Dec. 31, 2016 (State Water Resources Control Board 2011). DDW created an expert panel and advisory group in 2014 to assist its staff in investigating the feasibility of creating regulations for DPR. Based on the recommendation from the expert panel and advisory group, DDW issued a draft report on Sept. 8, 2016, concluding that it is feasible to develop and adopt regulations for using recycled water as drinking water, provided that certain research and key knowledge gaps are addressed. Following a 45-day public comment period with minor revisions, DDW finalized and submitted the report to the California legislature on Dec. 29, 2016. The final report lays the groundwork for creating regulations for a new and reliable source of potable water. However, the adoption of regulations related to DPR will not take place until the knowledge gaps are addressed and additional research is conducted related to specific public health issues (California State Water Resources Control Boards Media Release, Dec. 29, 2016). DDW does not have a schedule for adopting the DPR regulation yet.

Florida Regulations. The Florida Administrative Code (FAC), chapter 62-610, governs water reuse. Approved reuse applications are listed in Table 3-10.

For groundwater recharge and IPR applications, pilot testing is required for all projects that are required to provide full treatment and disinfection. The pilot testing program should accumulate 12 months of data to demonstrate that the reclaimed water will be of sufficient quality to protect public health and environmental quality. Local requirements may be more stringent than the state regulations, especially those that relate to nutrient concentrations.

The following additional approved uses are contingent on an engineering report and meeting all requirements of FAC chapter 62-610, part III:

- Water supply for commercial laundries;
- Vehicle washing, concrete mixing;
- Flushing sanitary sewers and reclaimed water lines;
- Manufacture of ice for ice rinks; and
- Cleaning roads, sidewalks, and outdoor work areas.

Table 3-10 Florida water reuse types described in Florida Administrative Code 62-610

Type of Reuse	Part in Chapter 62-610
Irrigation of pastures, trees, feed, fodder, fiber, or seed crops	II
Residential, golf course, and other landscape irrigation; toilet flushing; fire protection; dust control; aesthetic features (ponds and fountains); irrigation of edible crops (direct contact only with crops that will be peeled, skinned, cooked, or thermally processed)	III
Rapid infiltration basins; absorption fields	IV
Groundwater recharge and indirect potable reuse; salinity barrier; augmentation of surface waters	V
Overland flow systems	VI
Industrial uses: cooling water, wash water, process water (not to include food processing for human consumption	VII

Source: Florida Administrative Register and Florida Administrative Code, Chapter 62-210.

Nevada Regulations. Nevada Administrative Code (2012), chapter 445A, sections 274-280, and guidance documents WTS-1A, WTS-1B, WTS-3, and WTS-7 regulate water reuse in the state. Table 3-11 summarizes the five recycled water categories (A, B, C, D, and E) and approved end uses.

Texas Regulations. Texas Administrative Code, title 30, part 1, chapter 210, regulates water reuse in the state. Table 3-12 summarizes the two recycled water types and approved end uses.

El Paso, Texas, has practiced IPR by groundwater recharge for many years. Several DPR projects are either in operation (Big Spring and Wichita Falls) or in the planning process due to long-term drought, including those in Brownwood, Abilene, and Lubbock (White 2013). The Colorado River Municipal Water District in Big Spring obtained a permit for operation of an advanced treatment facility modeled after the California FAT-MF–RO–AOP process and typically produces 2.0 mgd of product water, meeting state and USEPA drinking water standards (WRRF 2011). Discharge is to a raw water main that crosses the project site where the product water is blended with lower-quality lake water and then treated at five conventional water treatment plants. Wichita Falls worked for four years to gain state approval for an advanced treatment facility for reclaimed water. The utility now provides 5 mgd to supplement the 12-mgd demand following significant community conservation in June 2014. Wichita Falls used an existing MF–RO treatment train to provide advanced treatment with discharge to a mixing basin, providing raw water to four existing conventional water treatment plant trains. Ultraviolet (UV) disinfection was added as an additional treatment barrier in February 2015. Heavy rains and flooding in May 2015 terminated the immediate need for DPR, and the city is converting to IPR for long-term lake augmentation (Nix 2015). Texas has not yet established regulations for IPR or DPR. These projects are approved on a case-by-case basis.

Table 3-11 Nevada water reuse types and classes

Type of Reuse	Minimum Class of Reclaimed Water Required
Spray irrigation of cemeteries, commercial lawns, golf courses, greenbelts or parks	A
Impoundments (with unrestricted access)	A
Spray irrigation of cemeteries, commercial lawns, golf courses, greenbelts or parks (with restricted access)	B
Subsurface irrigation of commercial lawns, greenbelts or parks	B
Cooling water in an industrial process	B
Firefighting operations in an urban area if approved by the local firefighting agency	B
Spray irrigation of cemeteries, golf courses, or greenbelts (with restricted access)	C
Watering of nursery stock (with restricted access)	C
Establishment, restoration, or maintenance of a wetland (with restricted access)	C
Washing of gravel used in concrete mixing; boiler feedwater	C
Impoundments (with restricted access)	C
Firefighting of wildland fires if approved by the local firefighting agency	C
Spray irrigation of agricultural land if public access is prohibited and a buffer zone of 400 ft minimum is maintained	D
Surface irrigation of greenbelts with prohibited public access, agriculture, cultivation of fruit-bearing and nut-bearing trees	D
Subsurface irrigation of land used for agricultural purposes if public access is controlled	D
Dust control, soil compaction, flushing sewer lines	D
Impoundments with prohibited public access and no body contact	D
Spray irrigation of agricultural land if public access is prohibited and a buffer zone of 800 ft minimum is maintained	E

Table 3-12 Texas water reuse types and classes

Type of Reuse	Minimum Class of Reclaimed Water Required
Residential irrigation including landscape irrigation at individual homes	I
Urban uses including irrigation of public parks, golf courses with unrestricted public access, school yards, or athletic fields	I
Fire protection, either in internal sprinkler systems or external fire hydrants	I
Irrigation of food crops where reclaimed water may have direct contact with the edible part of the crop, unless the food crop undergoes a pasteurization process	I
Irrigation of pastures for milking animals	I
Maintenance of impoundments or natural water bodies where recreational activities such as wading or fishing are anticipated even though the water body was not specifically designed for such a use	I
Toilet or urinal flush water	I
Other similar activities where the potential for unintentional human exposure may occur	I
Irrigation of sod farms, silviculture, limited access highway rights-of-way, and other restricted-access areas	II
Irrigation of food crops where the reclaimed water is not likely to have direct contact with the edible part of the crop or where the food crop undergoes pasteurization prior to distribution for consumption	II
Irrigation of animal feed crops other than pasture for milking animals	II
Maintenance of impoundments or natural water bodies where direct human contact is not likely	II
Soil compaction or dust control in construction areas where application procedures minimize aerosol drift to public areas	II
Cooling tower makeup water; use for cooling towers that produce significant aerosols adjacent to public access areas may have special requirements	II
Irrigation or other nonpotable uses of reclaimed water at a wastewater treatment facility	II

NOTE: Type I is for unrestricted public access, and type II is for restricted public access. Type I reclaimed water can also be used for any type II use.

Washington Regulations. Washington's water reuse program currently controls uses under chapter 90.46 of the Revised Code of Washington (2015). Table 3-13 summarizes the four recycled water types (A, B, C, and D) and approved end uses.

REUSE REQUIREMENTS

Regulatory Links

Table 3-14 shows links to water reuse regulations, guidance, and fact sheets in Arizona, California, Florida, Nevada, Texas, and Washington.

Table 3-13 Washington water reuse types and classes

Type of Reuse	Minimum Class of Reclaimed Water Required
Direct recharge into nonpotable supply or nonpotable groundwater aquifers	A
Spray or surface irrigation of food crops including root crops	A
Surface irrigation of food crops where there is no reclaimed water contact with the edible portion of the crop	B
Irrigation of nonfood crops (sod, ornamental plants for commercial use, and pasture access to dairy stock)	C
Orchards and vineyards, spray irrigation of food crops that undergo a physical/chemical process to destroy pathogens	D
Irrigation of nonfood crops (fiber, fodder, trees, seeds)	D
Nonrestricted areas (residential, golf courses, parks, playgrounds, recreational impoundments); decorative fountains	A
Restricted recreational impoundments, fish hatcheries	B
Landscape impoundments	C
Spray street washing	A
Fire protection (hydrants or sprinkler systems in buildings); toilet/urinal flushing	A
Industrial cooling that creates aerosols or mists	A
Industrial processes with worker exposure	A
Wash down of lots, sidewalks	B
Dust control, soil dampening, street sweeping, concrete mixing; boiler feed; industrial cooling with no mists created	C
Sanitary sewer flushing	D

Table 3-14 Links to water reuse regulations, guidance, and fact sheets

State	Regulation	Link to Regulation or Guidance	Alternate Link to Fact Sheet or Report
Arizona	Administrative code, title 18, environmental quality	http://www.azsos.gov/public_services/Title_18/18_table.htm	http://www.azdeq.gov/environ/water/permits/reclaimed.html
California	Title 22, code of regulations	http://www.cdph.ca.gov/certlic/drinkingwater/Pages/Lawbook.aspx http://www.cdph.ca.gov/HealthInfo/environhealth/water/Pages/Waterrecycling.aspx	http://www.waterboards.ca.gov/water_issues/programs/grants_loans/water_recycling/directory.shtml
Florida	Chapter 62-610, administrative code, reuse of reclaimed water and land application; section 403.064, Florida statutes	http://www.dep.state.fl.us/water/reuse/apprules.htm	
Nevada	Administrative code, chapter 445A, sections 274–280, WTS-1A, WTS-1B, WTS-3, WTS-7	http://www.leg.state.nv.us/nac/nac445a.html#NAC445ASec275	http://ndep.nv.gov/bwpc/fact01.htm
Texas		http://www.tceq.texas.gov/rules/indxpdf.html#210	
Washington	Chapter 90.46, revised code, reclaimed water use	http://apps.leg.wa.gov/rcw/default.aspx?cite=90.46&full=true	http://www.ecy.wa.gov/programs/wq/reclaim/index.html

Source: USEPA (2012).

GUIDELINES FOR OTHER COUNTRIES

Europe

Water recycling and reuse regulations are at the national level in European Union (EU) member countries and follow three EU policy documents:

- Urban Waste Water Treatment Directive (91/271/EEC, 5/19/1991) regarding municipal and industrial discharges and reuse where appropriate;
- Drinking Water Directive (98/83/EC, 11/3/1998) regarding drinking water quality; and
- Water Framework Directive (2000/60/EC, 10/23/2000) regarding water resources management.

Microbial water quality criteria for two representative EU members, France and Great Britain, are described below.

France. French water recycling criteria for nonpotable reuse cover the following four categories based on level of treatment (secondary treatment at a minimum) and water quality:

- Category A: *Escherichia coli* <1,000 cfu/100 mL, no Salmonella, no *Taenia* egg; for crops to be eaten raw and for irrigation for grazing, parks, sports areas, and golf courses;
- Category B: *E. coli* <1,000 cfu/100 mL; for crops to be eaten cooked or processed and for spray irrigation for flowers, nurseries, cereals, and fodder crops;
- Category C: *E. coli* <104 cfu/100 mL; for crops to be eaten cooked or processed and for nonspray irrigation for flowers, nurseries, cereals, and fodder crops; and
- Category D: No microbiological targets; for nonspray irrigation of restricted forests.

Great Britain. Great Britain established the following two standards for design, installation, maintenance, and compliance of domestic systems for nonpotable reuse: BS8515:2009 for rainwater harvesting systems and BS8525-1:2010 for graywater systems that use bath, sink, and laundry water. Both standards encompass standard risk management processes to protect human health.

Regular water quality testing and compliance criteria are not stipulated in standard BS8515:2009, as the end uses are of low direct human contact, such as for toilet flushing and laundry. Human health risks related to rainwater capture systems are minimized through professional design, construction, and installation and compliance with the standard. Minimum treatment, in the form of filtration prior to the collection tank, is mandated. UV disinfection may be required for applications with elevated human exposure.

In addition to requirements for professional design, construction, and installation of individual and communal domestic graywater systems, standard BS8525-1:2010 prescribes microbial compliance criteria. Table 3-15 summarizes the criteria.

Table 3-15 British microbial guidelines for graywater systems

Parameter	Spray Application Target	Nonspray Application Target			Analytical Test Method		System Type
	Pressure Washing, Car Washing, Spray Irrigation*	Toilet Flushing	Landscape Irrigation*	Laundry	Spray Application	Nonspray Application	
Escherichia coli (per 100 mL)	Not detected	250	250	Not detected	BS EN ISO 9308-1	BS EN ISO 9308-3	Single site and communal domestic systems
Intestinal enterococci	Not detected	100	100	Not detected	BS EN ISO 7899-1 or BS EN ISO 7899-2	BS EN ISO 7899-1	Single site and communal domestic systems
Legionella pneumophila (per 100 mL)	10	NA	NA	NA	BS 6068-4.12	NA	Where analysis is necessary as indicated by risk assessment
Total coliforms (per 100 mL)†	10	1,000	1,000	10	Blue Book 223, Method D (N2)	BS EN ISO 9308-3	Single site and communal domestic systems

Source: British Standards Institution (2010); graywater systems, part 1, code of practice.

* If graywater is used for food crops, notification regarding preparation of crops (for example, boiling, peeling, washing with potable water) is required for the user.

† Total coliforms are an indicator parameter for operational interpretation. The bacteriological guideline values for treated graywater reflect the need to control the quality of treated water for supply and use.

Singapore

Water recycling in Singapore arose out of insecure future prospects regarding long-term water purchases from neighboring Malaysia. NEWater is Singapore's Public Utilities Board branding for advanced-treated, reclaimed domestic wastewater used primarily for nonpotable industrial and commercial uses and also for IPR through raw water reservoir augmentation. Four wastewater plants treat influent through conventional treatment, UF or MF, RO, and UV for nonpotable and IPR end uses.

Monitoring and compliance targets for Singapore IPR projects generally follow WHO drinking water guidelines for disinfection residual, microbial, chemical, and radiological constituents. A summary of a few selected water quality standards is provided in Table 3-16.

Namibia

The first DPR (no retreating of advanced-treated recycled water in a water treatment plant) system in the world was developed in Windhoek, Namibia, at the Goreangab treatment facility. Advanced-treated municipal wastewater is blended with treated potable water to supplement scant and unreliable local supplies. The Goreangab facility began operations in 1968, making it the world's oldest DPR in the world.

Through four decades of operations, treatment processes were modified or added in response to evolving water quality issues and new treatment technologies. Expansion in 2002 included the following process modifications preozonation, ozonation, granular activated carbon, biological activated carbon, and UF. Addition of UF membranes in the 2002 expansion was the first use of membranes in this DPR treatment train (WRRF 2013). Up to 35 percent of treated wastewater is allowed to supplement surface and groundwater supplies. In response to a projected 4 percent per year population increase through 2040 and water quality issues that are straining the Goreangab facility, design of a parallel reuse treatment plant will begin in 2013 to increase much-needed capacity.

Extensive water quality monitoring accounts for approximately 20 percent of the Windhoek plant treatment costs. In addition to water quality testing, mortality rates and patterns are monitored in the service area. Internal water quality goals for selected parameters at the Goreangab facility effluent and drinking water quality standards based on WHO and USEPA guidelines are listed in Table 3-17.

Table 3-16 Singapore water quality standards, 2012

Parameter	World Health Organization and NEWater Limit	Unit
Escherichia coli	<1	cfu/100 mL
Turbidity	5	ntu
Chlorine	5	mg/L
Nitrate (as N)	11	mg/L
Nitrite (as N)	0.9	mg/L

Table 3-17 Windhoek direct potable reuse water quality criteria

Parameter	Unit	Internal Water Quality Limit	Regulatory Standard*
CCPP	mg/L as $CaCO_3$	NA	0–8
Chemical oxygen demand	mg/L	10	15
Color	mg/L platinum	10	15
Dissolved organic carbon	mg/L	3	5
Total dissolved solids	mg/L	1,000 (or 200 > raw water value)	1,200 (or 250 > raw water value)
Turbidity	ntu	0.1	0.2
$Ultraviolet_{254}$	Abs/cm	NA	0.06
Aluminum	mg/L	NA	0.15
Ammonia	mg/L	NA	0.1
Iron	mg/L	0.05	0.1
Manganese	mg/L	0.01	0.025
Heterotrophic plate count	per 1 mL	80	100
Total coliform	per 100 mL	NA	0
Fecal coliform	per 100 mL	NA	0
Escherichia coli	per 100 mL	NA	0
Coliphage	per 100 mL	NA	0
Enteric viruses	per 10 L	NA	0 (or 4-log removal)
Fecal *Streptococci*	per 100 mL	NA	0
Clostridium spores	per 100 mL	NA	0
Clostridium viable cells	per 100 mL	NA	0
Total trihalomethanes	µg/L	20	40
Chlorophyll-a	µg/L	NA	1
Giardia	per 10 L	0 (or 6-log removal)	0 (or 5-log removal)
Cryptosporidium	per 10 L	0 (or 6-log removal)	0 (or 5-log removal)

Source: Japan–United States (2009).

* Regulatory standards are from the World Health Organization, US Environmental Protection Agency, and Namibia.

REFERENCES

Anderson, P.; Denslow, N.; Drewes, J.; Olivieri, A.; Olivieri, D.; & Snyder, S. 2010. *Monitoring Strategies for Chemicals of Emerging Concern (CECs) in Recycled Water.* SWRCB, 6/25/2010. http://www.waterboards.ca.gov/water_issues/programs/water_recycling_policy/docs/cec_monitoring_rpt.pdf.

CDPH (California Department of Public Health). 2001. *California Health Laws Related to Recycled Water* ("The Purple Book"), June 2001 update. http://www.cdph.ca.gov/certlic/drinkingwater/Documents/Recharge/Purplebookupdate6-01.PDF

———. 2013. Groundwater Replenishment Reuse Draft Regulation. Title 22, California Code of Regulations, Division 4, Chapter 3, http://www.waterboards.ca.gov/drinking_water/certlic/drinkingwater/RecycledWater.shtml.

California State Senate. 2010. Senate Bill 918, amended June 1, 2010. Sacramento, California. (http://www.leginfo.ca.gov/pub/09-10/bill/sen/sb_0901-0950/sb_918_bill_20100601_amended_sen_v95.pdf).

Crook, J. 2010. *Regulatory Aspects of Direct Potable Reuse in California.* Fountain Valley, CA: National Water Research Institute.

EPA Victoria. 2003. *Guidelines for Environmental Management-Use of Reclaimed Water,* June. Retrieved from Horticultural Australia Limited, "Recycled Water in Australia." (http://www.recycledwater.com.au).

Florida, State of. Florida Administrative Register and Florida Administrative Code, Chapter 62-210, Reuse of Reclaimed Water and Land Application.https://www.flrules.org/gateway/ChapterHome.asp?Chapter=62-610.

Fulton, M. & Graf, C. 2011. Wastewater Treatment and Reclaimed Water Reuse in Arizona: Past and Present. http://www.flagstaff.az.gov/DocumentCenter/Home/View/15091.

Japan–United States. 2009. Fifth Japan–US Government Conference on Drinking Water Quality Management and Wastewater Control. March 4. http://www.niph.go.jp/soshiki/suido/pdf/h21JPUS/abstract/r7-1.pdf.

Nevada, State of. 2012. Administrative Code, March. Chapter 445A–Water Controls (Sec. 445A.274 through 445A.2771).http://www.leg.state.nv.us/nac/NAC-445A.html.

NHRMC (National Health and Medical Research Council) 2004. Australian Drinking Water Guidelines, 2004. http://www.nhmrc.gov.au/guidelines/search?keywords=drinking&subject=1101&status=rescinded.

Nix, D. 2015. Keep it Coming–Extending DPR in Wichita Falls. WateReuse Conference Proceedings, Texas.

NRMMC/EPHC/AHMC (Natural Resource Management Ministerial Council, Environment Protection and Heritage Council, and National Health and Medical Research Council). 2006. National Guidelines for Water Recycling: Managing Health and Environmental Risks (Phase 1). Canberra, Australia: NRMMC/EPHC/AHMC.

———. 2008. Australian Guidelines for Water Recycling: Managing Health and Environmental Risks (Phase 2). Augmentation of Drinking Water Supplies. Canberra, Australia: NRMMC/EPHC/AHMC.

———NRMMC/EPHC/NHMRC. 2009. Australian Guidelines for Water Recycling: Managing Health and Environmental Risks (Phase 2). Managed Aquifer Recharge. Canberra, Australia: NRMMC/EPHC/AHMC.

———NRMMC/EPHC/NHMRC. 2009. Australian Guidelines for Water Recycling: Managing Health and Environmental Risks (Phase 2). Stormwater Harvesting and Reuse. Canberra, Australia: NRMMC/EPHC/AHMC.

SWRCB (State Water Resources Control Board). 2011. California Water Code. Sacramento, CA: California SWRBC (http://www.leginfo.ca.gov/cgi-bin/calawquery?codesection=wat&codebody=&hits=20).

USEPA (US Environmental Protection Agency). 2004. *2004 EPA Guidelines for Water Reuse.* EPA/625/R-04/108. Cincinnati, OH: USEPA.

———. 2012. *2012 EPA Guidelines for Water Reuse.* EPA/600/R-12/618. Cincinnati, OH: USEPA.

Washington, State of. Revised Code, Chapter 90.46, Reclaimed Water Use. http://apps.leg.wa.gov/RCW/default.aspx?Cite=90.46

———. Department of Ecology. 1997. Water Reclamation and Reuse Standards. http://www.ecy.wa.gov/programs/wq/reclaim/advisorycommittee/standards.pdf.

———. 2005. Frequently Asked Questions About Reclaimed Water Use. https://fortress.wa.gov/ecy/publications/publications/0510012.pdf.

White, A. 2013. Water-Reuse Projects Move Forward, Despite Concerns. *Texas Tribune,* February 8, 2013.

WHO (World Health Organization). 2006a. *Guidelines for the Safe Use of Wastewater, Excreta and Greywater, vol. 1: Policy and Regulatory Aspects.* Geneva, Switzerland: WHO Press.

———. 2006b. WHO. 2006a *Guidelines for the Safe Use of Wastewater, Excreta and Greywater, vol. 2: Wastewater Use in Agriculture.* Geneva, Switzerland: WHO Press.

———. 2006c. *Guidelines for the Safe Use of Wastewater, Excreta and Greywater, vol. 3: Wastewater and Excreta Use in Aquaculture.* Geneva, Switzerland: WHO Press.

———. 2006d. *Guidelines for the Safe Use of Wastewater, Excreta and Greywater, vol. 4: Excreta and Greywater Use in Aquaculture.* Geneva, Switzerland: WHO Press.

WRRF (WateReuse Research Foundation). 2011. *Direct Potable Reuse: A Path Forward,* Alexandria, VA: WRRF.

———. 2013. *Potable Reuse: State of the Science Report and Equivalency Criteria for Treatment Trains.* Alexandria, VA: WRRF.

This page intentionally blank.

Chapter 4

Source and Treated Water Quality

The quality of municipal wastewater effluent varies and often depends on regional regulations, wastewater discharge category, and degree of treatment. Industrial effluent intended for reclamation is typically reused in the same or similar process and requires treatment to restore the water quality specific to the unit process. Membrane treatment is often selected to remove constituents that are added to the process water during its initial use. This chapter provides an overview of source water quality typically considered for reuse, with a focus on municipal and industrial reclamation using membranes for treatment.

WASTEWATER TREATMENT

In the United States, some wastewater must be treated to secondary levels, at a minimum. Wastewater treatment levels, which are described in state discharge permits, are based on receiving water quality, category of water reuse, and corresponding federal, state, and regional requirements. For applications that involve discharge to surface waters, states may choose to regulate them through the National Pollutant Discharge Elimination System's wastewater discharge permit program. The discharge streams must comply, at a minimum, with state surface water quality standards and any total maximum daily load requirements that would apply to the receiving water. Discharged water quality may also be regulated locally by terms of the Endangered Species Act or by specific water rights laws (USEPA 2012). For example, in California, the California Water Code (SWRCB 2011) was amended in 2010 to require the Division of Water and Environmental Management (DDWEM), formerly the California Department of Public Health, to develop and adopt uniform water reuse criteria for indirect potable reuse for groundwater recharge. The amended code also required DDWEM to develop and adopt uniform water reuse criteria for indirect potable reuse for surface water augmentation, if a specified expert panel convened pursuant to the bill found that the criteria would adequately protect public health. The California Water Code also requires DDWEM to investigate the feasibility of developing uniform water reuse criteria for direct potable reuse and provide a final report to the legislature.

WASTEWATER EFFLUENT QUALITY

Regardless of regional variations, wastewater treatment technologies, in general, involve similar physical, chemical, and biological principles in their design and operation. The 2012 US Environmental Protection Agency (USEPA) Water Reuse Guidelines included the concept of "fit-for-purpose" with application of the correct type and degree of treatment to meet the quality required for the intended use, as depicted in Figure 4-1.

For membrane applications, source water quality drives selection of the type and size of the pretreatment and membrane components for reuse purposes. Treatment processes for membrane applications are described in Chapter 5. Based on Figure 4-1, tertiary treatment could be selected solely to provide a higher-quality water for a specific use, but pretreatment could also be selected in preparation of membrane treatment in an advanced treatment process. Typical constituent ranges[H1] of water quality for untreated municipal wastewater, secondary treatment including conventional activated sludge, activated sludge with biological nutrient removal, and after membrane bioreactor treatment are listed in Table 4-1.

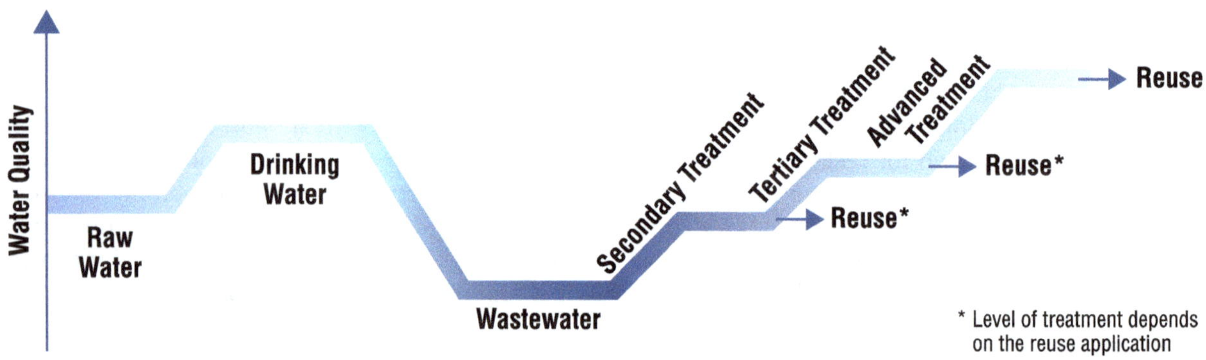

Source: USEPA (2012).

Figure 4-1 Treatment technologies to achieve desired levels of reclaimed water quality

Table 4-1 Typical ranges of municipal wastewater and secondary treatment effluent quality

Effluent			Range of Effluent Quality After Indicated Treatment		
Constituent	Unit	Untreated Wastewater	Conventional Activated Sludge	Activated Sludge With Biological Nutrient Removal	Membrane Bioreactor
Total suspended solids	mg/L	120–400	5–25	5–20	≤1
Biological oxygen demand	mg/L	110–350	5–25	5–15	1–5
Chemical oxygen demand	mg/L	250–800	40–80	20–40	<10–30
Total organic carbon	mg/L	80–260	10–40	8–20	0.5–5
Ammonia nitrogen	mg N/L	12–45	1–10	1–3.0	<1
Nitrate nitrogen	mg N/L	0–trace	10–30	2–8	<10
Nitrite nitrogen	mg N/L	0–trace	0–trace	0–trace	0–trace
Total nitrogen	mg N/L	20–70	15–35	3–8	<10
Total phosphorus	mg P/L	4–12	4–10	1–2.0	0.5–2.0
Turbidity	ntu		2–15	2–8	≤1
Volatile organic chemicals	μg/L	<100–>400	10–40	10–20	10–20
Metals	mg/L	1.5–2.5	1–1.5	1–1.5	trace
Surfactants	mg/L	4–10	0.5–2	0.1–1	0.1–0.5
Total dissolved solids	mg/L	270–860 (typical–500)	500–700	500–700	500–700
Trace constituents	μg/L	10–50	5–40	5–30	0.5–20
Total coliform	no./100 mL	10^6–10^9	10^4–10^5	10^4–10^5	<100
Protozoan cysts and oocysts	no./100 mL	10^1–10^4	10^1–10^2	0–10	0–1
Viruses	pfu/100 mL	10^1–10^4	10^1–10^3	10^1–10^3	10^0–<10^3

Source: Metcalf & Eddy/AECOM (2007; Table 3-14).

SPECIFIC WASTEWATER QUALITY PARAMETERS

Membrane performance is driven by the feedwater quality, operational conditions, and the complex interaction between feedwater and the membrane system. In addition to typical ranges of water quality parameters (for example, biological oxygen demand, chemical oxygen demand, total suspended solids [TSS]) in wastewater treatment plant effluent, attention must be paid to specific water quality parameters that could impact membrane performance over time. Low-pressure membrane processes such as microfiltration (MF) and ultrafiltration (UF) are typically used in pretreatment to remove TSS and reduce turbidity. Due to individual membrane pore size, these processes are used to remove some protozoan cysts and oocysts as well as some bacteria and viruses. High-pressure membrane processes such as nanofiltration (NF) and reverse osmosis (RO) are used to remove hardness ions, small to very small ions, viruses, and color. Water quality constituents that are important to the membrane application and are primarily removed by high-pressure membranes include the following:

- Major and minor cations and anions, metal hydroxides, sparingly soluble salts, silicates, and colloidal/particular material related to membrane scaling potential;
- Effluent organic matter from municipal wastewater treatment plants that includes natural organic matter, soluble microbial products, trace organic compounds and their compositions, relating to membrane organic fouling potential; and
- General physical and chemical parameters such as pH, total dissolved solids [TDS], ionic strength, and temperature.

A brief discussion of these classes of constituents and the corresponding potential influence on membrane applications is provided in Table 4-2.

Table 4-2 Water quality parameters and their impact on membranes for water reuse applications

Water Quality Parameter	Constituent Influence on Membrane Application	Impact on MF–UF	Impact on NF–RO
pH	pH is a means of expressing the hydrogen ion concentration. Membrane charge and hydrophobicity change as a function of pH. pH has a significant impact on solubility of sparingly soluble salts and/or dissolved gases. Low pH can foster shield charge on the membrane surface and neutralization of the charged macromolecular functional groups of NOM. The configuration of NOM is also affected by pH.	Medium	High Most NF and RO membranes are negatively charged at neutral pH due to the dissociation of acidic groups on the membrane surface. Membrane surface charge is also dependent on the presence of divalent ions and NOM that can shield or reduce the effective surface charge.
Ionic strength	Ionic strength is a measure of ionic interactions that cause deviation from ideal behavior. At high ionic strength, the foulant layer on the membrane surface could be more densely compacted. Ionic strength may also have direct impact on solubility of sparingly soluble salts and osmosis pressure across the membrane.	Low	High Often influenced by high Ca^{++} concentrations, which lead to a shielding of the surface membrane charge, increased rejection values for positively charged solutes, and decreased rejection values for negatively charged solutes.
Particulate and colloidal matter	TSS or turbidity limits are suggested as a measure of organic and inorganic particulate matter in reclaimed water that has received secondary treatment (USEPA 2012) to prevent clogging of membranes. In wastewater treatment, there is general agreement that particulate matter should be reduced to low levels, for example, 2 ntu or 5 mg/L TSS prior to disinfection to ensure reliable destruction of pathogenic microorganisms during the disinfection process. NF or RO should have influent turbidity levels less than 0.5 ntu.	Medium MF and UF are often used as pretreatment for NF and RO membranes to reduce turbidity to less than 0.5 ntu.	High Suspended matter in source water can quickly blind these membranes with a surface layer, preventing proper membrane function. Pretreatment to less than 0.5 ntu is typical. See Chapter 5.

(Continued)

Table 4-2 Water quality parameters and their impact on membranes for water reuse applications *(Continued)*

Water Quality Parameter	Constituent Influence on Membrane Application	Impact on MF–UF	Impact on NF–RO
Divalent cations	The presence of divalent cations such as calcium and magnesium has an influential role on membrane fouling. The presence of calcium ions can contribute to fouling in a number of ways: precipitation of calcium carbonate, precipitation of calcium–organic complexes, coprecipitation of organics and calcite, adsorption of organics onto calcite, and a coupling effect that may increase the colloidal fraction. The charge neutralization effect of the presence of the divalent cations also decreases the solubility of compounds such as humic acid.	Medium	High
EfOM	EfOM represents a large group of structurally complex, heterogeneous, and poorly defined organic compounds derived from raw wastewater and microbial metabolic activities in biological treatment systems. Researchers have attempted to determine whether membrane fouling could be attributed to any specific EfOM characteristic (Lozier et al. 2008; Zheng & Croue 2012). Heterotrophic plate count, adenosine triphosphate, total direct cell counts, and assimilable organic carbon can serve as indicators for potential biological fouling.	Medium Recent work indicates that UF membranes are most impacted by fouling from colloids at comparable dissolved organic carbon delivered concentrations.	High
Chlorine	Chlorination is the most common method of wastewater disinfection and is used worldwide for the removal of pathogens before discharge into receiving streams, rivers, and oceans. Some membranes are subject to membrane degradation induced by chlorination. In general, chlorine tolerances for low-pressure membranes (MF–UF) vary from 0 to 2 ppm during continuous feed and from 1,000 to 2,000 ppm while cleaning. Polypropylene and polyamide NF/RO membranes are not chlorine tolerant.	Medium Chloramination can be used if MF–UF membranes are not tolerant to free chlorine or in direct coupled to NF–RO trains.	High Chloramination can be provided for most NF–RO membranes to reduce or control biological fouling.

Abbreviations: EfOM, effluent organic matter; MF, microfiltration; NF, nanofiltration; NOM, natural organic matter; RO, reverse osmosis; TSS, total suspended solids; UF, ultrafiltration.

Table 4-3 lists typical feedwater quality for low-pressure membranes such as MF and UF membranes and for high-pressure membranes such as NF and RO membranes. Table 4-3 shows the representative range of water quality anticipated from secondary effluent from municipal wastewater facilities that can serve as a feed to MF–UF treatment and also shows feedwater quality when an RO system follows a membrane bioreactor or a conventional activated sludge system with MF–UF pretreatment.

Table 4-3 Typical membrane product water quality for low- and high-pressure membranes

		Low-Pressure Membranes	High-Pressure Membranes (Reverse Osmosis)	
Parameter	Unit	MF–UF	Conventional Activated Sludge Followed by MF–UF as Pretreatment	Membrane Bioreactor as Pretreatment
Total suspended solids	mg/L	5–30	Below Detection Limits	<1
Biological oxygen demand	mg/L	5–30	<5	<5
Turbidity	ntu	2–10	<0.1	<0.2
Total organic carbon	mg/L	5–30	5–25	5–25
pH	su	6–9	6–9	6–9
Temperature	°C	8–30	8–30	8–30
Total dissolved solids*	mg/L	500–1500	500–1500	500–1500
Total Kjeldahl nitrogen	mg N/L	10–20	5–30	<3
Bicarbonate	mg/L	150–350	150–350	150–350
Iron	mg/L	0–1	0–0.2	0–0.2
Total phosphorus	mg/L as P	1–8	0.1–8	<1[c]
Fats, oils, and grease	mg/L	0.1–10	0.1	0.1
Silt density index		Not applicable	<3	<3
Total residual chlorine	mg/L	0–2	2†	2†
Total coliforms‡	No./100 mL	50,000–2,000,000	Up to 6-log removal	Up to 6-log removal
Fecal coliforms‡	No./100 mL	10,000–1,600,000	Up to 6-log removal	Up to 6-log removal
Protozoan cysts and ocysts	No./100 mL	5–10	Up to 6-log removal	Up to 6-log removal
Viruses‡	pfu/100 mL	0.05–1000	Up to 3-log removal	Up to 3-log removal

Source: Water Environment Federation (2006).
* mg/L = ppm.
† Free chlorine is dechlorinated prior to RO; added chloramine concentration.
‡ Prior to disinfection.
Abbreviations: MF, microfiltration; UF, ultrafiltration.

INDUSTRIAL REUSE

Industrial reuse has increased substantially since the early 1990s for many of the same reasons municipal reuse has gained popularity. Reasons include water shortages, increased populations in drought areas, and legislation regarding water conservation and environmental compliance. To meet this increased demand, many utilities have increased the availability of reclaimed water to municipal and industrial users and have installed the necessary reclaimed water distribution lines. In the United States, many states such as California, Arizona, Texas, Florida, and Nevada have industrial facilities that use reclaimed water for cooling water and process/boiler-feed requirements. Utility power plants are ideal facilities for reuse due to their large water requirements for cooling, ash sluicing, and flue gas scrubber requirements. Petroleum refineries, chemical plants, and metal working facilities are other industrial facilities that benefit from reclaimed water not only for cooling but for process needs as well.

Cooling Water

Cooling water is the largest industrial use of reclaimed water because advancements in water treatment technologies have allowed industries to successfully use waters of lesser quality. Many municipal reclaimed water systems also have converted to use of reclaimed water for municipal cooling systems. Substitution of reclaimed water for current sources typically requires a change or increase in treatment process to maintain the water quality intended for the installed cooling equipment. These advancements have made it possible to better control deposits, corrosion, and biological problems that are often associated with use of reclaimed water in a concentrated cooling water system.

Two basic types of cooling water systems use reclaimed water:

- Once-through systems and
- Recirculating evaporative systems, which are the most common reclaimed water system due to their large water use and consumption by evaporation.

As implied by the name, once-through cooling water systems without cooling towers (typically) involve a single pass of cooling water through heat exchangers. There is no evaporation and, therefore, no consumption or concentration of the cooling water. Very few once-through cooling systems use reclaimed water. In most instances, these systems are confined to locations where reuse is convenient, such as where industries are located near a wastewater discharge pipe. Recirculating evaporative cooling water systems use water to absorb process heat and then transfer the heat by evaporation. As the cooling water is recirculated, makeup water is required to replace water lost through evaporation. Water must also be periodically removed from the cooling water system to prevent a buildup of dissolved solids in the cooling water.

Cooling Towers

Like all recirculating evaporative systems, cooling water towers are designed to take advantage of the absorption and transfer of heat through evaporation. Over the past 10 years, cooling towers have increased in efficiency so that in new units, only 1.5 to 1.75 percent of the recirculated water is evaporated for every 10°F (6°C) drop in process water heat, decreasing the need to supplement with makeup water (USEPA 2012). Because water is evaporated, the TDS and minerals remain in the recirculated water. TDS must be removed or treated to prevent accumulation on the cooling equipment as well as the cooling tower and are frequently reduced using NF or RO membranes. Removal is accomplished by discharging a portion of the cooling water, referred to as blowdown water. The blowdown water is usually treated using a chemical process and/or a filtration–softening–clarification process before disposal. Buildup of TDS can occur within the reclamation/industrial cooling system if the blowdown waste stream, with increased dissolved solids, is recirculated between the water reclamation plant and the cooling system.

The cycles of concentration (COC) are defined as the ratio of a given ion or compound in the cooling tower water to the identical ion or compound in the makeup water. For example, if the sodium chloride level in the cooling tower water is 200 mg/L and the same compound in the makeup water is 50 mg/L, then the COC is 200 divided by 50, or 4, often referred to as 4 cycles. Industries often operate their cooling towers at widely different cycles of concentration, as shown in Table 4-4. The reason for such variations is that the cooling water is used for different applications such as wash water, ash sluicing, and process water.

Cooling Water Quality Requirements

The most frequent water quality problems in cooling water systems are corrosion, biological growth, and scaling. These problems arise from contaminants in potable water as well as in reclaimed water, but the concentrations of some contaminants may be higher in reclaimed water than in potable water.

Corrosion concerns. The use of any water, including reclaimed water, as makeup in recirculating cooling tower systems results in concentration of dissolved solids in the heat exchange system. This concentration may or may not cause serious corrosion of components. Any contamination of the cooling water through process in-leakage, atmospheric conditions, or treatment chemicals impacts water quality and requires treatment changes to maintain desired water quality.

Biological concerns. Biological concerns are associated with the use of reclaimed water in cooling systems because increased nutrient concentrations in most reclaimed waters contribute to biological growth. Microbiological organisms contribute to the potential for deposits and microbiologically induced corrosion.

Microbiological organisms (bacteria, fungus, and algae) that contribute to deposits and corrosion are most often those that adhere to surfaces. Deposits can interfere with heat transfer and cause corrosion directly due to acid or corrosive by-products. Indirectly, deposits may shield metal surfaces from water treatment corrosion inhibitors and establish under deposit corrosion. Deposits can grow rapidly and plug heat exchangers, cooling tower film fill, and cooling tower water distribution nozzles and sprayers.

Table 4-4 Typical cycles of concentration

Industry	Cycle
Fossil power plant	5–8
Nuclear power plant	6–10
Petroleum refinery	6–8
Chemical plant	8–10
Steel mill	3–5
Heating, ventilation, and air conditioning	3–5
Paper mill	5–8

Source: Metcalf & Eddy/AECOM (2007).

Scaling concerns. The primary constituents for scale potential from reclaimed water are calcium, magnesium, sulfate, alkalinity, phosphate, silica, and fluoride. Combinations of these minerals that can produce scale in the concentrated cooling water generally include calcium phosphate (most common), silica (fairly common), calcium sulfate (fairly common), calcium carbonate (seldom found), calcium fluoride (seldom found), and magnesium silicate (seldom found). All constituents with the potential to form scale must be evaluated and controlled using chemical treatment and/or by adjusting the cycles of concentration. Reclaimed water quality must be evaluated, along with the scaling potential, to establish the use of specific scale inhibitors. Guidelines for selection and use of scale inhibitors are available as are scale predictive tools. Many of the same constituents that create scale potential for cooling systems also result in increased fouling of membrane systems. Selection of treatment system options is discussed in Chapter 5.

Boiler Makeup Water

The use of reclaimed water for boiler makeup water is similar to the use of conventional public water supply; both require extensive additional treatment. Quality requirements for boiler makeup water depend on the pressure at which the boiler is operated. Generally, the higher the pressure, the higher the quality of water required. Boilers with very high pressure (1,500 psi [10,340 kPa] and above) require high-quality makeup water. In general, both potable water and reclaimed water used for boiler water makeup must be treated to reduce the hardness and TDS of the boiler feedwater to close to zero. This may often include use of double-pass RO to control TDS concentrations. Removal or control of insoluble scales of calcium and magnesium and control of silica and alumina are required since these are the principal causes of scale buildup in boilers.

Industrial Process Water

The suitability of reclaimed water for use in industrial processes depends on the particular use. For example, the electronics industry requires water of almost distilled quality for washing circuit boards and other electronic components. The tanning industry can use relatively low-quality water. Requirements for textiles, pulp and paper, and metal fabricating are intermediate. Thus, in investigating the feasibility of industrial reuse with reclaimed water, potential users must be contacted to determine the specific requirements for their process water.

Water quality requirements for the chemical industry vary greatly according to production requirements. Generally, waters with near neutral pH (6.2 to 8.3) that are also moderately soft with low turbidity, suspended solids, and silica are required; dissolved solids and chloride content are generally not critical.

Waters used in textile manufacturing must be nonstaining; consequently, they must be low in turbidity, color, iron, and manganese. Hardness may cause curds to deposit on the textiles and may cause problems in some of the processes that use soap. Nitrates and nitrites may cause problems in dyeing. These water quality requirements often influence the selection of membrane treatment processes.

Petroleum and Coal Industries

Processes for the manufacture of petroleum and coal products can usually tolerate water of relatively low quality. Waters generally must be in the 6 to 9 pH range and have moderate suspended solids of no greater than 10 mg/L.

Typical water quality requirements for various industrial processes are shown in Tables 4-5 and 4-6. It is important to note that specific water quality requirements vary greatly with each industrial process.

Table 4-5 Typical reclaimed water quality requirements for various industrial processes

		Boiler Feedwater, psi		Cooling Water			
Parameter	Unit			Once-Through		Makeup for Recirculation	
		0–145	145–174	Fresh	Brackish	Fresh	Brackish
Silica	mg/L	30	10	50	25	50	25
Aluminum	mg/L	5	0.1	*	*	0.1	NA
Iron	mg/L	1	0.3	*	*	0.5	NA
Manganese	mg/L	0.3	0.1	*	*	0.5	NA
Copper	mg/L	0.5	0.05	*	*	NA	NA
Calcium	mg/L	*	*	200	520	50	420
Ammonia	mg/L	0.1	0.1	*	*	*	*
Bicarbonate	mg/L	170	120	600	N/A	25	N/A
Sulfate	mg/L	*	*	680	2700	200	2700
Chloride	mg/L	*	*	600		500	
Fluoride	mg/L	*	*	600	19,000	500	19,000
Dissolved solids	mg/L	700	500	1,000	35,000	500	35,000
Suspended solids	mg/L	10	5	5,000	2,500	100	100
Hardness	mg/L as $CaCO_3$	20	1.0	850	6,250	130	6250
Alkalinity	mg/L as $CaCO_3$	140	100	500	115	20	115
pH	SU	8–10	8–10	5.0–8.3	*	*	*
Chemical oxygen demand	mg/L	5	5	75	75	75	75
Turbidity	ntu	10	5	5,000	100	*	*

Source: Metcalf & Eddy/AECOM (2007).

* Specific water quality requirements vary greatly with each industrial process.

AGRICULTURAL REUSE

This section focuses on the following specific considerations for implementing a water reuse program for agricultural irrigation:

- Agricultural irrigation water quality demands,
- Reclaimed water quality, and
- Other system considerations.

Agricultural reuse is often included as a component in water reuse programs for the following reasons:

- Extremely high water demands for agricultural irrigation,
- Significant water conservation benefits associated with reuse in agriculture, and
- Ability to integrate agricultural reuse with other reuse applications.

Table 4-6 Typical reclaimed water quality requirements for various industrial processes

		Process Water by Industry					
Parameter	Unit	Textile	Pulp and Paper	Chemical	Petroleum and Coal Products	Primary Metal	Tanning
Silica	mg/L as SiO_3	25	50	50	60	*	*
Aluminum	mg/L as Al_2O_3	8		N/A		*	*
				N/A			
Iron	mg/L	0.1–0.3	0.3	0.1	1.0	*	50
Manganese	mg/L	0.01–0.05	0.1	0.1	*	*	0.2
Copper	mg/L	0.01–5		N/A	*	*	*
				N/A			
Calcium	mg/L	*	20	70	75	*	60
Magnesium	mg/L	*	12	20	30	*	*
Sodium	mg/L	*	*		230	*	*
Ammonia	mg/L	*	*		40	*	*
Bicarbonate	mg/L	*	*	130	480	*	*
Sulfate	mg/L	100	*	100	600	*	250
Chloride	mg/L	*	200	500	300	500	250
Fluoride	mg/L	*		5	1.2	*	*
Dissolved solids	mg/L	100–200	100	1,000	1,000	1,500	*
Suspended solids	mg/L	0–5	10	5	10	3,000	*
Hardness	mg/L as $CaCO_3$	0–50	475	250	350	1,000	150
Alkalinity	mg/L as $CaCO_3$	*	*	125	500	200	
pH	su	6–8	4.6–9.4	5.5–9.0	6-9	5-9	6-8
Color	Color units	0–5	10	20	25		5
Turbidity	ntu	0.3–5	*	*	*	*	*

Source: Metcalf & Eddy/AECOM (2007).

* Specific water quality requirements vary greatly with each industrial process

Reclaimed Water Quality

The chemical constituents of concern in reclaimed water used for agricultural irrigation are salinity, sodium, trace elements, excessive chlorine residual, and nutrients. Sensitivity is generally a function of a given plant's tolerance of constituents encountered in the root zone or deposited on the foliage. Reclaimed water tends to have higher concentrations of these constituents than the groundwater or surface water sources from which the water supply is drawn. Due to the cost associated with membrane treatment, the value of the agricultural crops cannot support the level of treatment required for this intended use.

WATER SUPPLY AUGMENTATION

Reclaimed water is increasingly being recognized as a viable source of water to augment the supply of surface and groundwater. A 2012 National Research Council report identified the long-term practice of "de facto reuse," that is, the discharge of treated wastewater to streams upstream of other community water supplies. Groundwater recharge has been practiced through infiltration from surface basins and, beginning with Water Factory 21, injection of membrane-treated reclaimed water as a saltwater intrusion barrier (USEPA 2012). Numerous systems have been developed in California and international locations for indirect potable reuse. Three facilities are currently practicing direct potable reuse (DPR; see Direct Potable Reuse below).

Health and Regulatory Considerations

Constraints on groundwater recharge are conditioned by the use of the extracted water and include health concerns, economic feasibility, physical limitations, legal restrictions, water quality constraints, and reclaimed water availability. Of these constraints, health concerns are the most important as they relate to most recharge projects. Where reclaimed water will be ingested, health effects due to prolonged exposure to low levels of contaminants must be considered as well as the acute health effects from pathogens or toxic substances. One problem with recharge is that boundaries between potable and nonpotable aquifers are not always well defined. Some risk of contaminating high-quality potable groundwater supplies is often incurred by recharging "nonpotable" aquifers. The recognized lack of knowledge about the fate and long-term health effects of contaminants found in reclaimed water necessitates a conservative approach in setting water quality standards and monitoring requirements for groundwater recharge. Because of these uncertainties, some agencies have set stringent water quality requirements and require high levels of treatment, in some cases, organic removal processes where groundwater recharge impacts potable aquifers.

AUGMENTATION OF INDIRECT POTABLE SUPPLIES

This section discusses indirect potable reuse via surface water augmentation, groundwater recharge, and DPR. Indirect potable reuse refers to the augmentation of a drinking water source (surface water or groundwater) with reclaimed water followed by an environmental buffer that precedes normal drinking water treatment. DPR refers to the introduction of reclaimed water directly into a potable water supply distribution system downstream of a water treatment plant or into the raw water supply immediately upstream of a water treatment plant.

Water Quality Objectives for Potable Reuse

Development of water quality requirements for either direct or indirect potable reuse is challenging. From a regulatory standpoint, there is a tendency to use the Safe Drinking Water Act National Primary Drinking Water Regulations (NPDWRs) as a starting point for defining potable water quality objectives. For years, water reuse advocates have argued that reclaimed water from municipal wastewater meets the requirements of the NPDWRs. However, the original NPDWRs were not intended to define potable water quality when the source is municipal wastewater.

Recently, there has been a dramatic increase in the ability to detect chemicals, for example, microconstituents such as endocrine disruptors, pharmaceuticals, and personal care products such as antibacterial soaps, sunscreen, and bath gels, that are present in municipal wastewaters at trace concentrations (nanograms per liter). Some of these constituents have been detected in surface and groundwater sources as well. Typically, none of these compounds are regulated or monitored. Some indirect potable reuse projects have started using toxicological assays to compare the drinking water source to the reclaimed water.

With a combination of advanced treatment barriers and added protection provided by an environmental buffer, the problem of defining water quality objectives for indirect potable reuse is manageable. By using treatment beyond typical disinfected tertiary treatment and an environmental buffer, indirect potable reuse projects can provide additional constituent removal and retention time for the reclaimed water, thereby furnishing added protection against the unknowns and uncertainty associated with trace organics.

DIRECT POTABLE REUSE

DPR is being practiced in Windhoek, Namibia, Big Spring, Texas, and Wichita Falls, Texas. A small project is near completion in Cloudcroft, N.M. The first DPR project came online in Namibia in 1969 but did not add UF membranes to an expanded treatment train until 2010 as an additional barrier for turbidity, microorganisms, and protozoa (du Pisani 2011). Planning for future treatment improvements include ozonation and RO. In the United States, the most extensive research focusing on DPR has been conducted in California, Colorado, and Florida. A considerable investment in potable reuse research has been made in Denver, Colo. This research included operation of a 1-mgd reclamation plant in many different process modes over a 10-year period (Lauer and Rogers 1998). The product water was reported to be of better quality than many potable water sources in the region. The WateReuse Research Foundation (WRRF) has solicited funding from numerous sources and is sponsoring $6 million in research on various aspects of potable reuse, with much of the research focused in southern California, beginning with a report authored by Tchobanoglous et al. (2011) on potable reuse potential. As of June 2015, WRRF had allocated $5.8 million to fund 34 DPR research projects. In addition, the California Department of Water Resources, San Diego County Water Authority, funded a $2.1 million DPR demonstration project in San Diego in September 2014 (WRRF 2015).

WATER QUALITY TREATMENT REQUIREMENTS

For each reuse application, the water quality requirements and regulations that are established by regulatory agencies vary depending on the regulatory agency that has primacy. Ranges of water quality for the general water reuse categories identified earlier are discussed in Chapter 3.

Constituents of emerging concern (CECs), including endocrine-disrupting compounds and pharmaceuticals and personal care products, have received increasing attention in the water and wastewater industries over the past several years, particularly in the area of water reuse. These include a broad range of compounds that potentially have adverse effects on human health and the environment but may not be currently regulated by the USEPA. Water and wastewater utilities have become increasingly concerned about CECs, but very little guidance exists regarding how to define the problem or develop action plans to improve the quality of public water supplies. Some states such as Florida, California, Arizona, and Massachusetts are using total organic carbon (TOC) as a surrogate parameter to monitor trace organics and are limiting the monitored water for reuse applications such as groundwater injection and rapid infiltration basins. As an example, the TOC limit in California for reclaimed water used for recharge of potable aquifers is <0.5 mg/L.

Multiple Barriers

Design inclusion of multiple barriers is an important consideration when selecting appropriate processes for potable reuse applications. The concept is based on establishing a series of barriers to prevent the passage of pathogens and harmful organic and inorganic contaminants into a water system. In the context of potable reuse, this section focuses on a combination of treatment processes wherein each process provides a specific level of constituent reduction.

The selection of technologies for potable reuse involves careful consideration and evaluation of numerous factors. Where there are several potential water reuse application scenarios, each user may have specific requirements for the quality of reclaimed water.

The use of multiple technologies may be appropriate for meeting the requirements in instances where:

- Multiple grades of product water are required that cannot be produced economically by a single process,
- Additional safeguards are needed for public health protection, and
- High levels of constituent removal are required to meet reclaimed water criteria.

Table 4-7 lists principal unit operations and processes considered in a multiple barrier approach along with the constituent classes for which they are used in water reuse applications. The removal of a contaminant is considered a relative measure of the unit process's ability to act as a barrier to that contaminant. Wastewater unit processes act as barriers to contaminants and are expected to result in at least 50 percent removal of the contaminant. Water reuse systems for potable reuse (direct or indirect) contain considerably more contaminant barriers than a conventional water treatment plant because of the direct link to wastewater origin.

Table 4-7 Unit processes used for the removal of classes of constituents found in wastewater for reuse applications

Unit Operation or Process	Suspended Solids	Colloidal Solids	Organic Matter (particulate)	Dissolved Organic Matter	Nitrogen	Phosphorus	Trace Constituents	Total Dissolved Solids	Bacteria	Protozoan Cysts and Oocysts	Viruses
Secondary treatment	x			x							
Secondary treatment with nutrient removal				x	x	x					
Depth filtration	x								x	x	
Surface filtration	x		x						x	x	
Microfiltration	x	x	x						x	x	
Ultrafiltration	x	x	x						x	x	x
Dissolved air flotation	x	x	x							x	x
Nanofiltration			x	x			x	x	x	x	x
Reverse osmosis				x	x	x	x	x	x	x	x
Electrodialysis		x						x			
Carbon adsorption				x			x				
Ion exchange					x		x	x			
Advanced oxidation			x	x			x		x	x	x
Disinfection				x					x	x	x

Source: Metcalf & Eddy (2006).

Role of Membranes

The removal of trace constituents may be required in certain reuse applications such as aquifer recharge, surface water augmentation, DPR, and in industrial applications. Advanced technologies currently being adopted for removal of trace constituents include membrane bioreactors, MF, UF coupled with NF and RO membranes, and advanced oxidation processes (AOPs) that allow treatment to the quality level fit for the intended use. Nonmembrane-based treatment processes typically include carbon and biologically activated carbon unit processes. The application of these technologies is discussed in more detail in Chapter 5.

MF and UF membranes typically reject turbidity, TSS, some bacteria, and some viruses with typical membrane product water (RO feedwater) quality as shown in Table 4-3. NF and RO membranes have the ability to remove both soluble organic and inorganic constituents. The potential impact of various water quality parameters is summarized in Table 4-2. Membranes are typically used in industrial applications to reduce TDS to desired process water quality levels and to control corrosion. Table 4-8 presents typical product (filtrate) water quality for high-pressure membranes (NF and RO) with single-pass and two-pass used for advanced post-treatment of tertiary wastewater effluent. Two-pass refers to a second pass of membrane treatment for the permeate produced in the first pass. It is important to note that any process variations upstream of the membranes or changes in feedwater quality can impact membrane performance, resulting in poor product water quality. Membrane processes combined with AOP and disinfection are becoming popular methods for extending water supplies through augmentation of surface and groundwater supplies, IPR, and DPR applications.

Current research and discussion in the water industry are focused on appropriate technologies for continuous monitoring of potable reuse membrane unit processes to ensure continuous maintenance of potable water quality. Current regulatory approaches limit the log-removal credits allowed for NF and RO membrane units due to limitations in direct integrity testing that would ensure higher log removal of microorganisms and virus. Several research projects are being conducted through the California Direct Potable Reuse Initiative (WRRF 2015) to identify appropriate monitoring approaches.

Table 4-8 Typical product water quality for reverse osmosis membranes

		High-Pressure Membranes Range of Values	
Parameter	Unit	Single-Pass	Two-Pass
Total dissolved solids (mg/L = ppm)	mg/L	10–60	1–10
Total suspended solids	mg/L	ND	ND
Turbidity	ntu	ND	ND
Total organic carbon	mg/L	0.2–1	0.1 0.5
Total Kjeldahl nitrogen	mg N/L	ND	ND
Bicarbonate	mg/L as $CaCO_3$	10–30	2–20
Iron*	mg/L	0–0.06	0–0.04

Source: Water Environment Federation (2006).

* Additional pretreatment may be required, depending on the form of iron present.
Abbreviation: ND, not detected.

REFERENCES

Bergman, R. 2007. *Reverse Osmosis and Nanofiltration*. Denver, CO: AWWA.

Du Pisani, P, et al. 2011. *Direct Potable Reclamation in Windhoek: A Critical Review of the Design Philosophy of New Goreangab Drinking Water Reclamation Plant*. IWA Water Reuse. *Journ. Of Water Reuse and Desalination*

Florida Department of Environmental Protection. 2002. *2001 Reuse Inventory*. Tallahassee, FL: Florida Department of Environmental Protection.

Lauer, W.C. & Rogers, S.E. 1998. Wastewater Reclamation and Reuse. *The Demonstration of Direct Potable Water Reuse: Denver's Landmark Project* (T. Asano, editor). Boca Raton, FL: CRC Press.

Lozier, J.; Amy, G.L.; Jacangelo, J.; & Mysore, C. 2008. *Natural Organic Matter Fouling of Low Pressure Membranes*. Denver, CO: AwwaRF.

Metcalf & Eddy–AECOM. 2007. *Water Reuse Issues, Technologies, and Applications*. New York: McGraw Hill.

———. 2003. *Wastewater Engineering Treatment and Reuse*. New York: McGraw Hill.

NRC (National Research Council). 2012. *Water Reuse: Potential for Expanding the Nation's Water Supply Through Reuse of Municipal Wastewater*. Washington, DC: National Academies Press.

Postel, S. 1999. *Pillar of Sand: Can the Irrigation Miracle Last?* New York: W.W. Norton and Company.

SWRCB (State Water Resources Control Board). 2002. *2002 Statewide Recycled Water Survey*. California SWRCB, Office of Water Recycling, Sacramento, California.

———. 2011. *California Water Code*. California SWRCB, Sacramento, California. (http://www.leginfo.ca.gov/cgi-bin/calawquery?codesection=wat&codebody=&hits=20) www.swrcb.ca.gov/recycling/recyfund/munirec/ index.html.

Tchobanoglous, G., et. al., 2011. *Potable Reuse: A Direct Path Forward*. WateReuse Research Foundation and WateReuse California.

USEPA (US Environmental Protection Agency). 2012. *2012 EPA Guidelines for Water Reuse*. EPA/600/R-12/618. Cincinnati, OH: USEPA.

van der Graaf, J.H.J.M., et al. 2005. Treatment Matrix for Reuse of Upgraded Wastewater. *Water Science and Technology: Water Supply*, 5:1:87–94.

Verliefde, A.R.D., E.R. Conclissen, S.G.J. Heijaman, G.L. Amy, B. Van der Bruggen, and J.C. van Dijk. 2008. The Role of Electrostatic Interactions on the Rejection of Organic Solutes in Aqueous Solution With Nanofiltration. *Journal of Membrane Science*, 322:1:52–66.

von Sperling, M., et al. 2002. Urban Wastewater Treatment Technologies and the Implementation of Discharge Standards in developing Countries. *Urban Water*, 4:105–114.

Water Environment Federation. 2006. *Membrane Systems for Wastewater Treatment.* New York: McGraw-Hill.

WRRF (WateReuse Research Foundation). 2015. *California Direct Potable Reuse Initiative, Reporting on Significant Progress.* WateReuse and Water Reuse California, Spring/Summer.

Zheng, X. & Croue, J.P. 2012. Contribution of Different Effluent Organic Matter Fractions to Membrane Fouling in Ultrafiltration of Treated Domestic Wastewater. *Journal of Water Reuse and Desalination,* 2:4:204–209.

This page intentionally blank.

Chapter 5

Membrane Process Treatment Facility Design

Membrane treatment processes are used in reuse to meet the water quality requirements of various reuse applications. This is done by removing and separating impurities and undesirable substances, which include colloidal particles, dissolved solids, organic and inorganic compounds, and trace substances. These substances are not effectively removed by conventional treatment processes such screening, granular media filtration, clarification, and ion exchange. In many water reuse facilities, membrane treatment is the final treatment process used to meet water reuse quality requirements.

PRETREATMENT OR FEEDWATER CONDITIONING FOR MEMBRANE TREATMENT PROCESSES

The membrane treatment processes used in reuse facilities are microfiltration (MF), ultrafiltration (UF), reverse osmosis (RO), nanofiltration (NF), electrodialysis (ED), and electrodialysis reversal (EDR). These processes share characteristics, operating conditions, and materials of construction and, therefore, require similar pretreatment and feed conditioning. The topics in this section follow the membrane treatment hierarchy shown in Figure 5-1 for MF–UF, RO–NF, and ED–EDR as well as specific pretreatment required for NF–RO. Discussion on the design for MF–UF and NF–RO as applied to water reuse are also included.

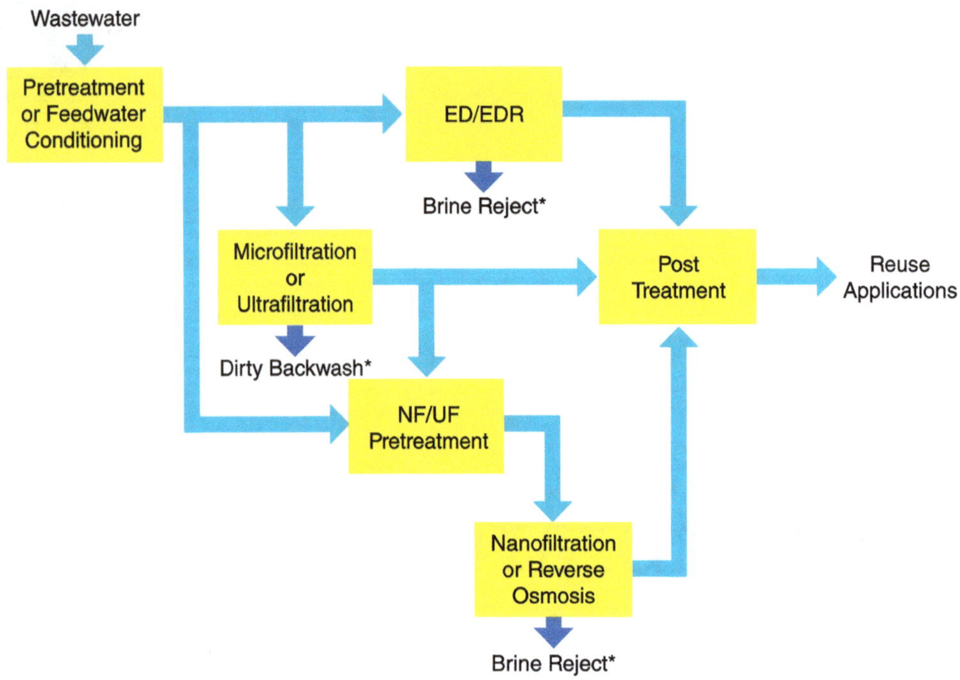

Figure 5-1 Water reuse membrane treatment hierarchy

Abbreviations: ED, electrodialysis; EDR, electrodialysis reversal; MF, microfiltration; NF, nanofiltration; RO, reverse osmosis; UF, ultrafiltration.

The feedwater used in reuse facilities comes from municipal wastewater treatment plants as well as industrial wastewater, including water produced from oil and gas exploration and other low-quality water from various sources. Other potential sources of feedwater are dirty backwash from conventional and membrane filtration plants and brine concentrate from NF, RO, ED, and EDR plants. This feedwater contains a variety of impurities and undesirable substances that often interfere with membrane treatment processes. If these materials and undesirable substances are not removed, they can lead to the following operational problems in the membrane treatment systems:

- Loss of membrane permeability or water passage, which leads to reduced production;
- Destruction of the membrane materials;
- Decrease in impurities retention (MF–UF) or decrease in rejection (increase in salt passage in NF–RO);
- Increase in operating pressure; and
- Short membrane life/increased replacement frequency.

In selecting pretreatment and feed conditioning for any membrane process, it is important to understand the following:

- Final water quality targets or acceptable ranges of constituents,
- Feedwater source(s),
- Physical and chemical characteristics of the feedwater,
- Physical and chemical properties of the membranes, and

- Operating conditions of the selected membrane (for example, temperature, flow rate, cleaning frequency).

Pretreatment and feed conditioning for membrane processes can be categorized into the following general treatment strategies and goals:

- Remove the materials and substances that may interfere with the membrane treatment processes. These materials and substances cause physical and chemical damage and alteration of the membrane materials that significantly affect membrane performance.
- Provide conditions that optimize the performance of the membrane treatment in terms of flux and rejection. This includes changing salt solubilities, slowing down crystallization, and oxidizing or reducing certain chemical species.
- Minimize accumulation of foulant on the surface of the membranes that can block the membrane pores and hinder water flow. Membrane pore sizes are shown in Table 5-1. MF and UF membranes have pores that range in size from 0.001 to 0.1 µm, and the removal of impurities is by size exclusion. NF and RO do not have pores, and the separation is by diffusion and separation of dissolved solids less than 0.001 µm in size. Membrane foulants come from two sources: the particulates present in the raw water and the foulant that is formed on the membrane surface as a result of biological and chemical activities. Scales are foulants that result from the concentration of salts and minerals.
- Change operating conditions to prolong the membrane's useful life.

The pretreatment and conditioning of feedwater must involve a total system approach for continuous and reliable operation. The pretreatment process involves one or more various processes to achieve treatment objectives. One membrane treatment process may be used to pretreat feedwater for another membrane process. It is common to use MF or UF to pretreat feed water for NF or RO. The pretreatment and conditioning processes are discussed in general terms in order for the process engineer or designer to evaluate and select the proper pretreatment. Detailed design procedures and parameters of each treatment process are not described in this manual.

Table 5-1 Filtration media pores sizes

Filtration Device	Pore Size	Common Materials Size
Media filter (sand) or bag filter	1,000 nm–0.001m (1µm–1mm)	Beach sand, granular activated carbon, ion-exchange resin, human hair, pollen, pin point, yeast cells, coal dust, red blood cells, bacteria, milled flour, *Cryptosporidium*, *Giardia* cyst
Microfiltration or ultrafiltration	100 nm–1,000 nm (0.1µm–1µm)	Asbestos, gelatin, latex/emulsion, tobacco smoke
Ultrafiltration	1 nm–100 nm (0.001µm–0.1µm)	Colloidal silica, virus, albumin protein, carbon black, endotoxin, pyrogen
Nanofiltration or reverse osmosis	< 1 nm (0.001µm); theoretical as there are no actual pores in these membranes	Aqueous salts, sugar, metal ion, synthetic dye, pesticides, herbicides, atomic radius

Adapted from *Osmonics, The Filtration Spectrum* (1998).

SCREENING FOR REMOVAL OF DEBRIS AND LARGE SOLIDS

Treated effluent from a wastewater treatment plant is free from large solids and debris. However, when discharged to open reservoirs, tanks, or ponds, it is important to note that these water bodies tend to pick up large solids and debris from aquatic plants, algae, land-based vegetation, trees, shrubs, dirt, rocks, garbage, and animals. Upsets in the wastewater treatment plant can result in carryover from clarifiers and filters. Often, the treatment plant has biological growths that generate debris. These materials can physically damage the membrane surface and clog pumps, pipes, fittings, and other equipment.

Fine screening is a crucial pretreatment component of MF–UF systems used for water reuse applications. Fine screening is used to remove large particles and debris before the membrane processes to protect the membranes from fiber damage and plugging. Systems in which the flow is from the inside to the outside generally have finer screens since they are more susceptible to plugging than an outside-in flow system. Typically screen size ranges from 100 to 1,000 µm. Inside-out as well as most outside-in MF–UF membrane manufacturers recommend fine screening with 100- to 300- µm openings. The preferred screen is the perforated plate type, although other types such as the wedge wire screen type are used. If the feed stream contains fibrous materials, a wedge wire screen is not recommended. Another screen commonly installed in front of MF and UF membranes is the disc type. The screens used as pretreatment for MF and UF are usually self-cleaning.

Automatic backwashing screens, as shown in Figure 5-2, typically require an inlet pressure of approximately 30 psi. The filters work by passing water through a perforated screen, weave wire, or wedge wire, which protects the cleaning mechanism and a fine multilevel screen (typically 300–500 µm for municipal water reuse applications) prior to discharge via an outlet pipe. The screens are programmed to automatically backwash based on preset values of filter run length or differential pressure.

Automatic backwashing disc filters (Figure 5-3, left) are comprised of polymeric, grooved discs (typically between 20 and 400 µm based on filtration requirements) stacked on a central assembly. During filtration, the discs are compressed using a preloaded spring; differential pressure forces water through the groves. These filters are programmed to automatically backwash based on preset values of filter run length or differential pressure. Another innovation is the use of wound microfiber in cassettes (Figure 5-3, right).

In drum screens, feedwater enters via an inlet pipe that distributes the water tangentially over the screen. As the drum rotates, solids that are too large to pass through the screen are collected near the inlet side of the screen and discharged to a hopper for disposal. The screens are typically equipped with spray nozzles and/or diverted plates, which are used to prevent accumulation of solids on the screen surface. The slots on the screens vary in shape and size. Typically, slots are wedge wire or perforated.

NF and RO systems typically include 5-µm cartridge filters that protect the membranes from debris and solids, although some designers argue that they are not required. The amount of dirt collected in the cartridge filter elements would justify their installation. A cartridge filter before NF–RO should not be designed for large-scale removal of solids but rather as protection of the NF and RO membranes from unforeseen events.

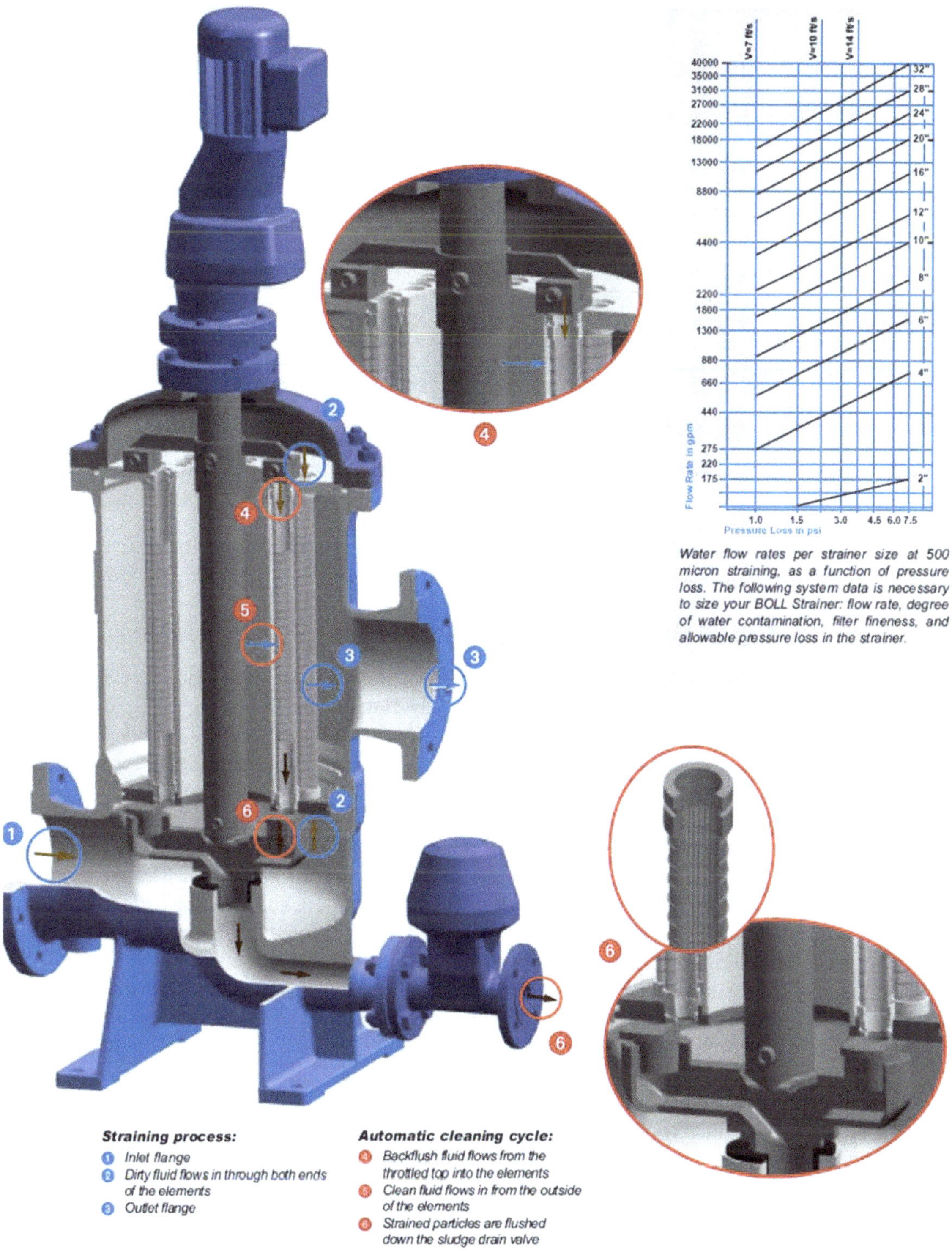

Courtesy of Boll Filter Corporation.

Figure 5-2 Typical automatic backwashing screen used in microfiltration–ultrafiltration systems

Courtesy of Amiad Water Systems.

Figure 5-3 Automatic backwashing disc (left) and wound microfiber cassette filters (right)

REMOVAL OF CHEMICALS INCOMPATIBLE WITH MEMBRANE MATERIALS

Certain chemicals permanently alter the chemical composition and physical integrity of membrane materials. Most commercially available membranes are made of organic polymeric materials, which are susceptible to chemical attacks from solvents, hydrocarbons, and oxidizing agents. The most common polymeric materials for MF, UF, NF, and RO are shown in Table 5-2.

Materials of membrane module components such as spacers, casing, channels, and glue must be checked for compatibility with the chemicals that may be present in the feedwater. In addition to the membrane module materials, it is critical to evaluate the compatibility of the associated materials and equipment such as pipes, valves and fittings, pumps, tanks, and instruments. When in doubt, the membrane or equipment manufacturer should be contacted.

As a rule of thumb, all oxidizing agents (for example, chlorine, permanganate, halogens) can harm the membrane and must be removed or eliminated, if possible. A polyamide membrane is stable against most chemicals at a pH of 2 to 11, providing these chemicals are dissolved and do not occur in an organic phase. Cellulose acetate membranes can tolerate most chemical compounds including oxidizing agent such as chlorine; however, they must operate at a pH range of 3 to 8 to prevent hydrolysis. In actual operation, cellulose acetate membranes are operated at a pH range of 5.5 to 6.

Table 5-2 Polymeric membrane materials

Filtration System	Polymeric Material
Microfiltration–ultrafiltration	Polysulfone, polyether sulfone, polyvinylidine difluoride, polypropylene, cellulose acetate, polyvinylpyrrolidone, polyacrylonitrile
Nanofiltration–reverse osmosis	Cellulose acetate, cellulose triacetate, polyamide
Electrodialysis–electrodialysis reversal	Polystyrene

Solvents such as ketones, aldehydes, alcohols, gasoline, some hydrocarbons, and other organic compounds that react directly with membrane materials should not be present in the feedwater at any time. Because their presence in feedwater may indicate accidental spillage from an industrial or commercial facility, permanent treatment capacity for these chemicals is not typically required. However, upstream monitoring is required to provide warning in the event a plant shutdown is needed. If there is a shutdown, solvents and other organic chemicals in the feedwater can be removed using conventional processes such as activated carbon absorption or using other specialty and proprietary processes.

Chlorine, which is a strong oxidizing agent, is normally present in municipal wastewater effluent for disinfection. Chlorine can be added as chlorine gas, generated on-site as NaOCl, commercial NaOCl, (sodium hypochlorite) or as $Ca(OCl)_2$ (calcium hypochlorite chlorine). The dosage of chlorine depends on the feedwater's chlorine demand and the required residual, which is normally 0.5 mg/L.

In MF–UF, chlorine is often dosed continuously or intermittently to prevent biological fouling. In NF and RO feed, free available chlorine must be reduced to less than 0.05 ppm as Cl_2 as a requirement for most polyamide membrane manufacturer warranties. The two most common pretreatment methods for reducing chlorine levels are absorption onto activated granular carbon filter media and use of a chemical-reducing agent such as sodium bisulfite (SBS).

Carbon filters are typically used on small systems (50 to 100 gpm or less) due to capital cost considerations. Typically the activated carbon is supplied in exchangeable bottles, while larger systems use high-pressure vessels. The advantage of a carbon filter is its ability to remove organics and small amounts of oil and grease, which could foul the NF–RO membranes, from the feedwater. Activated carbon is more reliable than a chemical feed system in treating feedwater. The disadvantage is that carbon filters are known for breeding bacteria that can result in biological fouling of the NF–RO membranes.

SBS is the chlorine-reducing agent of choice for larger membrane systems. Theoretically, 1.47 ppm of SBS (or 0.70 ppm of sodium metabisulfite) will stoichiometrically neutralize 1.0 ppm of chlorine. Design engineers have used a dosing rate of 1.8 to 3.0 ppm SBS per 1.0 ppm chlorine as an industrial safety factor for NF and brackish water RO systems. SBS must be added far enough upstream of the NF and RO elements to ensure at least 20 s of reaction time. Proper in-line mixing, which preferably includes a static mixer, is required. The advantage of SBS dechlorination is that it is less capital intensive than carbon filters for large systems. The reaction by-products are readily removed by the NF and RO membranes, and residual SBS is readily removed by RO. The disadvantages of SBS are the possible need to handle and mix a small volume of chemical; the increased risk of chlorine going to the membranes if sufficient monitoring and controls are not designed into the dechlorination system; and, in a few cases where sulfur-reducing bacteria are present in the feed supply, the functioning of bisulfite as a nutrient, which enhances growth of the bacteria. Another reducing agent, sulfur dioxide and its salts, is commonly used in wastewater treatment and can be applied in pretreatment for membrane processes.

Dechlorination can be monitored using a free available chlorine monitor or an oxidation–reduction potential (ORP) monitor or by monitoring for a residual bisulfite concentration. The ideal method is to monitor for a residual bisulfite concentration, as this ensures that sufficient bisulfite exists to neutralize all chlorine. Most commercial chlorine monitors measure accurately to 0.1 ppm, which is the upper limit for the polyamide membranes. The use of ORP monitors to indirectly measure residual bisulfite in the feedwater is unreliable due to variability of the baseline readings. It is common to use chloramine instead of chlorine in reuse plants.

Typical polyamide membranes have an estimated chlorine tolerance of 1,000 to 2,000 ppm-hour before a doubling in salt passage occurs. Polyamide membranes have higher tolerance that range from 50,000 to 200,000 ppm-hour before a noticeable increase in salt passage occurs. Dechlorination after chloramination may be required because chloramines are produced by mixing ammonia and chlorine; any free residual can be present if insufficient ammonia is used. Therefore, it is vital that there is always excess ammonia and that the system is monitored to ensure this.

NF and RO system designers are cautioned that, in a few applications, chlorine and chloramine tolerances, as described above, can be dramatically reduced by the catalytic effects of increasing temperature (90°F or higher), low and high pH, or the presence of transition metals on the membrane surface (for example, iron, manganese, zinc, copper, aluminum). Other halogen compounds should have a similar reaction with bisulfite as chlorine.

Oil and grease are not compatible with most membranes used in water treatment, and they should be avoided or removed from the feedwater. Most NF–RO membrane manufacturers limit oil and grease to less than 0.1 mg/L, effectively zero. There are specialized membranes that can separate oil from water. These oil-removal UF membranes can be used to pretreat feedwater for NF and RO. Specialty oil-removal MF–UF membranes are also used to pretreat feedwater for NF and RO. The membrane manufacturers' guidelines should be consulted in these situations.

Cationic polymers and other chemicals used in wastewater treatment can have detrimental effects on membranes and should be avoided. In addition, some esters can have detrimental effects on membranes. Aromatic hydrocarbon is detrimental to polysulfone membranes, and most MF–UF membranes have good resistance to aliphatic hydrocarbons. Some dissolved hydrocarbons do not affect MF–UF membranes; however, they can affect NF–RO membranes if not removed or eliminated in pretreatment. Activated carbon or a proprietary absorbent such as OptiPure are used to remove dissolved hydrocarbon.

Absorption of organic substances on membrane surfaces causes flux loss, and these substances must be removed or eliminated. Organic compounds present as an emulsion may form an organic film on the membrane surface. The adsorption process is favored by high molecular mass compounds that are hydrophobic or positively charge.

Ceramic, zirconium oxide, carbon, and sintered metal MF–UF membranes are resistant to most organic and inorganic chemicals and can operate at elevated temperatures. However, these membranes do have limitations. Ceramic membranes are damaged by hydrofluoric acid, and some inorganic acids attack some sintered metals.

FOULANT, SUSPENDED SOLIDS, AND PARTICULATES REMOVAL

Foulant is any material that physically blocks membrane pores or restricts the flow of water through the membrane. The major foulants are particulates, colloids, metal oxides, and biological materials. Feed to NF and RO membranes should be free from any foulants. MF and UF membranes, which are used to remove foulant, have limitations on how much foulant they can handle, depending on the configuration, membrane material, pore size, and pretreatment. It is important to refer to the membrane manufacturers' recommendations when removing foulant.

Suspended solids in feed negatively impact most membranes. Suspended solids plug the pores on the membrane surface and in water channels, increasing in pressure drop and hindering water flow, which reduces productivity. The increase in pressure drop requires increased feed pressure to maintain membrane productivity. Excessive pressure drop physically damages the membrane elements.

Other sources of fouling are chemicals added during pretreatment. These include coagulant, flocculant aid, some polymers (especially cationic), and dissolved matter present in the feedwater that may precipitate or solidify as a result of physical and chemical pretreatment processes.

Colloids are very small particles that range in size from 0.01 to 1.0 μm. Colloids are electrically charged and are in suspension because they repel one another. The sources of colloids in water reuse include the following:

- Feedwater that passes through or over clay geological structures and picks up aluminum silicate colloids;
- Corrosion products from carbon steel pumps, piping, and filters that occur as iron colloids;
- Feedwater pretreated with alum and improperly clarified water that contain aluminum hydroxide;
- Presence of colloidal sulfur from hydrogen sulfide; and
- Biological treatment process that precedes the membrane treatment.

Colloidal fouling, which is predominant in NF and RO membranes, is caused by coagulation of the colloids and deposition of the coagulated material on the membrane. For MF and UF membranes, fouling occurs when colloids pass through the membranes.

Aluminum fouling is caused by the precipitation of aluminum hydroxide. Wastewater treatment plants often use alum coagulation in advanced treatment of secondary wastewater. The aluminum ion is amphoteric, that is, capable of reacting chemically either as an acid or as a base, and has minimum solubility at a pH of approximately 6.5 to 6.7. If alum coagulation is used upstream, the pH should be 6.5 to 6.7 to minimize the solubility of aluminum.

The recommended pretreatment processes for suspended solids removal depend on the physical and chemical properties of the materials to be removed. Suspended solids or particulates in the feedwater can be settleable solids or colloids. Settleable solids can be removed by settling if the feedwater is allowed to stand for 3 min; colloidal solids require the addition of chemicals, coagulant, or flocculant for settling. Because colloids range in size from 0.3 to 1.0 μm and are charged, they can pass through MF and UF membranes. In order to remove the colloids, the feedwater is coagulated and flocculated. This causes the colloids to form large particles and also neutralizes their charge. The upper limit for suspended solids concentration for most MF and UF membranes operating in the dead-end filtration mode is 1,000 mg/L. The ultimate aim of suspended solids and particulate removal is to have pretreated raw NF and RO membrane feedwater that has a silt density index (SDI) of less than 3.0, turbidity less than 0.5 ntu, and total suspended solids (TSS) <10 mg/L.

COAGULATION–FLOCCULATION

Coagulation–flocculation is the most flexible pretreatment process when followed by clarification and/or filtration because it can work with variable suspended solids loads.

Common metallic salt coagulants include alum, ferric chloride, polyaluminum chloride, and aluminum chlorohydrate. Coagulation improves membrane performance by increasing membrane productivity, enhancing dissolved organic carbon (DOC) and/or nutrient (that is, phosphorus) removal, and reducing membrane cleaning requirements. The specific mechanisms responsible for these enhancements include the following:

- Reduction in the number of particles that enter the membrane pores,
- Conditioning of the cake layer, and
- Enhancement of back transport.

The type and dose of coagulant is dependent on several factors including feedwater quality, including temperature and pH, which is best optimized by performing jar testing.

Ferric salts are popular because they work well in water with a pH range of 5 to 8.5. The most common ferric salt used is ferric chloride ($FeCl_3$). The dosage for $FeCl_3$ depends on the quality of the feedwater, particularly as it relates to alkalinity and the suspended solids concentration (as confirmed by jar testing), which ranges from 1 to 30 mg/L as Fe^{3+}. Low-turbidity water requires sweep floc; this means that a relatively high concentration of Fe^{3+} is required. The hydrolysis reactions of ferric salts are very fast, requiring mixing times of less than 1 s, preferably 0.5 s. Ferric chloride is also effective in removing small amounts of oil and grease in feedwater. Alum and polyaluminum chloride are also used as coagulants in water reuse.

Coagulation to remove foulants and other contaminants from the feedwater can be performed before MF–UF systems to enhance membrane performance. Coagulation can be performed via conventional methods including coagulation–flocculation/sedimentation processes or in-line. Direct filtration, such as media filtration, can also be applied in MF–UF. Compared to conventional coagulation processes, direct filtration or in-line coagulation offers the benefit of a reduced footprint. However, high-rate tube plate settlers have been used upstream of MF–UF in advanced water treatment (AWT) applications. For example, the Bundamba AWT plant in Queensland, Australia, uses high-rate settlers upstream of the MF process to remove phosphorus; this prevents scaling (calcium phosphate) during the downstream RO processes (Freeman et al. 2008; Wallis-Lange et al. 2009) (also, see Chapter 9).

Coagulation is always followed by flocculation. Flocculation chambers with vertical-shaft turbines are preferred because a minimum area of metal is exposed to the water. The retention time in the flocculation chamber ranges from 5 to 30 min, with velocity gradients of 10 to 80/s. As an alternative, proprietary coagulation–flocculation basins, most of which are in circular configuration, are available from various manufacturers. Coagulation–flocculation is usually followed by clarification; however, the treated water stream can go directly to filtration. Coagulation–flocculation (followed by filtration) must be used when oils (hydrocarbon and silicone based) and greases are present in the NF or RO feedwater. For more details on coagulation–flocculation, see AWWA Manual M37 (2011).

CLARIFICATION

When the concentration of suspended solids after coagulation–flocculation is relatively high (>100 mg/L), it is recommended that coagulation–flocculation be followed by clarification. Clarification or sedimentation can be achieved in both circular and rectangular (square) tanks. Rectangular (or square) tanks with an inclined-plate separator or tilted parallel plates are recommended to minimize the clarifier footprint. When ferric salt is used as a coagulant, the resulting flocs or sludge settle very well in an inclined-plate separator or tilted parallel plate clarifiers and tube settlers. The typical surface loading rate (overflow rate) for a clarifier (inclined-plate separator or tilted parallel plate) is 0.5 to 1.0 gpm/ft^2, with a retention time of 1.5 to 4 hours. For horizontal-flow rectangular clarifier tanks with plate and tube settlers, the surface loading rate is 1.0 to 3.0 gpm/ft^2, with a retention time of 6 to 10 min for tube settlers and 15 to 25 min for plate settlers. For additional information on clarifier design, see AWWA–ASCE (2005).

DISSOLVED AIR FLOTATION

Dissolved air flotation (DAF) involves a clarifier that is used to separate suspended solids or oils that do not readily settle because they are lighter than water. Algae, aquatic organisms, plants, and solids with strong affinity to air float when DAF is used. One advantage of DAF is its ability to remove grease. Starting with TSS of 100 mg/L, DAF can reduce TSS to less than 10 mg/L. DAF can be a good pretreatment to include in MF–UF systems. For more information on the selection and design for DAF, see Edzwald (2012).

MEDIA FILTRATION

Media filtration in NF and RO membrane plants has two basic functions: suspended solids removal and improvement of the SDI value to less than 3.0. In most applications, these functions can be achieved using a single-stage filter when the colloidal suspended solids are less than 10 mg/L and TSS are less than 30 mg/L. This limitation is based on the assumption that most of the suspended solids are settleable (that is, discrete and will settle within 3 min). When colloidal suspended solids are greater than 10 mg/L and TSS are greater than 30 mg/L, two-stage filtration is recommended. In two-stage filtration, the first-stage filter is for suspended solids removal (settleable and colloidal solids) and the second-stage, or polishing, filter is for improvement of SDI.

The selection of the filter configuration depends on water quality, mainly suspended solids concentration and SDI values. Some plants do have single-stage filters but with different functions. For example, a plant draws feedwater from a source with a measured SDI of less than 3 (sample taken from a large tank with more than 3 min retention), TSS 20 mg/L, and colloidal solids less than 10 mg/L. Another plant takes feedwater from a source with suspended solids of 10 mg/L and SDI values greater than 5. In the first case, a single suspended solids removal filter is recommended; for the second, a single-stage SDI improvement filter is recommended. The SDI improvement filter can handle suspended solids to some extent and it can reduce the filtration cycle.

Filtration rates for removal of suspended solids range from 3 to 6 gpm/ft^2 (7.33 to 14.6 m/h) and from 2 to 3.6 gpm/ft^2 (4.88 to 8.78 m/h). The backwash rate depends on the selection of the filter media but can typically range from 10 to 20 gpm/ft^2 (24.4 to 48.8 m/h) for polishing or SDI improvement.

Common filter media include sand, anthracite, and garnet. For a single-media filter, sand is typically used. Most dual-media filters use sand and anthracite, while multimedia filters use sand, anthracite, and garnet. Filter media depth depends on the media size and density run. Most RO membrane manufacturers recommended sand media filter depth of at least 20 in. (500 mm). The typical filter run is 24 to 72 hours.

The selection of gravity or pressure filter depends on economics and regulations. Some existing regulations require daily inspection of the filter surface, which is done only with gravity filters. Gravity filters are made from concrete, steel, or fiberglass reinforced pipe (FRP). Pressure filters can be vertical or horizontal. If horizontal filters are selected, the minimum diameter is 8 ft (2.43 m), but 10 ft (3.0 m) is preferred due to the minimum bed depth required.

The recovery for media filtration is greater than 90 percent. The design of media filters is not covered here. A common mistake that engineers make in designing media filtration for NF or RO pretreatment is to use the same data that is used for potable water filtration, which are based on turbidity. For NF and RO applications, SDI is the criteria.

Granular media filters are also used in MF–UF installations, although they are not common. Potential benefits of adding granular media filtration upstream of MF–UF include reduced biological and particulate fouling, which can lead to operation at higher flux rates, and reduced time between cleanings. Testing at the San Diego, Calif., advanced water purification system showed that MF and UF systems that operate on tertiary filtered effluent had cleaning frequencies of 6 months at flux of 30 gfd (City of San Diego 2013). For more information on media filtration, see *Water Quality and Treatment* and AWWA Manual M37 (2011).

MICROFILTRATION–ULTRAFILTRATION AS PRETREATMENT

Using microfiltration and ultrafiltration as pretreatment for NF–RO is preferred over two-stage media filtration, especially where there is limited space and the construction cost is high. MF or UF can be used in surface water treatment is usually used as pretreatment for NF and RO for water reuse. One advantage of MF–UF is that these processes can consistently produce filtered water with less than 0.5 ntu and SDI of less than 3. The typical flux for MF–UF ranges from 30 to 45 gfd (50 to 76 l/h × m^2). Most NF and RO manufacturers recommend operation at higher flux when the feed is pretreated with MF–UF.

OTHER PRETREATMENT PROCESSES FOR SUSPENDED SOLIDS REMOVAL

The other pretreatment processes used to remove suspended solids include the following:

- Hydrocyclones,
- Stationary or rotating screens,
- Degritters,
- Strainers,
- Centrifuges, and
- Settling basins.

CONTROLLING TEMPERATURE

Most polymeric membranes used in water treatment operate at temperatures that range from 0 to 45°C. There are special polymeric membranes that can operate at up to 80°C. Inorganic membranes such as ceramic and sintered metal types can operate at greater than 100°C. Temperature significantly affects NF and RO membrane treatment processes. Increased temperature results in a decrease in salt rejection (increased salt passage), an increase in permeability, and relatively lower operating pressure. In MF and UF, the effect of temperature is not as significant. However, temperature does affect the water's viscosity and operating pressure; it may also affect the passage of some organic macromolecules. When the water temperature is outside of the operating range required by the process, the feedwater stream must be either cooled or heated. Cooling and heating can be accomplished through heat exchangers, and cooling can also be achieved using an evaporative cooling tower or by dilution.

pH

pH plays a very important role in membrane treatment, especially in NF and RO because a change in pH can significantly affect salt passage and salt rejection. pH also affects the solubility of many salts and the speciation of many ions, which can trigger other problems such scaling, corrosion, and fouling.

DISSOLVED IONIC SPECIES THAT REQUIRE PRECONDITIONING

Some dissolved ionic species are effectively removed or separated by membrane processes if the feedwater is preconditioned. Preconditioning includes pH adjustment, addition of oxidizing or reducing agents, and use of chemical additives. Common examples of these ionic species are arsenic, chromium, selenium, silica, boron, carbon dioxide, and hydrogen sulfide.

OXIDATION AND REDUCTION PROCESSES

Some ionic species present in feedwater are effectively removed if they are oxidized. Ferrous ions that are dissolved in feedwater can pass through MF or UF membranes and, to some extent, can pass through NF and RO membranes. However, in oxidized form, ferric iron precipitates and can be rejected by MF or UF membranes. Oxidation of ferrous ions can be achieved by aeration, ozonation, or use of an oxidizing filter such as manganese dioxide or a zeolite catalyst called BIRM. Other oxidizing agents include chlorine dioxide, hydrogen peroxide, and potassium permanganate.

Precipitation of metal salts dissolved in feedwater can occur in membranes. This is not a serious concern for MF and UF membranes. However, for NF and RO membranes, metal precipitation can lead to serious problems such as fouling and binding. The most common metal precipitants are iron oxides and aluminum compounds.

Oxidation of ferrous ion on a membrane can be a problem. Ferrous ion alone does not affect the performance of NF and RO membranes and is rejected by these membranes. The presence of dissolved oxygen (DO) or an oxidizing agent can cause the ferrous iron to oxidize to ferric oxide, especially at high pH, and form deposits on the membrane. Controlling the feed pH and eliminating oxygen and oxidizing agents in the feed can prevent this form of iron precipitation.

SOFTENING

Lime–Soda Ash Softening

Lime softening is used to remove calcium and magnesium. This is done by adding hydrated lime in a lime reactor–clarifier and follows the chemical reaction shown below:

$$Ca(HCO_3)_2 + Ca(OH)_2 \rightarrow 2CaCO_3 + 2H_2O \qquad \text{(Eq 5-1)}$$

$$Mg(HCO_3)_2 + 2Ca(OH)_2 \rightarrow Mg(OH)_2 + 2CaCO_3 + 2H_2O \qquad \text{(Eq 5-2)}$$

With the addition of soda ash or sodium bicarbonate:

$$CaCl_2 + Na_2CO_3 \rightarrow 2NaCl + CaCO_3 \qquad \text{(Eq 5-3)}$$

The lime–soda ash process can also be used to reduce the silica concentration. The addition of sodium aluminate and ferric chloride forms precipitate that includes calcium carbonate and a complex with silicic acid, aluminum oxide, and iron. The ultra-high lime process (pH 11.3) also precipitates silica. With hot lime, silicic acid is removed at 60 to 70°C, resulting in 1 mg/L silica. This is done by adding a mixture of lime and porous magnesium hydroxide. Another variation of lime softening is a pellet softener, which produces crystalline pellets instead of sludge.

A variation of lime softening used in water reuse is the fluidized bed reactor. These softeners are often referred to as pellet softeners because crystalline pellets are generated instead of sludge. Figure 5-4 shows a pellet softener installation in Hillsboro Beach, Fla. One potential application of the pellet softener is in the recovery of brine reject.

Another variation of lime softening is the integrated precipitative membrane process, which uses lime or lime–soda ash to soften wastewater (Figure 5-5). A slurry consisting of lime and calcium carbonate is continuously fed directly into a MF or UF tubular or flat sheet membrane system in cross-flow filtration mode. This process has been proven effective in removing silica from water.

Courtesy of the Town of Hillsboro Beach Water Department.

Figure 5-4 Hillsboro Beach, Fla., pellet lime softener installation

Strong Acid Cation Softening

Scale-forming cations such as calcium (Ca^{2+}), barium (Ba^{2+}), and strontium (Sr^{2+}) can be removed by ion-exchange and replaced with sodium ions. When exhausted, the ion-exchange resin is regenerated with sodium chloride. The pH of the softened water is not changed. The carbon dioxide present in the water can pass through the membrane. By adding caustic soda to the softened water before feeding to an NF or RO membrane, the permeate conductivity can be reduced. The carbon dioxide is converted to bicarbonate, which is rejected by the NF or RO membrane.

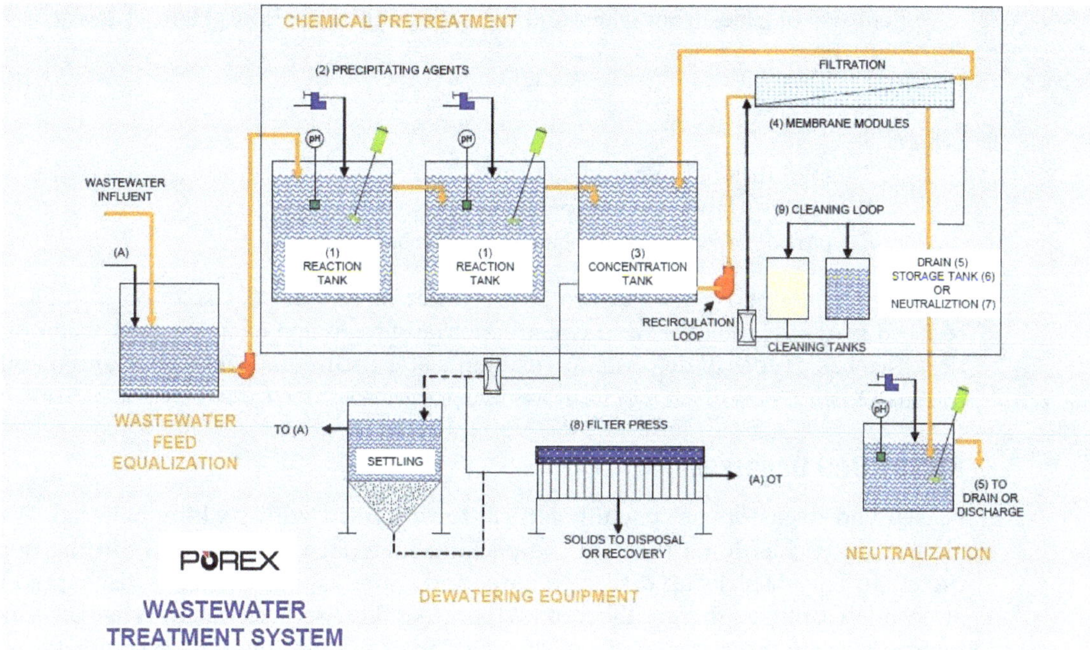

Courtesy of RWL Water and Porex.

Figure 5-5 Integrated precipitative membrane treatment

Weak Acid Cation Dealkalization

In dealkalization that uses weak acid cation-exchange resin, only Ca^{2+}, Ba^{2+}, and Sr^{2+} linked to bicarbonate (temporary hardness) are removed and replaced by H^+, which lowers the pH to 4 to 5. The acidic groups of the resin are carboxylic groups. The ion exchange stops when the pH reaches 4.2; at that point the carboxylic groups are no longer dissociated. This process is ideal for water with high bicarbonate content. The bicarbonate is converted into carbon dioxide as shown in the following:

$$\rightarrow HCO_3^- + H^+ \; H_2O + CO_2 \qquad \text{(Eq 5-4)}$$

DEGASSER

Some wastewater may contain significant concentrations of undesirable dissolved gases that must be removed as part of treatment. Some dissolved gases may interfere with the membrane treatment process, while others can pose hazards to plant operations.

Dissolved gases can be found in the treated effluent that feeds the membrane treatment system. In most cases, these dissolved gases do not have any effect on the membrane treatment processes because they are not rejected by the membranes. In some cases, as with carbon dioxide, the gases change the water's pH and can affect the treatment process. Hydrogen sulfide, in the presence of air or other oxidant and when oxidized to elemental sulfur, can cause membrane fouling. Flammable gases such as methane must be removed because they are a fire hazard.

Removal of gases from water and aqueous solutions follows Henry's law:

$$S = kP \quad \text{(Eq 5-5)}$$

Where:

S = solubility of the gas in solution
K = Henry's law constant for the gas in solution
P = partial pressure of the gas above the solution

When the space above the surface of the water or an aqueous solution is subjected to reduced pressure, the gases and water vapor tend to escape from the bulk water or solution. Two basic types of degasser are used in water treatment: packed bed and membrane or contactor.

Packed Bed Degasser

Packed bed degassers are usually vertical towers filled with packing material. Water is allowed to trickle down through the packing material while air flows in the opposite direction. The air picks up dissolved gases and water vapor, which are discharged to the atmosphere or through a scrubber to further treat the dissolved gases. When air-sensitive materials are in the water or when introduction of oxygen into the treated water is to be avoided, a vacuum degasser is used.

Membrane Degasser

A membrane degasser is an alternative to a packed bed degasser when the footprint is limited, direct contact of air and water is not allowed, or in sterile conditions where atmospheric air must be avoided due to possible introduction of airborne microorganism. In a membrane degasser, water or aqueous solution flows through bundles of membrane tubes. This allows the gases to permeate through the membrane walls, but not the water, when reduced pressure is applied outside the membrane bundle. In a membrane degasser, the removal of dissolved gases from the bulk water or aqueous solution also follows Henry's law.

CONTROLLING BIOLOGICAL FOULING

The presence of assimilable organic matter (AOM) and microorganisms in feedwater creates the potential for deposition of microbial by-products on the membrane surface. This is known as biofouling. Bacterial activity can result in both membrane fouling and growth of bacterial slime, which can plug the feed channel of the membrane element and result in deterioration of system performance and increased cleaning frequency. In severe cases, membrane replacement frequency also increases. The strategies for preventing biofouling include the following:

- Removal or elimination of the AOM, which is food for the microorganism;
- Removal and/or destruction of the microorganisms;
- Prevention or minimization of the formation of biofouling on the membrane; and
- Removal of the biofoulant after it is deposited on the membrane surface.

Biological activity within the membrane treatment system can occur in several forms, each requiring unique solutions. Algae can enter a system with the feedwater and/or grow within the pretreatment system. Bacteria can enter with the feedwater and grow within the pretreatment system or the membrane elements. Many methods for controlling biological activity can be used. Although some techniques are based on seemingly logical theories, they do not always solve the problem due to unanticipated or unforeseen interactions.

CHLORINATION

Ideally, maintaining a biocide such as chlorine throughout the membrane system would be the way to control biological fouling. However, as described previously, most membranes have limited chlorine tolerance. The tolerance of most polyamide membranes is approximately 1,000 ppm-hour, which can be greatly reduced if iron is present (for example, corrosion products, coagulant carryover).

To prevent deterioration of membrane performance due to biological fouling, many systems use chlorination and dechlorination (sodium bisulfite addition) in pretreatment. The goal is to eliminate all biological activity with chlorine and then dechlorinate for membrane protection. However, in some cases, rather than reducing the membrane element fouling rate, chlorination and dechlorination result in an increased rate of fouling due to the breakdown of organic matter into AOMs.

Bacterial regrowth following dechlorination is often more problematic than bacteria in the raw feed. Unless the feedwater is severely contaminated (for example, from a sewage outfall), chlorination and dechlorination are not required or desirable. The preferred method is to periodically use chlorination and dechlorination equipment only for shock treatment to control growth within the inlet structure and pretreatment equipment.

If there is significant biological activity in the raw feed, chlorination and dechlorination may be the best option. However, alternative biocides for periodic shock treatment of the RO membrane should be considered to augment the chlorination and dechlorination processes. Several nonoxidizing biocides that are compatible with membranes are used periodically for shock treatment to arrest biological activity.

If dechlorination is necessary, sodium bisulfite (sodium metabisulfite addition) is the preferred method for seawater systems. Generally a 3:1 weight ratio of bisulfite to chlorine residual is adequate. If chlorine residual is expected to be erratic, more bisulfite can be added. The sodium metabisulfite should not be a cobalt-activated grade. When an analytical method is used to verify dechlorination, it is most effective to measure residual sulfite or reduction potential, not "zero" chlorine.

Chloramine

Chloramines include monochloramine, dichloramine, and trichloramine. Chloramine is of interest in water reuse because ammonia is present in most treated effluent used as feedwater. Monochloramine is the desired disinfectant because dichloramine and trichloramine produce a sharp taste and odor. Chloramine is usually used in place of chlorine to reduce specific disinfection by-products (DBPs), that is, trihalomethanes and haloacetic acids. However, there are potential disadvantages to using chloramines, for example, they can produce a variety of nonregulated DBPs such as *N*-nitrosodimethylamine (NDMA), degrade rubber and elastomeric compounds used in various water treatment equipment, and interfere with aquaculture applications.

Chloramines are generated by combining chlorine and ammonia at specific ratios to form various types of chloramine. At a ratio of 5:1 (milligrams per liter of residual free chlorine to milligrams per liter ammonia as N) or less, monochloramine is formed; at a ratio of 5:1 to 7:1, dichloramine is formed; and at a ratio of 7:1 to 10:1, trichloramine is formed. At ratios above 10:1, all of the ammonia is reacted and free chlorine is again formed. The ratios change with pH and temperature.

The use of chloramines before MF–UF is an effective way to reduce organic and biological fouling in water reuse applications. Typically, chloramines are dosed to achieve a 3-mg/L residual in the filtrate; for RO systems, this reduces fouling of the downstream membranes. An important consideration when using chloramines in potable reuse applications is to minimize NDMA formation. Research has shown that NDMA formation is linked directly to a specific water matrix of chlorine–chloramine addition, pH, and order of reagent addition (Schreiber & Mitch 2005). Chloramines can form by the sequential addition of ammonia and chlorine into the feedwater or by a sidestream process in which chloramines are preformed prior to addition to the feedwater. The latter method is used at the Gibson Island AWT, Brisbane, Australia, and has proven to reduce NDMA formation by adding only monochloramine and allowing time for the monochloramine to react completely before it reacts with nitrogenous precursors (Crittenden et al. 2012). Farre et al. (2010) reported that the most significant factor that contributes to NDMA formation in purified recycled water facilities is the contact time of the wastewater with the disinfectant.

Ozone

Ozone is often use to disinfect membrane feedwater when there is a specific treatment objective such as oxidation of iron or manganese, taste, odor, color, or high levels of natural organic matter or when ozonation is part of a biological treatment process. Ozone is a very strong oxidizing gas that must be generated on site. It is an effective disinfectant for bacteria, viruses, and parasites.

The use of ozone as pretreatment for MF–UF systems that are part of a reuse treatment train is a relatively new approach and the topic of a recent research study funded by the WateReuse Research Foundation (*Ozone Pretreatment of a Non-Nitrified Secondary Effluent Before Microfiltration*). The study showed that ozone generally improved MF performance by chemically oxidizing wastewater effluent organic matter. The study also identified several challenges including increased potential for manganese fouling (based on pilot testing of polyvinylidene difluoride membranes), NDMA formation, and bromate formation. Based on this research, the West Basin Municipal Water District recently implemented ozone upstream of their MF system at the Edward C. Little Water Recycling Facility, Carson, Calif. At the time this manual was prepared, San Diego, Calif., was testing the use of ozone followed by biologically active filtration upstream of MF as part of a potential direct potable reuse train.

Chlorine Dioxide

Chlorine dioxide is gaining popularity for use in membrane treatment plants. Studies and real-world experiences showed the following benefits: high biocidal effectiveness, limited production of harmful by-products, and relatively mild effects on polymeric membrane structures. In addition to disinfection, chlorine dioxide is also used to oxidize iron and manganese, hydrogen sulfide, and phenolic compounds and to control taste, odor, and color. Like ozone, chlorine dioxide is generated on site. When generating chlorine dioxide from chlorine or other chlorine compounds, it must be ensured that there is no chlorine residual in the chlorine dioxide solution. Chlorine dioxide does not react, in general, with organics in the same way and to the same degree as chlorine.

Ultraviolet Disinfection

Ultraviolet (UV) light is a form of electromagnetic radiation with a wavelength that is shorter than that of visible light, in the range of 10 to 400 nm. Water can be disinfected using UV lights within the germicidal range (UV-C wavelengths from 200 to 280 nm and UV-B wavelengths from 280 to 300 nm). UV light within these germicidal ranges damages nucleic acid, preventing microbes from replicating. In addition to wavelength, the effectiveness of UV light as a sterilizer is determined by dose, which is calculated as UV intensity multiplied by the time of exposure, usually expressed as millijoule per square centimeter. UV light's effectiveness as a sterilizer is also affected by the UV transmittance in water, which is influenced by the presence of UV-absorbing organic chemicals. It is important to keep the water free from suspended solids, which can form deposits in the UV lamp sleeves. UV disinfection is not effective for virus inactivation. For more details on UV sterilization, refer to *Ultraviolet Disinfection–Guidelines for Drinking Water and Water Reuse* published by the National Water Research Institute in collaboration with Water Research Foundation (2012).

Algicide

Algae are often present in feedwater that comes from open reservoir or storage systems. They are removed by pretreatment and can be a problem if allowed to grow within the pretreatment system. The pretreatment system should be designed to avoid storage of feedwater, especially in open tanks. Exposure to air and sunlight should be avoided. Copper sulfate is an effective algicide at 0.5 mg/L. Although not effective against bacteria, copper sulfate can be used to inhibit algal growth within intake structures and media filters. Copper sulfate cannot be used to remove algae from raw water. Some regions, including in the United States, have restrictions on the discharge of copper into brine, which would eliminate copper sulfate as a treatment option. Copper is very well rejected by NF–RO membranes and should not be an issue for meeting potable permeate standards.

MICELLAR-ENHANCED MEMBRANE SEPARATION

Low-molecular-weight organic compounds that are normally permeable to MF–UF and NF membranes can be separated if they are complexed with larger impermeable molecules. Surfactants can form complexes with organic molecules and metals under certain conditions. Surfactants contain a hydrophobic tail and a hydrophilic head and consist of anionic, cationic, or nonionic groups, as designed. These characteristics make surfactants an effective cleaning agent, where the hydrophobic interior of the micelle serves as a solvent for organic matter and the polar hydrophilic group orients itself outward toward the water. Micellar-enhanced membrane separation takes advantage of a surfactant's ability to associate with itself and form micelles above a certain concentration, known as the critical micelle concentration. Micelles can be 20 to 200 molecules in size. If solute–micelle interactions take place, it is possible to separate the micelles with membranes, as they are complex low-molecular-weight solutes.

Heavy metals can be separated from wastewater if macromolecules such as sodium polystyrene sulfonate, polyacrylic acid, polyethylenimine, or polyvinyl alcohol are added to bind the metals, thus forming large complexes that can be separated by membranes.

OPTIMIZATION OF UPSTREAM BIOLOGICAL WASTEWATER TREATMENT

Optimization of upstream wastewater treatment processes is an effective pretreatment for MF–UF used for water reuse applications. The longer solids retention time that is associated with the upstream wastewater treatment process can lead to lower dissolved organic constituents in the MF–UF feed, which can reduce membrane fouling. Bradshaw (2013) compared the performance of MF systems that operate with and without upstream nitrification–denitrification (NdN) and found that systems with NdN operated at higher flux and recovery and required less frequent membrane cleaning.

MF–UF DESIGN CONSIDERATIONS

As noted previously, MF–UF membranes have pores in the surface (Figure 5-6), and the separation of impurities is by size exclusion. It is for this reason that water analysis alone cannot be used when selecting MF–UF membranes and designing MF–UF systems. The information used in designing MF–UF systems is usually based on information collected by membrane manufacturers, similar feed streams, prior projects, actual plant operating logs, and data from bench-scale and/or pilot test results.

The two basic operating modes in MF–UF are dead-end filtration and cross-flow filtration (Figure 5-7). In dead-end filtration, all of the feed goes through the membrane (100 percent filtrate recovery), and the suspended solids are continuously deposited on the membrane surface, causing an increase in deferential pressure across the membrane. At the set differential pressure, the membrane is backwashed by reversing the flow, and the solids are removed from the membrane surface. In cross-flow mode, the feed flow across the membrane surface is at a specific velocity that causes some of the filtrate to pass through the membrane; the suspended solids remain in the bulk stream, which concentrates the solids. The feed is added to the bulk stream, which recirculates through the membrane. To control the concentration of solids in the bulk stream, part of the bulk stream is wasted.

Courtesy of Porex.

Figure 5-6 Ultrafiltration membrane surface showing pores

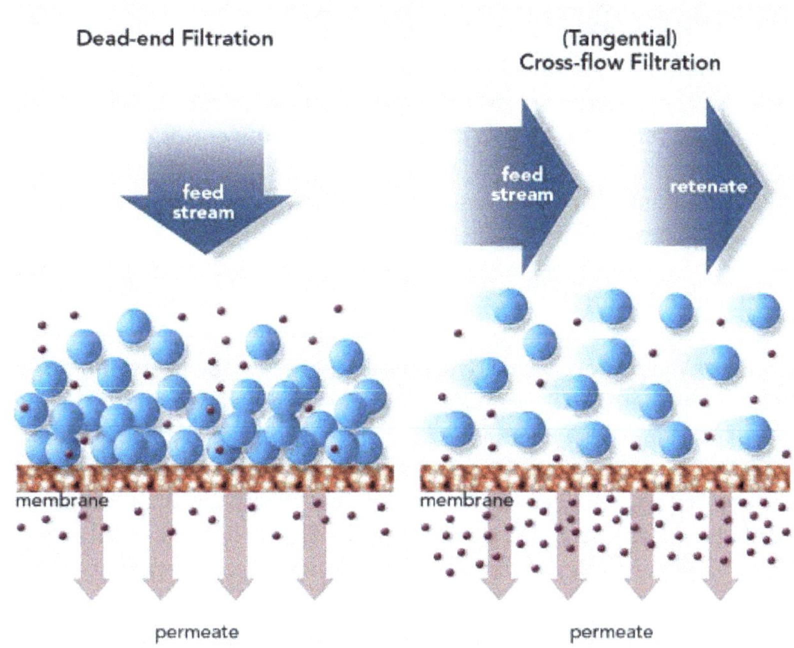

Courtesy of Pentair.

Figure 5-7 Microfiltration–ultrafiltration filtration modes

MF/UF DESIGN CONSIDERATIONS

The following section provides information on basic design considerations related to MF and UF membrane systems used for water reuse applications. The intent is to provide general information related to various aspects of MF–UF, including pretreatment, configuration, suppliers, design parameters, filtrate quality, and monitoring. For additional information, see AWWA Manual M53 (AWWA 2005).

MF–UF system design varies based on specific site and project conditions and on the selected membrane product. Consequently, it is crucial to engage the membrane manufacturers early in the design phase to ensure that the products being considered are capable of treating the specific feedwater. The products must also meet all performance requirements, including the quality and quantity of filtrate, and operating requirements, such as cleaning interval and pretreatment.

Some membrane system suppliers and consulting engineers have tried to implement a universal MF–UF train frame design that is based on a vertical membrane module configuration to allow interchangeability of membrane modules. Currently, MF–UF modules are not standardized, unlike NF–RO modules. Outside of the United States, most large MF–UF trains are horizontal. For very large MF–UF trains with capacities that exceed 5 mgd, a horizontal configuration offers a more economical footprint.

MF–UF FILTRATE QUALITY REQUIREMENTS

The following below is based on current design practices in use at the time this manual was written. Because the number of MF–UF systems used for water reuse is increasing rapidly, it is anticipated that this information will be further refined and changed in the future due to increased experience, improvements in membrane products and related technology, and further development of relevant regulations.

MF–UF Filtrate Quality

MF–UF membranes remove contaminants present in the feedwater via physical sieving. As such, these systems consistently produce low-turbidity filtered water and achieve high microbial log removals. Unlike operations with conventional granular media, these performance criteria are provided by MF–UF independent of chemical preconditioning, such as with a coagulant, since the removal mechanism is physical sieving. AWWA Manual M53 (2005) includes a review of 122 studies that showed that, "MF and UF membranes produce an extremely high-quality water regardless of influent turbidity and that there is no apparent difference in turbidity removal between membrane type, manufacturer, or whether a coagulant was used." Mean filtrate values reported in all of the studies had a median of 0.06 ntu and the maximum values had a median of 0.08 ntu. Jacangelo, Adham, and Laine (1991, 1995), Coffey et al. (1993), and others have shown that MF and UF also consistently remove *Giardia* cysts and *Cryptosporidium* oocysts, with removals greater than 4 log and ranging from 6 to 7 log. Cysts are not detected in the filtrate of intact MF–UF membrane systems; therefore, reported values are limited by the sensitivity of the measurements. While there is no observed difference between MF and UF regarding removal of protozoan cysts (which are 2 to 15 μm in size), UF provides greater removal of viruses (which are 1 to 2 orders of magnitude smaller). Generally, a virus log-removal credit of 0 to 0.5 log is granted for MF and 2 to 4 log is granted for UF, depending on the specific membrane, level of testing sensitivity, and the regulatory agency. There is no definition that differentiates MF from UF, other than the pore size ranges and virus removal ranges indicated above.

Membranes used for water treatment have pore sizes that range from 0.1 to 0.5 μm for MF and from <0.01 to 0.1 μm for UF. Some UF membranes with pores smaller than 0.01 μm are rated based on the molecular weight cutoff (MWCO); nominal MWCO values of 100,000 to 200,000 Daltons are typical. Neither MF nor UF achieve substantial removal of dissolved iron and manganese, hardness, true color, or DBP precursors such as DOC. To augment removal of dissolved contaminants, MF–UF is integrated with processes that convert the dissolved contaminants into particulate form, such as coagulant or powered activate carbon, in addition to adsorb dissolved organics, or oxidants to precipitate manganese and iron.

MF–UF FILTRATE QUALITY REQUIREMENTS FOR WATER REUSE

In general, the primary water quality requirement for nonpotable water reuse is low-level filtrate turbidity. For example, the California Code of Regulations, Title 22, Water Recycling Criteria (Title 22), requires that turbidity for membranes not exceed 0.2 NTU "more than 5 percent of the time within 24-hour period and 0.5 NTU any time." MF–UF membranes are capable of exceeding this requirement consistently and in a verifiable manner. Information on specific MF–UF technologies that have received conditional acceptance to meet the filtration requirements for Title 22 is provided elsewhere (CDPH 2013).

For reuse applications where MF–UF is used to provide pretreatment for downstream pressure membrane systems (that is, NF–RO), including agricultural reuse (for example, total dissolved solids [TDS]/chloride removal) and potable reuse (via groundwater recharge or surface water augmentation), additional filtrate water quality requirements may apply. Typical targeted MF–UF filtrate water quality requirements to promote effective NF–RO operation include the following: SDI, <3; suspended solids, undetectable; turbidity, 95th percentile <0.1 NTU; *Giardia*, undetectable; and *Cryptosporidium*, undetectable. In addition, some reuse applications target removal of dissolved materials including inorganic material (for example, manganese, iron, phosphorus) and organic materials (for example, DOC) by MF–UF. Because these materials are smaller than the membrane pore size, additional pretreatment is required to facilitate removal.

For potable reuse applications in California, MF–UF can also be used to achieve required log removals for *Giardia*, *Cryptosporidium*, and viruses (CDPH 2014). For such applications, integrity monitoring and membrane repair (as needed) are essential for compliance. In addition, the specific log reduction value (LRV) allotted for pathogens must be proven via demonstration testing or documented performance from other facilities that use the same membrane product. The California Division of Drinking Water and Environmental Management typically grants MF and UF LRVs of 0.5 and 2.5 to 4 for virus removal, respectively, and an LRV of 4 for *Giardia* and *Cryptosporidium* removal for both MF and UF (CDPH 2001, 2011).

IMPACT OF FEEDWATER QUALITY ON PRETREATMENT AND MF–UF SYSTEM DESIGN

Pretreatment requirements and design parameters for MF–UF systems are largely impacted by feedwater quality. Typically, in water reuse applications, MF–UF are used as part of a membrane bioreactor (MBR) system or downstream of secondary or tertiary wastewater treatment processes. Details of MBR system design are not provided in this manual but can be found elsewhere (WEF 2006; Judd 2006, 2011). Typical design parameters that are applicable to MF–UF systems used for municipal water reuse applications that treat secondary or tertiary wastewater are provided in Table 5-3. Note these values can vary and are site specific. It is recommended that the membrane manufacturer and the membrane system supplier be consulted.

Table 5-3 Typical microfiltration–ultrafiltration design criteria for water reuse facilities

Design Parameter	Unit	Typical Value	Notes
Maximum trans-membrane pressure	Pounds per square inch	35–40 (pressurized) 14.4 (submerged)	Typical manufacturer recommend maximum values.
Instantaneous flux	Gallons per square foot per day (liters per square meter per hour)	20–35 (34–60)	This is the most important design parameter because it is used to determine the number of membrane modules required to produce design filtrate capacity. Typically source water with poor quality (for example, high suspended solids, organic matter) is designed with lower flux.
Minimum recovery	Percent	90–95	Defined as percentage of feedwater converted to filtrate over a set period of time (typically a day or interval between two maintenance washes). It takes into account water used for cleaning purposes and is dependent on flux, frequency, and duration of cleaning and the amount of water used for cleaning purposes.
Backwash/air-scour duration	Seconds	60–100	Deterioration in feedwater quality may require increasing the backwash/air-scour duration.
Backwash frequency	Minutes	15–40	Deterioration in feedwater quality may require reduction in the time interval between backwashing.
Maintenance wash frequency Chemical-enhanced backwash	Hours	12–72	Typically practiced in municipal water reuse to mitigate organic/inorganic fouling.
Clean-in-place frequency	Days	30–90	Required to mitigate fouling for source waters high in organic/inorganic constituents.

Even though particle removal and turbidity control with MF–UF are independent of raw water quality and pretreatment conditions, membrane productivity can be affected by feedwater quality. For example, if raw water turbidity increases, the operator may need to reduce the membrane flux rate or decrease the time interval between backwashing and chemical cleaning operations to avoid undesirable levels of fouling. However, such changes could impact the overall net filtrate capacity, thereby impacting downstream processes such as RO. Chemical dosing may also need to be adjusted. Currently there are no guidelines on when these strategies should be used. Consequently, the operator should be aware of the impact of raw water and pretreatment changes at the facility, as well as the options available for addressing these changes. For municipal water reuse applications, changes in upstream wastewater processes can impact MF–UF performance. For example, upsets in the biological secondary process can lead to increased organic fouling. Degradation of upstream tertiary filter performance and changes in coagulant and polymer dosing are other factors that may impact MF–UF performance.

MF–UF EQUIPMENT

Membrane Process Configuration

MF–UF systems have two process configurations—pressurized (or encased) and submerged (or immersed). A brief description of the configurations is provided below.

Pressurized membrane configuration. In this configuration, the membrane fibers are mounted inside pressure vessels. In direct flow systems, 100 percent of the feedwater is driven through the membrane, and solids accumulate on the fiber as a cake during the filtration mode. The backwash/air scour/chemical-enhanced backwash/clean-in-place (BW/AS/CEB/CIP) cycles are the only way to remove this cake from the fiber. Therefore, it is important that these steps be executed properly. The filtered air or water that flows backward through the fiber wall is used to free accumulated material from the fiber surface. Consequently, the velocities or pressures must be within the manufacturer's recommendations. In addition, the BW water and air need to be very clean to prevent membrane damage or fouling. Pressurized membrane systems can also be operated with crossflow to reduce cake formation on the fiber. In addition to maintaining proper BW/AS/EB/CIP conditions, it is important that crossflow velocities be maintained within the proper range. Crossflow is typically used to obtain higher unit productivity, especially during periods of elevated feed turbidity. Therefore, failures can have a dramatic impact on productivity. The down side of this configuration is that it operates at lower recovery.

Submerged membrane configuration. In this configuration, the membrane modules are mounted in tanks that are open to the atmosphere. Submerged systems are usually equipped with air systems for agitation and with bleed systems to control solids buildup within the basin. The air may be intermittent or continuous depending on the application. It is important to maintain proper air flow within the basin to prevent solids from building up in the bottom of the basin and on the fiber. The appropriate bleed rate should also be maintained to prevent excessive solids buildup within the basin. These systems also include BW/AS/CEB/CIP cycles, which should be maintained properly for the reasons listed for other systems.

Comparison of membrane configurations. Several factors should be considered when comparing submerged and pressurized membrane systems based on specific site conditions. Factors to consider include capacity, hydraulic profile, building and excavation limitations, time frame for construction, availability, condition and configuration of using existing tanks, and cost. Larger capacities can favor submerged systems because of the smaller footprint.

Membrane Flow Configurations

The majority of MF–UF membrane products are configured with an outside–in flow pattern. Membranes with inside–out configurations are pressurized and generally use BW/AS/CEB to control fouling. It is important that pretreatment systems operate properly to prevent large debris from entering the system and plugging the inside lumen or entrance to the fibers. Membrane fibers that are small in diameter are more susceptible to plugging and require more sophisticated straining or pretreatment. The use of crossflow to control fouling of inside–out systems is not common due to the high energy requirements.

Membrane Material

Commercially available MF–UF membrane products are manufactured from a variety of polymers including polypropylene (PP), polyvinylidene difluoride (PVDF), cellulose derivatives, polyethersulfone (PES)–polyvinylpyrrolidione (PVP) blends, polysulfone, and polyacrylonitrile.

These materials vary in mechanical strength, fouling resistance, hydrophobicity, hydrophilicity, and chemical resistance. An important consideration when selecting MF–UF membranes for reuse applications is tolerance of oxidants allowing high concentration chlorine cleaning or continuous chloramines dosing which are shown to control organic and microbial fouling. Of the materials listed above, PES–PVP blends and PVDF offer the highest oxidant tolerance, while PP is the least oxidant tolerant. Typically, PP membranes are not subjected to free chlorine, but in several reuse plants, chloramines are used to control biofouling.

In addition to polymeric-based membranes, several ceramic MF–UF membrane products are commercially available. While ceramic membranes offer several advantages including high durability (chemical, mechanical, and thermal stability), high production (permeability and flux), long service life, and microbial fouling resistance, their production cost is significantly higher than for polymeric-based membranes.

Module Geometries

For water reuse applications, there are several commercially available membrane geometries. The most common include the following:

- Hollow fiber,
- Tubular, and
- Flat-sheet.

Hollow-fiber and tubular configurations allow the membranes to be backwashed, which controls particulate and organic fouling. Flat-sheet configurations are typically used for MBR applications where the membranes are operated at a lower flux and air scour is continuously or intermittently applied to mitigate membrane fouling.

Pilot Testing

Regardless of the membrane product being considered, it is important to conduct pilot testing using the source water to confirm the manufacturer's projected performance and operating conditions prior to final selection. One benefit of pilot testing is that the required cleaning regime can be determined. The cost of pilot testing is minimal compared to a full scale commitment.

Large-Capacity Water Reuse Projects That Use MF–UF

The number of membrane-based water reclamation projects is increasing. Table 5-4 lists examples of large-capacity membrane-based water projects under construction or operational at the time this manual was prepared.

Filtrate Water Quality and Monitoring

Performance monitoring is a key aspect of operating a MF–UF membrane system that is used for water reuse applications. Table 5-5 provides information on key monitoring parameters categorized as water quality, membrane integrity, and operational performance. It has been shown that MF–UF membrane filtration provides greater than 4-log removal of cysts that are 2 to 15 µm in size as long as the physical barrier is intact. When the barrier is intact, MF–UF does not exhibit particle or turbidity breakthrough during filtration. An important aspect of maintaining water quality performance is integrity monitoring and fiber repair, as required. Several methods exist for monitoring the integrity of MF–UF membranes (AwwaRF 2004). Direct methods, including pressure decay integrity testing (PDIT), provide the most sensitive measure of membrane integrity but require that the system be taken off line. Typically in reuse applications (such as potable reuse) where MF–UF must meet microbial LRV, PDIT would be done every 24 hours. The pressure decay rate measured over a 5-min period can be directly correlated to *Cryptosporidium* removal (USEPA 2005). Indirect methods such as online filtrate monitoring (that is, particle counting and turbidity) provide a less-sensitive degree of integrity verification but provide real-time feedback and can be used in concert with direct methods as part of a robust integrity monitoring plan. Another key component of the integrity monitoring plan is to establish acceptable limits and an action plan to follow when the limits are approached or exceeded.

Process monitoring is essential when determining optimal operating conditions and cleaning strategies. Routine monitoring of filtrate flow, feed and filtrate pressure, and temperature is required to calculate and trend temperature-corrected specific flux and transmembrane pressure (TMP). A decrease in specific flux with time indicates that membrane fouling is occurring. TMP is used to assess membrane fouling; as foulants plug the membranes, TMP increases. Membrane fouling can be reversible or irreversible. Irreversible fouling can be assessed by comparing temperature-corrected results after each clean-in-place (CIP) cycle.

Table 5-4 Example large-capacity water reuse projects that use microfiltration–ultrafiltration

Location	Project	Total Capacity, mgd
Scottsdale, Arizona	Scottsdale Water Campus	12
Orange County, California	OCWD Groundwater Replenishment System	100
Carson, California	West Basin Municipal Water District Project	17
Santa Clara, California		
SVAWPC is for total dissolved solids removal; not a permitted reuse plant		
Sulaibiya, Kuwait	Sulaibiya Wastewater Treatment and Reclamation Plant	100
Singapore	NEWater projects	54
South East Queensland, Australia	Western Corridor Recycled Water Project	61–79

Table 5-5 Monitoring methods for microfiltration–ultrafiltration systems

Monitoring Parameter	Method	Purpose
	Water quality	
Filtrate turbidity	Online analyzer	Compliance monitoring, gross indirect measure of membrane integrity
Microbial	Grab samples for coliform, coliphage	Compliance monitoring, gross measure of membrane integrity
Organic	Grab samples for total organic carbon, dissolved organic carbon	Assess pretreatment performance
	Membrane integrity	
Pressure decay	Pressure decay integrity testing	Sensitive direct measure of membrane integrity
Particle counting	Online analyzer	Sensitive indirect measure of membrane integrity, often done in concert with pressure decay integrity testing
	Process performance	
Filtrate flow	Online meter	Assess membrane fouling, used to determine normalized specific flux
Feed/filtrate pressure	Online pressure sensors	Assess membrane fouling, used to determine transmembrane pressure and normalized specific flux
Filtrate temperature	Online temperature sensors	Used to determine temperature-corrected specific flux

REVERSE OSMOSIS–NANOFILTRATION

For some reclaimed wastewater applications, removal of particulate material, using either granular media filtration or MF–UF membrane filtration (discussed in an earlier section), is adequate. However, with increased need for use of reclaimed wastewater in applications that require high-quality water, there is a need for a reliable and efficient treatment process that removes dissolved constituents. These applications include industrial applications that require high-purity water, aquifer recharge for withdrawal at a later time for potable consumption, surface water augmentation for indirect potable reuse, and direct potable reuse applications. Dissolved constituents are typically removed using one of two types of membranes processes, each characterized by different driving force. These processes are reverse osmosis and nanofiltration.

In RO and NF processes, water is forced under a positive hydrostatic pressure gradient through semipermeable membranes that allow the passage of water while limiting the passage of salt. These membranes are capable of high rejection of most dissolved constituents and are discussed in this section. A detailed description of RO and NF processes is available in AWWA Manual M46 (2007). The discussion here includes a brief overview, followed by specific guidelines for reuse applications.

RO and NF membranes are semipermeable, allowing passage of water across them while restricting passage of dissolved constituents. When these membranes separate low-salinity water from high-salinity water, pure water tends to diffuse across the membrane from the low-salinity side to high-salinity side to equalize the salinity on both sides, a process referred to as osmosis (Figure 5-8A and 5-8B). The solutions on both sides of the membrane then reach equilibrium; the static head difference between the two sides of the membrane is referred to as the osmotic potential (Figure 5-8C). By applying a pressure greater than the osmotic potential on the high-salinity side, natural osmotic diffusion can be reversed. This causes pure water to diffuse from the high-salinity side to the low-salinity side; this process is referred to as reverse osmosis (Figure 5-8D). Through this process, both RO and NF membranes remove dissolved constituents from water that is to be treated.

Types of Membranes

The membranes that operate on the principle of RO are broadly categorized as RO membranes or NF membranes. The primary difference between these two membranes is their rejection ability and the operating pressure required. Rejection, in general, follows a lyotrophic series, that is, increasing rejection with increasing hydrated radius. This predicts the following rejection order for cations and anions: Mg^{2+} > Ca2+ > Sr^{2+} > Ba^{2+} > Ra^{2+} > Li^{2+} > Na^{+} > K^{+}; SO_4^{2-} > Cl^{-} > Br^{-} > NO_3^{-}. While this is a generalization, the rejection is based on specific properties. For example, ions with higher charge are better rejected, for example, divalent ions are better rejected than monovalent ions; ions with larger hydrated size are better removed; charged species are better removed than weakly or uncharged species; and higher-molecular-weight species are better removed than lower-molecular-weight species. Typically gases are poorly removed, if at all.

Compared to NF membranes, RO membranes are capable of high rejection of both monovalent and divalent ions. Within each category of RO and NF, there are additional classifications that are based on rejection ability. For instance, some RO membranes achieve high rejection and require higher operating pressure; others achieve slightly lower rejection and also require lower operating pressure. In reuse applications, RO membranes are used to meet the treatment goals because of this high rejection. Similar types of membranes are available for NF, which are broadly classified as tight and loose membranes. Some rejection properties of these membranes and the required operating pressures are shown in Figure 5-9.

In general, NF membranes allow a higher degree of salt passage compared to RO membranes and typically require significantly less driving pressure. Also, NF membranes achieve relatively higher removal of divalent ions compared to monovalent ions and are typically used to remove hardness-causing ions and organic material. Because of their lower rejection capability, NF membranes are not used in municipal water reuse applications and are not discussed further.

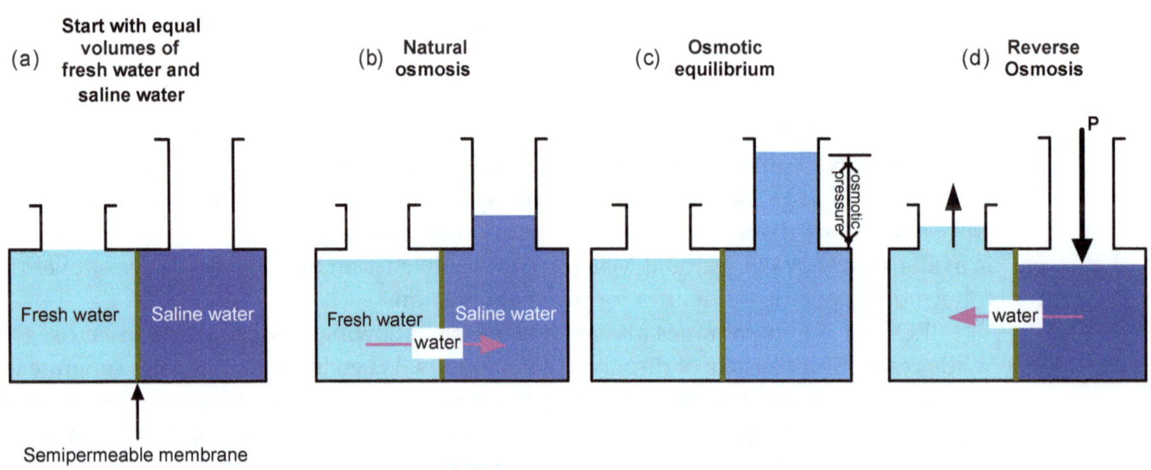

Figure 5-8 Principle of reverse osmosis

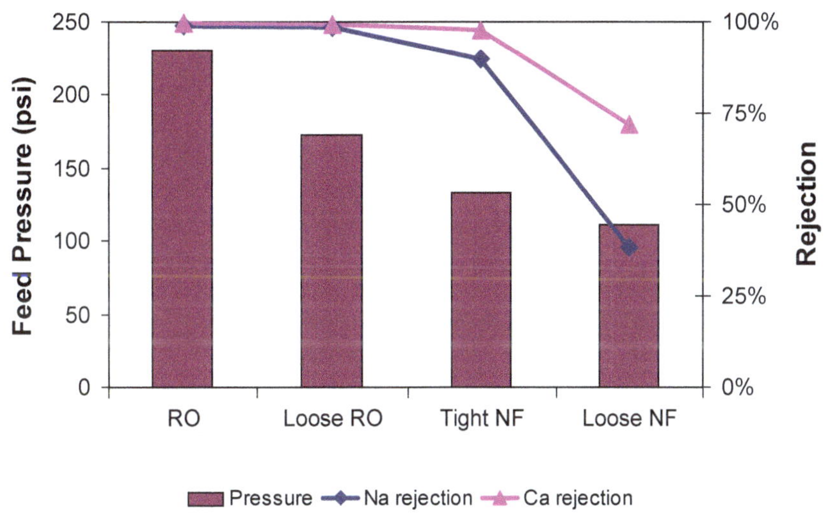

Figure 5-9 Rejection capabilities of various membranes and required operating pressure

Abbreviations: NF, nanofiltration; RO, reverse osmosis.

RO Membrane Material and Element Configurations

The most commonly used RO and NF membranes are made of polyamide material and are spiral wound. A thin layer of the polyamide membrane layer is mounted on a support layer made of polysulfone, which is supported by a reinforcing fabric layer. Two layers of this membrane are wrapped around a permeate tube to form a membrane element. Specific details can be found in AWWA Manual M46 (2007) and references therein. Hollow-fiber RO membranes are also available, wherein the membrane is cast as hollow fiber, and feedwater flows from outside to inside of the membrane fiber. These are not commonly used in large-scale municipal applications. Plate and frame RO membrane elements are also available for very small-scale applications.

RO membranes made of polyamide have pH tolerance that ranges from 2 to 12 without hydrolysis. The membranes made of polyamide have extremely limited tolerance to strong oxidants such as chlorine but their tolerance for chloramines is much higher. For instance, the typical tolerance of an RO membrane is approximately 1000 ppm-hours for free chlorine at pH 8 and 25°C (77°F) and approximately 60,000 ppm-hours for chloramines at the same pH and temperature. These limits could be significantly reduced if metals such as iron are present and if pH and temperature are different. In wastewater applications, it is common to use chloramines to control biofouling, as discussed later.

NF–RO Permeate Quality Requirements

NF–RO permeate quality. NF–RO is often considered the final membrane treatment in water reuse plants because it removes most dissolved salts and organic and inorganic substances; however, it does not remove dissolved gases. Since it allows passage of gases, the RO permeate can be saturated with carbon dioxide and oxygen. The NF–RO permeate is aggressive at low pH, requiring stabilization.

NF–RO permeate quality requirements for water reuse. The requirement for nonpotable water reuse is low turbidity. The California Code of Regulations, Title 22, Water Recycling Criteria, requires that turbidity for membranes not exceed 0.2 ntu "more than 5 percent of the time within 24-hour period and 0.5 ntu any time." This requirement can be consistently exceeded when NF–RO membranes are used. Reuse applications for some agricultural irrigation require a boron concentration of less than 0.5 mg/L and a sodium absorption ratio of less than 13. Typically the NF–RO permeate requires post-treatment to achieve the required value.

Treated water for storage in an aquifer storage and recovery (ASR) system requires removal of DO from the water before injection. Any microcrystalline pyrite (iron sulfide) that is present in the aquifer is oxidized by DO in the treated water, releasing arsenic. The membrane contactors tested during pilot testing and demonstration at a facility in Clearwater, Fla., reduced the DO from 6 to 9 mg/L (100 percent saturated) to 100 ppb. With the use of sweep gas, DO can be reduced to 1 ppb.

IMPACT OF FEEDWATER QUALITY ON PRETREATMENT AND NF–RO SYSTEM DESIGN

Potential Foulants and Pretreatment Requirements

The most commonly used NF–RO membrane elements are thin-film composite membranes that are made of a polyamide membrane layer. The membrane element configuration that is commonly used is spiral wound in which alternating layers of membrane and feed and permeate spacers are wound around a product-water collection tube. By design, it is not possible to backwash these membranes. Also, because the membranes recover water from the feedwater, several sparingly soluble salts can exceed their solubility product, and precipitation may occur. These issues should be addressed with pretreatment.

Fouling Indicator

Several indicators are used to determine the fouling potential of the feedwater that goes to NF–RO. While water quality parameters such as turbidity, suspended solids, and iron and manganese concentrations are good indicators of fouling potential, one indicator that is commonly used to characterize NF–RO fouling is the SDI. While variations of this test method have been developed, such as the modified fouling index and the mini-plugging factor index, SDI is the most commonly used fouling indicator.

The test is conducted in accordance with ASTM D4189-07, Standard Test Method for Silt Density Index (SDI) of Water. The test is performed by filtering water through 0.45-μm filter paper at a pressure of 30 psi. The time it takes to filter 500 mL at the beginning of the test and at the end of the test is recorded. The SDI is calculated as follows:

$$SDI_t = \frac{\%P_{30}}{t} = \frac{\left[1 - \frac{t_i}{t_f}\right] \times 100}{t} \qquad \text{(Eq 5-6)}$$

Where:

SDI_t = SDI measured at time interval t
$\%P_{30}$ = percent plugging at 30 psi feed pressure
t_i = time to collect initial 500-mL sample
t_f = time to collect 500-mL sample after time t elapsed since the beginning of test

Typically, an SDI test is performed for 15 minutes, which results in values that range from 0 to 6.67. SDI. Values that are less than 3 indicate lower fouling potential, while values greater than 5 suggest high fouling potential. For values in between, the fouling potential is dependent on the nature of applications and other water quality parameters. The impact of all potential fouling constituents is not captured by an SDI test. For instance, organics that may be present in the water at sufficient concentrations to foul an RO membrane may pass through the SDI filter pad and not be reflected in the SDI value. Similarly, biofouling potential is not reflected in the SDI value. Even when SDI values are low, it is possible for a high rate of fouling to occur on the RO membranes.

NF–RO Typical Foulants and Pretreatment Requirements

Suspended solids. Suspended solids that are partly reflected by turbidity measurements can deposit on the membrane surface and clog the feed channel. This results in increased head loss along the length of the membrane element and higher operating pressure to meet the treatment goals. The passage of salt may also increase as the foulant layer can increase the thickness of the concentration polarization (CP) layer. As the permeate passes through the membrane, it leaves behind rejected ions at the surface of the RO membrane. This results in a higher concentration near the membrane surface. While the RO membrane element is designed for high turbulence near the surface of the membrane element to minimize this CP layer, foulant deposits may decrease the turbulence. Thicker CP layers result in higher salt passage through the membrane. Consequently, pretreatment should remove suspended solids in the RO feedwater to minimize fouling. Typical pretreatment options include use of a robust filtration process upstream of RO and cartridge filtration as a safety barrier. Use of MF–UF has become a standard for reuse applications, even if the wastewater treatment has tertiary filtration using media filters.

Inorganic colloids. Inorganic colloids such as iron, manganese, and aluminum can foul the RO membrane. Typically, the concentration of these constituents in the wastewater is low. High concentrations of iron and aluminum could exist if these salts are used as coagulants in the wastewater treatment plant that is providing water for the reuse facility. The oxidized forms of metals are colloidal and have the potential to deposit on the membrane's surface. Manganese typically occurs as a contaminant in ferric salts that are used as a coagulant. Of particular concern is the formation of aluminum or iron silicates, which depends on the pH and aluminum, iron, and silica concentrations. Typically, these coagulants are used in a wastewater plant to remove phosphorus. In general, the optimal pH range for alum coagulant is very narrow. This may result in carryover of aluminum to the downstream RO process. Often, polyaluminum chloride or aluminum chlorohydrate have lower carryover due to the broader optimal pH range. Similarly, iron salts have a broader optimal pH range. Consequently, the preferred coagulant in a wastewater treatment plant is polyaluminum chloride–aluminum chlorohydrate or iron. Use of alum should be avoided.

Preventing Scaling From Sparingly Soluble Salts

Scale formation is expected to happen on most membranes as the salts in the reject or brine stream become concentrated. Scales deposited on membrane surfaces can severely reduce the flow or flux of permeate and can irreversibly damage the membrane. These salts include the calcium salts of carbonate, sulfate, and fluoride; the sulfate salts of barium and strontium; calcium phosphate; and silica. In most wastewater applications, the most common scalants are calcium carbonate, calcium phosphate, and silica (particularly if aluminum is present). Since wastewater quality changes considerably depending on the source water and industrial discharges, it is highly recommended that a detailed historical water quality analysis be performed. The precipitation of these salts can often be prevented or minimized through the addition of acid and antiscalant. Acid is used to control the carbonate scale precipitation potential by converting the bicarbonate to carbonic acid.

Antiscalant typically controls the precipitation of all other sparingly soluble salts. As discussed earlier, the only way to prevent the formation of aluminum silicate is by reducing the aluminum concentration in the feedwater or by lowering the recovery of the NF–RO system. Antiscalants do not prevent precipitation of salts but they do prevent the formation of strong chemical bonds in salts, forming hard crystalline deposits. Other scales of concern are $CaSO_4$, which can damage membranes, $BaSO_4$, $SrSO_4$, CaF_2, silica, and $Ca_3(PO_4)_2$.

In the absence of pH adjustment, nearly all NF and RO systems that operate at high recovery have a calcium carbonate scaling potential; however, this scaling potential can be prevented by most antiscalants and dispersants. In recommending acidification, antiscalant, or both, the following factors should be considered:

- Scaling potential versus capabilities of acids and antiscalants,
- Local cost and availability of acids and antiscalants,
- Presence of suspended solids in the NF and RO feedwater,
- Coagulation and filter aid polymer used in pretreatment,
- Personnel safety,
- Post-treatment requirements, and
- Regulations on concentrate disposal.

Acid Addition for Scale Control

Acid addition is an effective way to control calcium carbonate scale. All NF–RO membrane manufacturers' have design programs that can predict the concentration of various constituents in the concentrate stream. These programs use the values to calculate the Langelier saturation index (LSI) or the Stiff-Davis saturation index (S-DSI), which quantifies the scaling potential of calcium carbonate and can be used to optimize acid addition. A positive LSI or S-DSI in the concentrate stream indicates that there is a calcium carbonate scaling potential, while a negative value indicates no scaling potential. LSI is not a valid index for very high TDS brines and indicates a false scaling potential; in these situations, S-DSI should be used. Acidification will not prevent precipitation of other sparingly soluble salts (for example, $CaSO_4$, $SrSO_4$, $BaSO_4$, SiO_2), and there are safety and corrosion considerations.

Acidification can be beneficial if the permeate must be stabilized or remineralized. Without acidification, NF and RO permeate generally lacks adequate alkalinity to provide the buffering capacity for a stable permeate. Feed acidification increases the permeate carbon dioxide concentration. If this were the only objective, chemical addition to the permeate would be the most cost-effective process to use. However, when acid is also considered for scale control, the combined benefits can make it an appropriate choice. Some NF and RO manufacturers have cost analysis software that can be used to determine the most cost-effective chemical addition program, including pretreatment and post-treatment.

Significant reduction in the feedwater pH increases the corrosion rate. Depending on the salinity and materials of construction, this may be a serious issue.

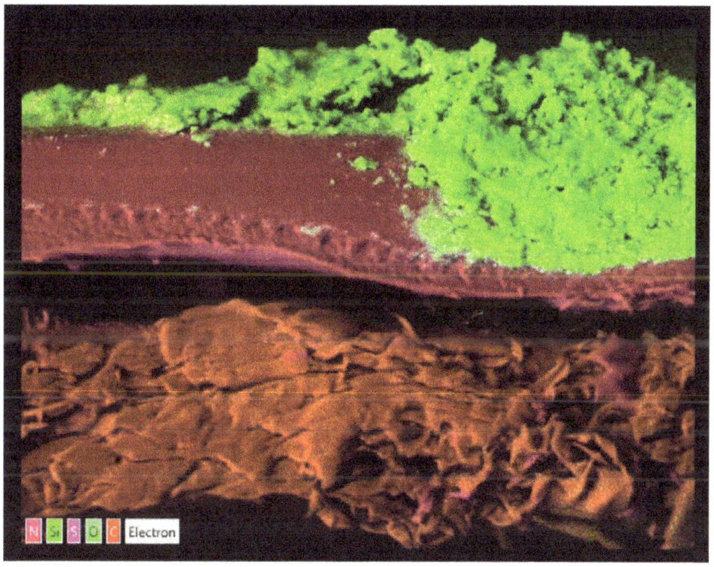

Figure 5-10 Scanning electron microscope image of a fouled membrane used to treat secondary effluent. Severe fouling by calcium aluminum silicates is seen.

Antiscalant and Dispersant Addition

There is a variety of antiscalants with various capabilities. Sodium hexametaphosphate (SHMP) has limited capability in inhibiting calcium carbonate and calcium sulfate precipitation. The advantages of SHMP are low cost and good availability worldwide. Disadvantages include the following:

- Limited capability for calcium carbonate up to LSI of +1.0 in the concentrate;
- When dissolved in water, SHMP converts to orthophosphate at a moderate rate, which can promote precipitation. The SHMP day tank must be emptied and refilled every two to seven days depending on temperature and pH; and
- Environmental concerns regarding disposal of concentrate that contains phosphates.

A variety of organic polymers (often polyacrylates) that inhibit scale formation are available. Some are a blended formulation and include particle dispersants. Major advantages over SHMP are that these polymers are better able to inhibit scale formation and to disperse particles that were not removed in pretreatment. Disadvantages include cost, potential to foul the membrane if overdosed, and potential interaction with other polymers used in pretreatment.

When calcium carbonate scale inhibition is the only consideration, it is generally recommended that acidification or antiscalants be used. The choice is dependent on the site-specific factors listed above (for example, cost, safety, corrosion).

When sulfate scales are also a consideration, antiscalants are needed. Examples of when both acid and antiscalants are used include the following:

- When carbon dioxide in the permeate is desirable and the antiscalant capabilities are also required; and

- When an extremely high calcium carbonate scaling potential exists (due to feedwater chemistry or recovery). Most membrane manufacturers provide a list of antiscalants and dispersants that are compatible with their membrane. This information should be reviewed carefully. Assuming the design results in a brine concentration that is within the manufacturer's published safety limits, antiscalants can be added.

In some instances, acidification is the preferred method of calcium carbonate scale control. Because antiscalants slow the precipitation process and do not stop it, systems that use an antiscalant must be flushed at every shutdown to prevent precipitation. If a particular site cannot ensure that this flush will occur (for example, frequent shutdowns, power interruptions, minimal staffing), then acidification is a "safer" way to control calcium carbonate scale.

Many manufacturers claim that their more advanced antiscalant–dispersant blends disperse suspended particles. Use of antiscalant–dispersant for this purpose should be performed with caution. Many applications include coagulants and filter aids in pretreatment, and interaction between excess pretreatment polymers and the RO antiscalants is common. The result is a viscous, slime-like membrane foulant. Minimizing use of a coagulant and antiscalants is the best way to prevent this foulant.

Improving and/or Enhancing the Performance of NF and RO Membranes

The performance of NF and RO membranes is improved and/or enhanced by the addition of acid or caustic soda. These chemicals change the characteristic of specific constituents in the feedwater, which affects their passage or rejection by the NF–RO membranes.

Acid addition. Acid can be added to promote the passage of dissolved gases through the membrane and to prevent precipitation of iron and aluminum. Acid addition is a common pretreatment in NF and RO plants for controlling scale formation. When present in the feedwater, hydrogen sulfide can precipitate in the membrane as elemental sulfur. When deposited in the membrane surface, elemental sulfur can cause irreversible damage. At lower pH (<7.0), hydrogen sulfide predominantly exists as a gas and can pass through the membrane without damaging it. In some instances, it is necessary to convert bicarbonate into carbon dioxide in order to have dissolved carbon dioxide in the permeate, which is required for remineralization.

Caustic soda addition. For some process applications, caustic soda can be injected to increase the NF and RO feed pH. Typically, the only caustic injected is sodium hydroxide, which is based on cost, availability, and its solubility in water. A high pH value helps to prevent fouling because both the membrane and many organic substances assume a negative charge at pH >9.

Caution is required when raising the feed pH because it can decrease the solubility of calcium carbonate, as measured by increasing LSI, iron, and manganese.

The most common process application that uses a caustic feed involves second-pass RO systems. In second-pass RO, the permeate from a first-pass RO becomes the feed to the second pass. The second pass RO "polishes" the first pass permeate to quality levels as high as 4 $M\Omega$-cm. Caustic is introduced into the second-pass feed for the following reasons:

- At a pH of 8.2 and higher, all carbon dioxide gas is converted into bicarbonate ion, which is rejected by the RO membrane. As a gas, carbon dioxide passes through the RO membrane into the permeate stream unimpeded. The carbon dioxide becomes an undesirable load to downstream polishing mixed beds.

- Certain total organic carbon constituents are better rejected at a higher pH.
- Silica rejection and solubility are higher at a higher pH (particularly above pH 9).
- Boron rejection is higher at a higher pH (particularly above pH 9).

One unique process application, frequently referred to as the HERO process, involves a caustic feed to raise pH to 9 or 10. The RO operates in a first-pass mode and uses brackish feedwater. This water can contain potential foulants, for example, hardness, alkalinity, iron, manganese, that can cause problems at higher pH levels. Pretreatment frequently involves a weak-acid cation-exchange system and a degassifier to remove these foulants

Organic constituents. Organic constituents in the NF–RO feedwater include natural organic matter that is typically present in the water supply and particulates that are generated during the wastewater treatment processes. The secondary effluent may contain high concentrations of organic macromolecular compounds such as polysaccharides, proteins, amino-sugars, nucleic acids, humic acids, fulvic acids, and organic acids. These compounds can foul membranes alone or by aggravating biofouling. Typically, the only way to reduce organic fouling is by optimizing the wastewater treatment process to reduce their concentration. Organic fouling can be minimized, to a limited extent, by adjusting the pH of the RO feedwater. However, an adjustment may adversely impact the scaling potential.

Biological activity. Effluent from a wastewater treatment plant has high biological activity, which may continue through MF–UF and RO. This is primarily because the feedwater has high concentrations of organic material and nutrients and because the RO membranes provide large surface areas for growth. Typically, colonization of the microorganisms forms a biofilm on the membrane surface due to the production of extracellular polymers that stick to the membranes. Other foulants get imbedded within this biofilm, increasing the severity of the fouling. Biofouling is controlled by adequately pretreating the feedwater, reducing stagnant feedwater in the membranes, using oxidants that do not damage the membranes, and performing a periodic chemical cleaning of the membranes.

Process chemicals. Often, wastewater treatment plants use polymers in their processes, such as in coagulation or during residual treatment. Polymers can adhere to the RO membranes, causing irreversible fouling. Consequently, polymers should be avoided in wastewater treatment. If absolutely necessary, its concentration should be well controlled to minimize the potential for carryover. Since it is difficult to ensure that no carryover occurs, it is preferable to not use polymers.

Phosphate scaling. High concentrations of phosphate in the RO feedwater could result in potential scaling of RO membranes. While the potential for phosphate scale formation can be mitigated by lower operating recovery and pH adjustment, additional pretreatment could be included to reduce phosphate concentration. Often, this involves coagulation and clarification upstream of MF–UF.

NF–RO Equipment

After pretreatment or conditioning, wastewater may or may not be treated with MF–UF and goes to the NF–RO train. The main equipment in a NF–RO system includes cartridge filters, chemical dosing systems, high-pressure pumps, NF–RO membranes, high- and low-pressure manifolds, and a brine control valve.

Cartridge Filtration

The quality of NF–RO feedwater obtained through filtration by MF–UF membranes may be very high with respect to colloidal material and can be treated with NF–RO directly. However, an interprocess tank is usually included between the MF–UF and RO processes for better hydraulic control. Contamination of the water in this tank is possible, and there is the potential for equipment failures such as tank lining failure and corrosion of valves and other equipment. To address such concerns, cartridge filtration is usually included between MF–UF and RO. The typical pore size of the cartridges ranges from 5 to 20 μm, with 5 μm being the most common size. Larger sizes can also be used such as in OCWD, Orange County. The cartridge filter should not be considered part of the pretreatment process but rather as a safeguard in the event of accidental introduction of large particulates in the NF–RO feed.

Chemical Addition

In water reuse, the chemicals that are most commonly added upstream of NF–RO are chloramines, acid, and antiscalant. Chloramines are typically added upstream of MF–UF at a concentration necessary to maintain a residual through the NF–RO process. The residual concentration typically ranges from 1 to 3 mg/L through the NF–RO process; the concentration to be maintained is dependent on the biological fouling potential.

Acid is typically added to control calcium carbonate scale formation and, in some instances, to control calcium phosphate scale formation. Sulfuric acid is the most commonly used acid, although hydrochloric acid may be considered if sulfate scale formation potential is very high. Use of hydrochloric acid requires larger storage volumes than sulfuric acid and additional safety measures.

Antiscalant is used to prevent precipitation of sulfate salts of calcium, barium, and strontium; calcium fluoride; calcium phosphate; calcium carbonate; and silica. Some antiscalants contain dispersants that stabilize colloids (including metal colloids such as oxide particles of iron and manganese) and prevent them from sticking to the membrane surface. Because antiscalants only slow down the kinetics of precipitation, concentrated feedwater should not be left in the membrane system for extended periods when not in operation.

High-Pressure Pump

The selection of a high-pressure pump depends on the feed flow and pressure requirement of the NF–RO membrane. Centrifugal single or multistage pumps are commonly used, although positive-displacement pumps are used for relatively low-flow, high-pressure applications. High-pressure pumps can be installed vertically or horizontally. The high-pressure pumps with high efficiency are selected. The materials of construction can be stainless steel type 304, stainless steel type 316, or duplex (for feedwater with a high chlorine concentration). The pumps that are used as interstage booster pumps have the same characteristics as the high-pressure pumps. The pump drive can be an electric motor or an energy recovery device (ERD). The ERD is a turbine or reverse-running pump driven by the high-pressure brine reject.

NF–RO Train Design

The most important design considerations for an RO system in water reuse applications are recovery and permeate flux. Additional design parameters, some of which are affected by these two primary design parameters, include array design (number of stages, number of elements per pressure vessel, cross-flow velocity) and number of passes, which are discussed below.

Courtesy of Black & Veatch.

Figure 5-11 Microfiltration (foreground) and reverse osmosis (RO; background) installation at the Bundamba advanced water reclamation plant in Australia. The RO membranes are 18 in. in diameter.

NF–RO System Nomenclature Typically

The nomenclature used for NF–RO systems varies considerably. The terms used in this manual are defined here.

Feed flow: The water that enters the NF–RO system to be treated. Also referred to as feedwater.

Permeate flow: The water that passes through the membrane is referred to as permeate.

Concentrate flow: The water that leaves the system untreated (and also carries all constituents rejected by the NF–RO stream) is referred to as concentrate, reject, or brine. In this manual, this stream is referred to as concentrate.

Stage: A collection of NF–RO pressure vessels that have common headers for feed flow, permeate flow, and concentrate flow. Sometimes also referred to as bank. A typical NF–RO train has two or three stages. Stage also refers to a complete NF–RO system that treats the brine reject from another NF–RO system, usually from the first-stage NF–RO system.

Pass: A complete NF–RO system that treats water. If two NF–RO systems treat water, with the second NF–RO system treating permeate from the first NF–RO system, it is called a two-pass NF–RO system. The first pass is the first NF–RO that treats the feedwater, and the second pass is the second NF–RO that treats permeate from the first-pass NF–RO.

NF–RO Recovery

Recovery is defined as the percentage of feedwater that is converted to permeate within the NF–RO system. It is calculated as follows:

$$Recovery\ (R) = Q_p / Q_f \qquad \text{(Eq 5-7)}$$

Where:

Q_p = permeate flow
Q_f = feed flow

It is primarily determined by the concentrations of sparingly soluble salts. As recovery increases, more permeate is extracted from the feedwater, resulting in higher concentrations of all salts in the feed stream. As the concentrations of some sparingly soluble salts exceed their solubility product, precipitation of these salts occurs. Through the use of antiscalants, operation at higher recoveries are possible resulting in salt concentrations several times higher than that of their thermodynamic saturation limit.

The concentrations of sparingly soluble salts that set the recovery limit are not the concentrations in the bulk NF–RO concentrate stream but rather the concentration at the membrane surface. As the feedwater passes through the membrane, it leaves behind the rejected salts at the membrane surface because the diffusion of salts back into the bulk flow is smaller than the convective flow of permeate through the membrane. This is referred to as concentration polarization. Its effect is typically minimized by improved mixing through use of turbulence created by increasing spacers on membrane surface and by maintaining sufficient cross-flow velocity along the membrane surface. The concentration at the membrane surface is expected to be 50 to 100 percent higher than that in the bulk solution, and the recovery is limited by this concentration.

In addition to limitations on recovery due to scale formation, increased concentrations of foulants such as organic material may also limit recovery. This cannot be modeled and should be determined through pilot testing or based on the experience of similar facilities that use the same source water. Typically, in wastewater applications, the overall recovery ranges from 75 to 85 percent for a two-stage system, with most facilities operating in the range of 80 to 85 percent.

NF–RO Permeate Flux

Permeate flux is the permeate flow rate per unit area of the membrane. It is commonly expressed as gallons per square foot per day or liters per square meter per hour. Higher flux results in use of less membrane area and potentially lower capital costs. However, it also results in higher operating pressures due to the need for higher pressure to push more permeate through the same membrane area. There is also the potential for increased risk of fouling. Through extensive experience at both pilot and full-scale facilities, a permeate flux that ranges from 10 to 12 gallons per foot per day has been found to be practical in reuse applications.

The range of 10 to 12 gfd (16.9 to 20.33 lmh) permeate flux is for overall system flux. However, the permeate flux of individual stages is often overlooked; this should also be optimized. In basic design, a high-pressure pump pressurizes the feedwater to the first stage, and the concentrate from the first stage is feedwater for the second stage, and so on. The first-stage feedwater has the highest pressure and also the lowest average salinity compared to feedwater from the other stages. This results in higher flux through the first stage and lower flux through the other stages. This could result in fouling of the first stage at a rate higher than of the other stages. Since the first stage typically produces around 50 percent of the total permeate flow, fouling of this stage could result in higher energy consumption. Consequently, ways to balance the flux between various stages should be evaluated.

NF–RO Train Design

The NF–RO train design is dependent on several parameters including recovery, flux, and end user preferences such as train capacity and height.

The NF–RO train size should be selected taking into consideration the minimum treatment capacity requirements, spare train requirements, and capital costs. Larger trains typically result in lower capital costs. However, if a spare train is necessary, the overall costs will be higher due to the higher cost of a larger train that is typically not in operation. The minimum treatment capacity also dictates the train size; often, optimal design means

Courtesy of Black & Veatch.

Figure 5-12 Reverse osmosis trains at Changi NEWater Facility, Singapore

that the minimum treatment capacity is met by operating multiple trains (typically one or two). NF–RO trains should not be operated at capacities that are significantly below their design capacity because cross-flow velocity across the membrane surface could be lower than that required to minimize concentrate polarization. Similarly, NF–RO trains should not be operated above design capacity because operation at higher fluxes can increase the fouling potential.

Number of Stages

The number of stages in a NF–RO train is dependent solely on the recovery. Since pressure vessels that house the NF–RO membrane elements typically house between six and eight elements, a fixed range of membrane area is available in each stage. Coupled with flux limitations and the need to maintain a minimum concentrate flow through each membrane, the amount of water that can be produced by one stage is limited. To increase recovery, the concentrate from each stage should be further treated by another stage. The number of pressure vessels and, therefore, the number of membrane elements typically varies by a factor of 2. For example, for a three-stage system, the ratio of elements in the first through third stages is 4:2:1.

The number of stages required with respect to recovery is as follows:

- One stage: approximately 50 percent,
- Two stages: approximately 50 percent to approximately 75 percent, and
- Three stages: 75 percent to 85 percent.

Through use of pressure vessels that house seven or eight elements, an NF–RO system can be designed to achieve 85 percent recovery with two stages or 90 percent recovery with three stages. However, such designs should be carefully evaluated to ensure that no single element is overstressed with respect to permeate flux and recovery and that the minimum cross-flow velocity requirements are met.

Courtesy Black & Veatch.

Figure 5-13 A three-stage reverse osmosis train

The required number of NF–RO passes is solely dependent on the water quality required. For most applications, a single-pass NF–RO system is adequate. For specialized reuse applications such as feedwater for a high-pressure boiler, a two-pass system is required. The design of the second-pass system is different from the design of the first-pass system, as the feedwater to the second pass is of high quality. Often, the flux of the second pass is 20 gfd (33.8 lmh) and recovery is 90 percent, achieved with only two stages due to the high-quality feedwater.

Flux Balancing

As discussed above, even when overall system flux is within the acceptable range, consideration should be given to balancing the fluxes of individual stages and to ensuring that individual element design parameters are met. These parameters include the minimum concentrate flow from each pressure vessel, individual element recovery, and flux through each element.

Often, flux can be balanced in several ways including the following:

- Throttling of the permeate in the first stage reduces the flux through the first stage and increase it through other stages. This is not the preferred solution as it results in energy waste.
- An interstage booster pump between the first and second stages or between the second and third stages helps to balance the fluxes. If coupled to an energy-recovery turbine, the interstage booster pump results in overall energy savings.
- An energy-recovery device can be used to recover energy from the concentrate and to provide an interstage boost between the first and second stages or between the second and third stages.

Courtesy of Black & Veatch.

Figure 5-14 Energy-recovery device in a reverse osmosis system that recovers the energy in the concentrate and reuses it to boost the second-stage feed pressure

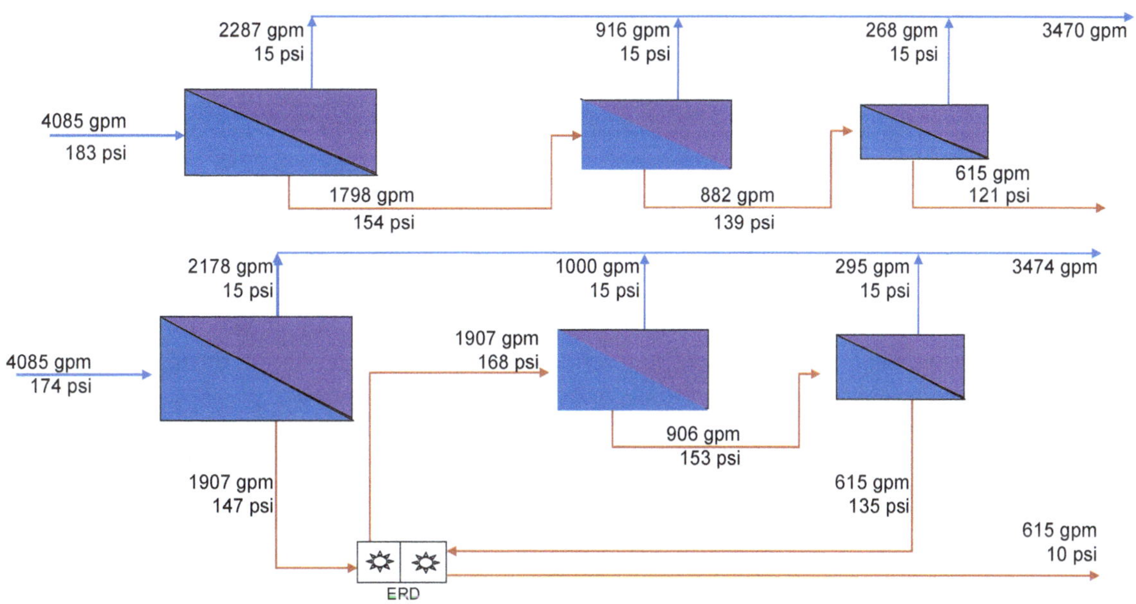

Figure 5-15 Flows and pressures through each stage of a reverse osmosis train, with and without an energy-recovery device

Abbreviation: ERD, energy-recovery device.

Fouling Allowance

Even with optimal design of the pretreatment and NF–RO systems to minimize fouling, in wastewater treatment applications, fouling of the NF–RO membranes is inevitable. Consequently, in reuse applications, the NF–RO system should be designed taking into consideration a higher rate of fouling than in other applications. Typically, an NF–RO system is designed to account for a 50 to 100 percent increase in operating pressures over the life of the membrane.

Ancillary System Designs

Typical ancillary systems needed for NF–RO trains include CIP and flushing systems. A detailed description of ancillary systems can be found in AWWA Manual M46 (2007).

CIP system. Over the course of operation, RO membranes become fouled and require periodic cleaning. Membranes are cleaned with either low pH or high pH solution, with or without surfactants. If the foulants are inorganic or colloidal in nature, low pH solution is often used. For organic or biological foulants, high pH solution is used. The order of cleaning is also dependent on the suspected nature of the foulants. If organic or biological foulants are the likely cause of reduced performance, cleaning first with high pH solution is often recommended. The cleaning system consists of cleaning solution tanks, pump, cartridge filter, and a neutralization system.

While one cleaning tank is sufficient, two tanks, each dedicated to a type of cleaning chemical, are desirable. This allows for heating of the second cleaning solution in the second tank, while the NF–RO train is being cleaned with the first cleaning solution. The system should also have a method for heating the cleaning solution within a reasonable time period to a temperature of 30 to 40°C (86–104°F). The cleaning pump should be sized to convey up to 50 gpm of water per pressure vessel for the largest section of the RO train to be cleaned. This is the entire first stage or half of the first stage, if the CIP system is so designed.

The spent cleaning solution should be neutralized in a separate tank, rather than in the CIP tanks, to prevent precipitation of foulants and the potential for introducing any foulants into the RO system during subsequent cleaning.

In reuse applications, often the foulants are biological or organic in nature, and cleaning with a high pH solution is more frequent than with a low pH solution. However, when inorganic foulants such as iron and aluminum are present in the feedwater, frequent cleaning with low pH solution is to be expected.

Flushing system. Whenever the NF–RO system is off line in reuse applications, the NF–RO system should be flushed with feedwater and then with permeate to remove the feedwater from the system. This ensures that the supersaturated, sparingly soluble salts are flushed out of the system and that organic and biological foulants from the feedwater do not result in continued fouling of the NF–RO membranes during downtime.

Electrodialysis

ED and other related processes are gaining popularity for treating wastewater for reuse, specifically for increasing recovery. One advantage of ED is that it does not require pretreatment, which is normally required for pressure-driven membrane processes. ED is not affected by silica scaling or particulate fouling. The ED membrane process is electrically driven. Water passes between pairs of ion-selective membranes, and an electrical potential is applied across the membranes. As water passes through the system, ions migrate toward the electrode with an opposite charge, resulting in lower-salinity water. This process does not remove uncharged or weakly charged particles including suspended solids, organic molecules, and pathogens. Because of this limitation, ED membranes have limited application as the primary treatment process in reuse applications and are be discussed further. Readers interested in learning more about this process for industrial reuse applications can refer to AWWA Manual M38 (AWWA 2007).

POST-TREATMENT, STABILIZATION, AND DISINFECTION

Post-treatment requirements for water reuse supplies vary depending on the upstream treatment technologies used and the desired finished water goals and/or requirements. Filtered effluents may require only disinfection prior to final distribution to end users. Demineralized supplies from NF and RO processes may require partial remineralization to ensure a corrosion-free product. In some cases, particularly for supplies destined for industrial reuse, pretreatment requirements may be limited in order to preserve the purity achieved by upstream treatment processes.

It is therefore important to first define the post-treatment and disinfection requirements for a given supply and end use. State regulators typically establish disinfection requirements based on the targeted end use and can apply specific constituent targets for parameters such as nitrogen, total organic carbon, and priority pollutants. Requirements are generally more stringent for supplies destined for human contact or eventual consumption and less so for irrigation and industrial reuse applications.

Additional post-treatment is based on the supply's end-use requirement. For high-purity uses, it is often sufficient to remove excess carbon dioxide prior to final distribution. For supplies destined for integration with other reuse supplies or contact with existing infrastructure and pipelines, requirements related to these supplies must be considered to ensure a compatible, noncorrosive supply.

Disinfection

Treated wastewater for reuse is often disinfected to destroy microorganism that may be present after treatment. Membrane treatment such as MF–UF or NF–RO provides a barrier that prevents microorganism from passing through the membranes. However, it is still possible for the microorganism to enter the permeate or filtrate side through broken fibers, leaks in the brine seals, and cuts in the membrane surface. The most common disinfectants are chlorine and chloramines. Ozone, treatment with hydrogen peroxide, and exposure to UV light, especially for destruction of organic compounds, can also provide some degree of disinfection. When evaluating the disinfection options for membrane-treated wastewater, it is important to consider the following:

- Reliability and effectiveness of the disinfection process;
- Capital expenditures and operating expenditures;
- Safety, transportation, and handling of the disinfectant;
- Availability of disinfectants;
- DBP issues; and
- Change in water quality after disinfection.

When chlorinating membrane-treated wastewater, it is important that the treated water be chemically stabilized, as the efficiency of chlorine is pH dependent. Substances such as organic compounds interfere and exert a chlorine demand. Turbidity caused by particulate matter prevents chlorine from acting on the microorganism. Ammonia, which can be present in the treated wastewater, reacts with chlorine to form chloramine, which is a less powerful than free chlorine. To ensure complete disinfection, free chlorine residual concentration should be achieved.

The impact of disinfection processes on final finished water quality should be carefully investigated, as there will be a change in pH, alkalinity, and the calcium carbonate precipitation potential. Addition of chlorine gas decreases pH and alkalinity, while addition of sodium hypochlorite and calcium hypochlorite increases pH and alkalinity.

There are cases where chlorinated-treated wastewater must be dechlorinated, as required in some industrial applications. The most common dechlorinating agents are sodium bisulfite and sulfur dioxide.

Ozone, a powerful chemical oxidant, destroys bacteria and viruses by rapidly oxidating their protein mass in a very short time. Ozone is very unstable and should be generated on site. When membrane treatment is incorporated into an advanced wastewater treatment plant, ozone is used to remove tastes and color and to destroy organic compounds.

UV light at a wavelength of 254 nm penetrates bacterial cell walls and interrupts the action of cellular nucleic acids, preventing replication and causing instant death. Suspended solids and particulates tend to shield the UV light and should be removed. Turbidity of less than 2 ntu enhances the performance of UV light.

The disinfection practice used to treat wastewater is identical to pretreatment disinfection as described in this chapter. AWWA Manual M20 (AWWA 2006), a USEPA design manual (USEPA 1986), and Connell (2002) provide guidance in evaluating disinfection options.

STABILIZATION

Gauging Stability

Reuse water supplies can have scaling or corrosion, depending on water quality and the nature of the receiving systems and equipment. Whether water is scaling or corrosive is gauged by the saturation level (SL) of its various compounds in solution. SL is the ratio of ion activity product (IAP) to the thermodynamic solubility product (K_{sp}) for a specific compound in a given water supply. For example, if calcium carbonate is the compound of concern, its SL is defined by the following equation:

$$SL = \frac{a_{ca^{2+}} \times a_{co_3^{2-}}}{K_{sp}} \qquad \text{(Eq 5-8)}$$

The product of the individual ion activity coefficients is the IAP. K_{sp} is a measure of ionic concentration when both dissolved and undissolved ions are in equilibrium at a given temperature, such that SL values greater than 1 indicate a scaling tendency and values less than 1 are associated with ion dissolution or corrosion. A stability index that uses SL values directly is difficult to calculate due the dependency on temperature and solute concentrations within a given solution. Some computer models, such as the Rothberg, Tamburini, and Winsor model, are used to calculate a parameter that is based on SL known as the calcium carbonate precipitation potential. It is a reliable index but difficult to calculate by hand.

Table 5-6 Values of calcium carbonate precipitation potential

Stability of Water	Calcium Carbonate Precipitation Potential Value, mg/L $CaCO_3$
Scaling (protective)	>0
Passive	0 to −5
Mildly corrosive	−5 to −10
Corrosive (aggressive)	< −10

Indices that permit direct calculation include the LSI, the Ryznar stability index (RSI), and the Larson–Skold index. Each is discussed separately below.

Langelier saturation index. The LSI is widely used, particularly in the industrial water industry. It is defined as follows:

$$LSI = pH - pH_s \quad \text{(Eq 5-9)}$$

Its basic definition is the difference between the actual pH of a given water supply and the theoretical pH at which a solution of calcium carbonate would be in equilibrium (pH_s). Calculating pH_s requires that values for TDS, pH, temperature, calcium, and alkalinity be known or estimated. pH_s can then be calculated using the following equations:

$$pH_s = (9.3 + A + B) - (C + D) \quad \text{(Eq 5-10)}$$

Where:

$$A = \frac{(\text{Log}_{10}[\text{TDS}] - 1)}{10} \quad \text{(Eq 5-11)}$$

$$B = -13.2 \times \text{Log}_{10}(°C + 273) + 34.55 \quad \text{(Eq 5-12)}$$

$$C = \text{Log}_{10}[\text{Ca2}^{+} \text{ as } CaCO_3] - 0.4 \quad \text{(Eq 5-13)}$$

$$D = \text{Log}_{10}[\text{Alkalinity as } CaCO_3] \quad \text{(Eq 5-14)}$$

For calculated values of LSI, negative values indicate a potential to dissolve or corrode, while positive values indicate a potential to scale. In terms of stability in recycled water supplies, slightly positive values that range from +0.2 to +0.4 are generally consistent with a stable water.

Ryznar stability index. The RSI uses the calculated pH_s from Eq. 5-10 in the following equation:

$$RSI = 2\,(pH_s - pH) \quad \text{(Eq 5-15)}$$

The index was developed based on a correlation established between an empirical database of scale thickness observed in municipal water systems and associated water quality characteristics of the supply.

Results are evaluated as follows:

- <6: oversaturated and likely scaling,
- >7: mild to slightly corrosive, and
- >8: aggressive or corrosive.

Larson–Skold index. The Larson–Skold index was developed based on observations of steel pipe corrosion conditions in the Great Lakes region compared to transported water quality, making it similar to the RSI. However, the Larson–Skold index looks at the relative levels of chloride, sulfate, carbonate, and bicarbonate according to the following equation (epm = equivalents per million):

$$\text{Larson–Skold Index} = \frac{(\text{epm } Cl^- + \text{epm } SO_4^{2-})}{(\text{epm } HCO_3^- + \text{epm } CO_3^{2-})} \qquad \text{(Eq 5-16)}$$

Larson–Skold index values are generally gauged by the ability to interfere with natural protective film development in pipelines. The following guidelines apply:

- <0.8: chloride and sulfate levels should not impact film formation,
- >0.8 to <1.2: chloride and sulfate levels may interfere with film formation and lead to higher rates of corrosion, and
- >1.2: film formation will likely be impeded, and accelerated corrosion in localized areas can occur.

TREATMENT REQUIREMENTS

If the indices described in the previous section indicate that the proposed recycled water is stable, no further treatment is required. However, if the indices indicate the supply is corrosive, additional treatment is needed.

In determining treatment requirements, Eq. 5-9 along with the succeeding equations are illustrative. Stable water is associated with a positive LSI value, which requires the water's pH to be slightly greater than the calculated pH_s. A useful calculation for determining the pH of a given water with negligible amounts of carbonate alkalinity (essentially water with a pH value <8.2 and with some residual carbon dioxide present) is as follows:

$$pH = \text{Log}_{10}\left(\frac{[\text{Alkalinity as } CaCO_3]}{[CO_2 \text{ as } CaCO_3]}\right) + 6.3 \qquad \text{(Eq 5-17)}$$

Using Eq. 5-17, the water's pH can be calculated and compared to the pH_s from Eq. 5-10. If the resultant LSI value is negative, adjustments need to be made to the water. One approach is to add caustic or remove carbon dioxide and raise the pH above the calculated pH_s. If the negative value is only a few tenths, this can be an acceptable approach. However, if the calculated LSI ranges from −4 to −5, which is often the case with product water from an RO system, simply adjusting the pH will not solve the problem. In these cases, the water needs to be treated to decrease the pH_s value.

As noted in Eq. 5-10, the calculation of pH_s, parameters A and B are additive, while parameters C and D are subtractive. Thus, if the goal is to decrease pH_s, the contributions of the A and B parameters need to be reduced and those of C and D need to be increased. The parameters show the following relationships:

- A is proportional to TDS,
- B is proportional to temperature,
- C is proportional to calcium ion concentration, and
- D is proportional to alkalinity.

Because it is often not desirable or cost effective to reduce TDS or temperature further, most stabilization treatments focus on increasing levels of calcium and alkalinity.

TREATMENT PROCESSES

Decarbonation

Excess carbon dioxide, which often occurs with RO and NF product supplies that use feed acidification, can be economically removed to increase pH using a decarbonator. A decarbonator is a packed tower that uses countercurrent air and water flow to induce the transfer of carbon dioxide from the water to the air. The packing media is inert thermoplastic with a high surface area that accelerates the transfer rate.

Water is introduced into the top of the tower, where a liquid distributor spreads it over the tower cross-section to flow down through the packing. Liquid distributors are composed of nozzles that spray the water over the packing or have a tray or weir style. The main difference between the two types of distributors is the operating pressure requirement. Nozzle-based distributors typically require a minimum of 5 psig pressure to operate, while weir-type distributors can operated a low head (<2 ft). Air is drawn from the atmosphere and induced or forced through the tower. Typically, centrifugal forced draft blowers located at the tower base are used. Entry air is often filtered to prevent the entrance of contaminants or debris.

The air-to-water flow ratio for a decarbonator is usually 3:1 (cfm:gpm). The tower cross-section should limit the flow of water to 25 gpm/ft^2. A packing depth of 8 ft is generally sufficient for achieving a carbon dioxide effluent concentration of 8 mg/L of less. Under these conditions, the air pressure drop across the tower is 2.5 in-wc or less, provided the media stays clean. When treating RO-treated product water directly, the bed can be expected to remain clean. If the water has not been RO treated, partially disinfected supplies are introduced to the tower. Connections that allow the circulation of cleaning solution through the bed are recommended.

Parameters that affect decarbonator sizing include flow, pH, temperature, carbon dioxide concentration, TDS, alkalinity, and total hardness. Manufacturers of packing media can model the decarbonator's performance with these parameters defined. This is a good check, particularly if the amount of residual carbon dioxide required is important.

Deoxygenation

Treated wastewater stored in an ASR needs to be deoxygenated to remove DO. Water in an ASR tends to have an arsenic concentration that exceeds regulatory limits. While oxygenated water is stored in an ASR, naturally occurring arsenic leaches out of the limestone deposits. The mobilization of arsenic is caused by a chain of oxygen-reduction reactions that occur when DO is present in water. By removing oxygen prior to pumping the water to an ASR, arsenic leaching can be minimized to the point where drinking water standards can be met.

Deoxygenation is carried out using a membrane gas contactor (MGC; Figure 5-16). The MGC is a shell–tube device that contains microporous hydrophilic polypropylene hollow fibers. Water does not pass through the pores. Instead, the membrane surface acts as an inert support that allows water to come in direct contact with a gas phase without dispersion. The partial pressure of the gas can be adjusted to control the solubility of gas in water. The MGC is operated in a vacuum to initiate the partial pressure gradient or a sweep gas such as nitrogen is used. By adjusting the gas concentration on the inside of the fiber, the concentration of dissolved gases in the aqueous phase outside of the fiber can be controlled.

Alkalinity Recovery

Alkalinity can be recovered by converting residual carbon dioxide to bicarbonate alkalinity, according to the following relationship:

$$NaOH + CO_2 \rightarrow HCO^-_3 + Na^+ \quad \text{(Eq 5-18)}$$

This can be an effective way to stabilize RO product supplies that are blended with other nondemineralized supply sources. If the final product water that requires stabilization is straight RO product water, then the final product water is typically not sufficient.

Increasing Calcium Hardness

Calcium hardness can be increased using the following:

- Quicklime,
- Hydrated lime,
- Lime slurry,
- Lime saturator,
- Limestone contactor, and
- Calcium chloride.

Each is discussed below.

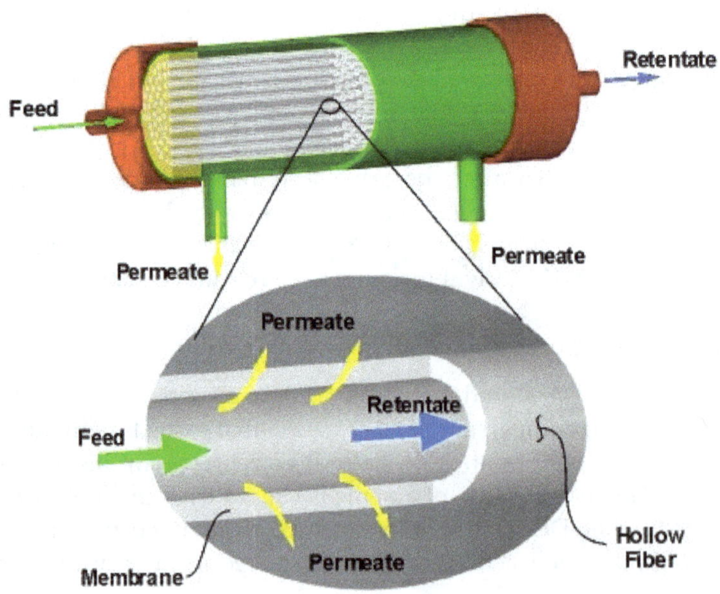

Courtesy of 3M.

Figure 5-16 Liqui-cel membrane contactor

Quicklime. Quicklime is a dry form of calcium oxide that has not been in contact with water. Quicklime requires wetting using a slaker, typically in a batch process. A gravimetric feeder is used to transfer dry chemical from storage to the slaker. The reaction of quicklime with water forms calcium hydroxide. This reaction creates heat, and the mixture needs to age before it can be used. Raw quicklime typically contains significant inert materials that must be removed effectively to prevent carryover to the downstream processes where it can accumulate in pipelines and reservoirs. Once aged, the hydrated lime is transferred to the point of application using hose pumps. It is good practice to use large, redundant transfer lines with multiple cleanouts to assist with long-term maintenance.

Hydrated lime. Hydrated lime (calcium hydroxide) is a dry material that is produced from quicklime and water prior to shipment to the site. Hydrated lime readily forms a slurry when contacted with water since it has already been hydrated and does not require aging. Therefore, continuous slurry production and transfer is possible. Metering of the process is similar to that for quicklime.

Lime slurry system. Proprietary lime slurry products with concentrations of 45 percent are commercially available. The advantage of this type of lime slurry is that it can be handled and dosed directly to the feed point, eliminating the need to handle solid lime and the need for a mixing tank.

Lime saturator. In a lime saturator, slurries of hydrated lime are transferred to a storage tank or saturator, where additional demineralized water is added. The saturator is essentially a solids contact clarifier designed to produce a saturated lime solution at a concentration of around 1,650 mg/L. The saturator eliminates undissolved calcium hydroxide in the solution that is transferred to the process. Saturated solutions can be stored in a tank prior to final metering. Centrifugal pumps can be used for metering.

Limestone contactor system. Limestone contactors use calcite chips (calcium carbonate). Water flows through the contactor and dissolves/absorbs the calcite. Contactors can be gravity fed or pressure type. An air–water backwash system is used to keep the bed uncompacted and free-flowing. As the media dissolves, new calcite must be added to the contactor(s). It is recommended that a high-capacity drain be placed at the base of each contactor to allow the removal and flushing of undissolved materials. The transfer efficiency of a limestone contactor is less than what is achieved by direct transfer of a hydrated lime solution. However, a limestone contactor is sometimes preferred because it eliminates the need to store and handle lime, along with its aesthetic and maintenance issues.

Calcium chloride. Calcium chloride can be delivered, stored, and metered as a 45 percent solution by weight. However, it does require an increased amount of solution in comparison to lime-based systems; this can result in a higher total cost. It also adds chloride ion to the treated water, which can increase TDS and potentially exceed specific chloride goals for the product supply.

ADVANCED OXIDATION SYSTEMS

An advanced oxidation system (AOS) has been integrated into many membrane water reuse treatment schemes in the United States and around the world. Figure 5-17 shows the Clearwater, Fla., replenishment process flow diagram that incorporates an AOS. Most systems treat permeate from the RO system; this is considered a polishing step to provide disinfection and remove trace organics that pass through the RO membrane. Advanced oxidation includes processes that produce highly reactive hydroxyl radicals ($OH^{\bullet}$). These processes include Fenton's reagent (hydrogen peroxide [H_2O_2] and iron) and combinations of ozone (O_3), UV irradiation, and hydrogen peroxide (O_3/UV, O_3/H_2O_2, and UV/H_2O_2). The hydroxyl radical is a powerful oxidant that can eliminate contaminants from water by chemically transforming them through oxidation. Complete oxidation of carbon-based compounds converts the contaminant to carbon dioxide and water. While they are very powerful, hydroxyl radicals have a very short half-life, on the order of seconds, and they do not last long in natural waters.

While there are many ways to generate hydroxyl radicals, the most common methods in water reuse are use of UV/H_2O_2 or ozone with or without the addition of hydrogen peroxide. While both ozone and UV irradiation are effective disinfectants, both can be used for advanced oxidation, especially when combined with H_2O_2. An AOS has been shown to be very effective in the degradation of organic pollutants and endocrine-disrupting chemicals (EDCs) through the formation of the hydroxyl radical. Use of ozone in this application has gained interest because highly treated wastewater requires lower ozone doses, making it cost effective in some cases. When ozone is fed with hydrogen peroxide, it reacts with the deprotonated form of hydrogen peroxide (HO^{2-}) to form hydroxyl radicals. While the O_3/H_2O_2 reactions are complex, involving initiation and propagation steps, a simplified version of the overall reaction can be shown as:

$$H_2O_2 = 2O_3 \rightarrow 2OH \times + 3O_2 \qquad \text{(Eq 5-19)}$$

To be effective, H_2O_2 does not always need to be fed with ozone. Many compounds directly react with ozone and be oxidized. Ozone can also react with organics and other constituents to generate hydroxyl radicals that can react with the contaminants. The need for peroxide depends on the application, water quality, and the target contaminants.

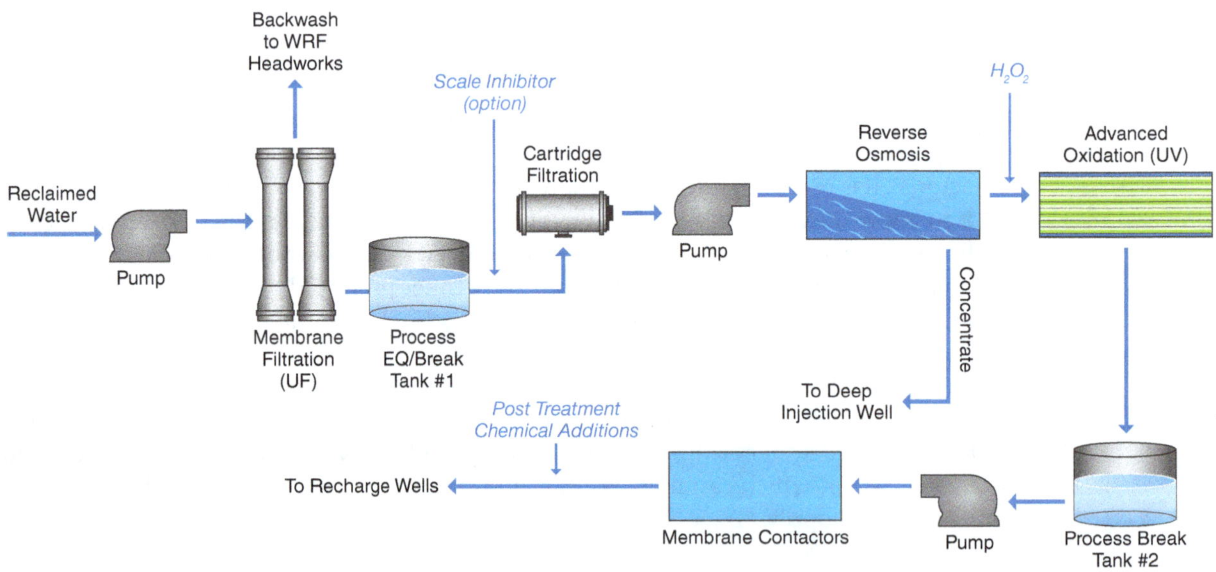

Source: Mercer et al. (2015).

Figure 5-17 Clearwater, Fla., groundwater replenishment process flow diagram

Abbreviations: EQ equalization; UF, ultrafiltration; UV, ultraviolet; WRF, water recovery facility.

UV/H_2O_2 oxidation treatment has been used in full-scale applications for many years. When H_2O_2 is present in water and is irradiated with UV light, hydroxyl radicals are formed. The UV light, represented as *hv* in Eq 5-20, causes the dissociation of H_2O_2 into two hydroxyl radicals. The mechanism is represented by the following equation

$$H_2O_2 + hv \rightarrow 2OH \times \quad \text{(Eq 5-20)}$$

In water reuse, many utilities use UV/H_2O_2 for their AOS because it forms hydroxyl radicals for oxidation of contaminants and because UV can also remove NDMA, a by-product found in wastewater effluent, through direct photolysis. The systems are typically designed to provide high-level disinfection and 1-log removal of NDMA and 0.5-log removal of 1,4-dioxane. NDMA and 1,4-dioxane are used to size the system because they represent the full spectrum for the compounds being removed. NDMA represents compounds that only react directly with UV energy and 1,4-dioxane represents compounds that only react with hydroxyl radicals. Most compounds fall somewhere in between because they react partially through direct photolysis and with hydroxyl radicals.

Whether the oxidation is generated by UV/H_2O_2 or ozone, both behave similarly. The most complete study of microconstituent removal by a variety of technologies is by Snyder and Wert (2007). A summary of their findings is provided in Table 5-7. Their results demonstrate that while UV alone did not provide a high level of removal of many of the constituents, when UV was combined with hydrogen peroxide, it provided a high level of removal for many of the same compounds as ozone alone did.

UV With Hydrogen Peroxide

The most common AOS for water reuse is UV/H_2O_2. It has been shown to be very effective for disinfection and for of organic contaminants. In disinfection applications, the two options for lamp technology—low pressure high output (LPHO) and medium pressure (MP)—have a fairly equal market share, and the selected technology depends on project conditions. However, in water reuse, LPHO lamp technology seems to dominate the market and applications. This is most likely due to the high doses (irradiance) that are required and the fact that the lamps are always on; the power draw of the system can range from 10 to 20 kW/mil gal treated. The high power draw tends to favor LPHO lamps because they use less power and therefore have a lower operating cost.

An advantage of a UV/H_2O_2 system is that the doses of UV and hydrogen peroxide can be varied to achieve a set of objectives and to optimize operations cost. The balance of the two depends on power costs, hydrogen peroxide costs, the water's UV transmittance, and the target contaminant concentration. One issue is that the conversion of peroxide to hydroxyl radicals is only 15 to 20 percent efficient. Consequently, there is a hydrogen peroxide residual that, in some cases, must be quenched. With hydrogen peroxide doses typically in the range of 5 to 10 mg/L, there is still 4 to 8 mg/L of peroxide residual. Quenching methods include use of chlorine, which must be dosed at 2 mg/L of chlorine per 1 mg/L of hydrogen peroxide, or granular activated carbon contactors.

Ozone and Ozone With Hydrogen Peroxide

Because recent research has shown that ozone is very effective in removing microconstituents, such as EDCs and pharmaceutical and personal care products, there has been a renewed interest in ozone's use in wastewater and water reuse. Current research on ozone in water reuse shows that ozone is very effective at inactivating viruses and coliforms, which are indicator organisms that are commonly found water reuse. Testing and full-scale sampling has shown that it can meet certain disinfection requirements for irrigation and agricultural reuse. However, to meet more stringent disinfection requirements, as required in California's Title 22 guidelines, ozone needs to be coupled with filtration. Some reuse applications that use ozone follow ozonation with some form of biological filtration because most organic by-products that are formed during ozonation are biodegradable. Other by-products formed during ozonation, such as bromate and NDMA, have been documented, but their formation depends on many factors including water quality and ozone dosage. Research suggests that the DOC concentration is one of the most important water quality parameters to consider when selecting an ozone dose for design or operation.

Table 5-7 Removal of microconstituents by ultraviolet (UV) irradiation, UV irradiation/hydrogen peroxide, and ozone

	Percent Removal			
Microconstituent	UV 40 mJ/cm^2	UV 439 mJ/cm^2	UV 372 mJ/cm^2 5 mg/L H_2O_2	Ozone Drinking Water Doses
Acetaminophen	20–50	>80	>80	>95
Androstenedione	<20	<20	50–80	>80
Atrazine	<20	50–80	50–80	20–50
Caffeine	<20	<20	50–80	>80
Carbamazepine	<20	20–50	>80	>95
DEET	<20	<20	50–80	50–80
Diazepam	<20	<20	50–80	50–80
Diclofenac	50–80	>80	>80	>95
Dilantin	<20	50–80	>80	50–80
Erythromycin	<20	50–80	50–80	>95
Estradiol	<20	>80	>80	>95
Estriol	<20	>80	>80	>95
Estrone	<20	>80	>80	>95
Ethynyl estradiol	<20	>80	>80	>95
Fluoxetine	<20	>80	>80	>95
Gemfibrozil	<20	20–50	>80	>95
Hydrocodone	<20	>80	>80	>95
Ibuprofen	<20	20–50	>80	50–80
Iopromide	<20	50–80	50–80	20–50
Meprobamate	<20	<20	20–50	20–50
Naproxen	<20	>80	>80	>95
Oxybenzone	<20	>80	>80	>95
Pentoxifylline	<20	20–50	50–80	>80
Progesterone	<20	20–50	>80	>80
Sulfamethoxazole	50–80	>80	>80	>95
Tris(2-chloroethyl) phosphate	<20	<20	<20	<20
Testosterone	<20	20–50	>80	>80
Triclosan	50–80	>80	>80	>95
Trimethoprim	<20	20–50	>80	>95

Source: Adapted from Snyder and Wert (2007).

To date, use of ozone for reuse in the United States has been limited. In a recent survey of ozone use in wastewater, only three facilities were identified as using ozone as part of an advanced treatment process for indirect potable reuse. The three facilities were the Fred Hervey Water Reclamation Plant in El Paso, Texas, the F. Wayne Hill Water Reclamation Center in Gwinnett County, Ga., and the Clark County Water Reclamation District in Las Vegas, Nev. A survey of ozone use in Japan showed that at the end of 2005, there were 51 facilities using ozone for wastewater treatment. Of those facilities, 90 percent were for water reuse and the remaining 10 percent were for final discharge. Ozone dosages for reuse ranged from 10 to 15 mg/L and from 5 to 10 mg/L for final discharge. Of the facilities with ozone, 90 percent had some form of pretreatment such as filtration, and more than half had chlorination as post-treatment.

DESIGN CONSIDERATIONS

An AOS that uses UV/H_2O_2 includes a UV reactor(s) with lamps and quartz sleeves, UV control panels, UV intensity sensors, an on-line UV transmittance monitor, and a hydrogen peroxide storage and feed system. Design considerations for an AOS include selection of lamp type, optimization of peroxide use versus energy, selection of a fouling factor, selection of an end of lamp-life factor, meeting a process guarantee, and conducting full-scale performance testing. Lamp type, that is, LPHO versus MP, should be selected by evaluating the present cost of the system. An MP system would typically be smaller and have lower capital cost; however, a LHPO system would use less power, which could result in lower operating cost. Based on a present cost analysis, the system should be based on design. If the system is designed without actual water quality data, models should be used to project the water quality at the AOS. When designing a system using assumed water quality, challenges include selecting a fouling factor, end of lamp-life factor, and design UV transmittance value that are conservative enough to ensure that the system meets the performance guarantee but not too conservative so as to oversize the system and drive up the capital cost.

One of the biggest challenges of using ozone for water reuse treatment is selecting an ozone dose that meets the treatment goal. Ozone doses required to treat wastewater are highly variable because the effluent quality is highly variable. It is best to perform bench-scale or pilot-scale testing when selecting an ozone dose.

WATER QUALITY IMPACTS

Influent water quality has a direct impact on a system's efficiency in removing organic contaminants. Oxidation is a nonselective process. Therefore, other compounds in the water can react in addition to the contaminants of concern. This increases the oxidant demand, thus increasing costs. These other reactions can also prevent complete oxidation, resulting in by-products other than carbon dioxide and water. Examples of how water quality can affect the AOS include the following:

- Total organic carbon (organic matter) exerts an oxidant demand.
- Other oxidizable species, for example, arsenics (III) and reduced forms of iron and manganese, exert an oxidant demand, especially if they are more easily oxidizable than the target organics. Some can form scale in the presence of UV light.
- Alkalinity scavenges hydroxyl free radicals and may require pH reduction to minimize the scavenging effect.

- Nitrate absorbs UV light and is converted to nitrite, which exerts an oxidant demand. This may only be a concern for some MP systems.
- Bromide, if present in sufficient concentrations, can be converted to bromate (which has a maximum contaminant level) by ozone. However, there are numerous bromate control methods available.
- Turbidity, which is not generally a problem with reuse after RO, lowers transmittance of UV light.

Because of these water quality impacts, it is desirable to locate the AOS near the end of the process.

By-Product Formation

One concern with most disinfectants and oxidants is the formation of by-products. Ozone can form various acids, aldehydes, and aldo- and ketoacids and, in the presence of bromide, can form bromate and other brominated by-products. Recent research has also shown that ozone can form NDMA in reuse applications. In testing done on a full-scale reactor used to treat secondary effluent, the results showed that there was rapid formation of NDMA in the first part of the ozone contactor, followed by a slight decrease through the ozone contactor and another decrease through sand filtration. In a pilot-scale study of ozone before microfiltration, ozonation increased the average NDMA concentration from 11 ppt to 114 ppt, well above the California action level. The formation in NF–RO permeate should be less of a concern. Most research has shown that there is minimal by-product formation when NF–RO permeate is treated with UV/H_2O_2.

REFERENCES

Asano, T., F, Burton, and H. Leverenz. 2007. *Water Reuse: Issues, Technologies, and Applications.* New York: McGraw-Hill.

AWWA. 1995. M38, *Electrodialysis and Electrodialysis Reversal.* Denver, CO: AWWA.

———. 2000. M37, *Operational Control of Coagulation and Filtration Processes.* Denver, CO: AWWA.

———. 2005. M53, *Microfiltration and Ultrafiltration Membrane for Drinking Water.* Denver, CO: AWWA.

———. 2006. M20, *Water Chlorination/Chloramination Practices and Principles.* Denver, CO: AWWA.

———. 2007. M46, *Reverse Osmosis and Nanofiltration.* Denver, CO: AWWA.

AWWA/ASCE. 2005. *Water Treatment Plant Design,* 4th ed. New York: McGraw-Hill.

AwwaRF. 2004. *Assessment and Development of Low Pressure Membrane Integrity Monitoring Tools.* Denver, CO: AWWA Research Foundation.

———. 2005. *Development of a Microfiltration and Ultrafiltration Knowledge Base.* Denver, CO: AWWA Research Foundation.

Bradshaw, G. 2013. Role of WWTPs for DPR. WateReuse DPR Conference, November 7, 2013, Newport Beach, CA.

CDPH (California Department of Public Health). 2001. California Surface Water Treatment Alternative Filtration Technology (AFT) Report (Draft), June. Sacramento, CA: CDPH

———. 2011. Alternative Filtration Technology (AFT) Report, August. Sacramento, CA: CDPH (http://www.cdph.ca.gov/certlic/drinkingwater/Pages/Publications.aspx.

———. 2013. Alternative Treatment Technology Report for Recycled Water, October. Sacramento, CA: CDPH

———. 2014. Recycled Water Regulations, Article 5.2 Indirect Potable Reuse: Groundwater Replenishment–Subsurface Application, updated June 18, 2014, from Title 22 and 17 California Code of Regulations. Sacramento, CA: CDPH

City of San Diego. 2013. Indirect Potable Reuse/Reservoir Augmentation Demonstration Project Advanced Water Purification Facility, Advanced Water Purification Facility Study Report. City of San Diego. 2013. Indirect Potable Reuse/Reservoir Augmentation Demonstration Project Advanced Water Purification Facility, Advanced Water.

Cheryan, T. 1998. *Ultrafiltration and Microfiltration*, 2nd ed. Boca Raton, FL: CRC Press.

Coffey, B.M. 1993. Evaluation of Microfiltration for Metropolitan's Small Domestic Water Systems. Proceedings 1993 AWWA Membrane Technology Conference. Baltimore, Md. Denver, CO: AWWA.

Connell, G.F. 2002. *Chlorination/Dechlorination Handbook*. Alexandria, VA: Water Environment Federation.

Crittenden, J.C.; Trussell, R.R.; Hand, D.W.; Howe, K.J.; & Tchobanoglous, G. 2012. *Water Treatment Principles and Design*, 3rd ed. New York: John Wiley & Sons, Inc.

De, S. & Modal, S. 2012. *Micellar Enhanced Ultrafiltration*. Boca Raton, FL: CRC Press.

Edzwald, J.K. 2012. *Dissolved Air Flotation for Water Clarification*. New York: McGraw Hill

Farre, J.M.; Doderer, K.; Keller, J.; Poussade, Y.; Mueller, J.; & Gernjak, W. 2010. *N*-nitrosodimethylamine (NDMA) in Purified Recycled Water. Collaboration Between Urban Water Security Research Alliance, Veolia Water Australia, and WaterSecure.

Freeman, S.; Bates, J.; Wallis-Lange, C.; & Mcevoy, J. 2008. Drought Relief in South East, Queensland, Australia, Provided by Membrane-Reclaimed Water. *Journal of the American Water Works* Association, 100:2.

Gerrity, Daniel, et al. 2011 Pilot-Scale Evaluation of Ozone and Biological Activated Carbon for Trace Organic Contaminant Mitigation and Disinfection. *Water Research*, 45: 2155–2165.

Hashimoto, T.; Nakazawa, H.; & Murakami, T. 2009. State of Ozonation to Municipal Wastewater Treatment in Japan. 19th Ozone World Congress, Tokyo, Japan. International Ozone Association.

Jacangelo, J.G.; Adham, S.S.; & Laine, J-M. 1991. Low-Pressure Membrane Filtration for Removing *Giardia* and Microbial Indicators. *Journal of the American Water Works Association* 83:9:97.

———. 1995. Mechanism of *Cryptosporidium, Giardia,* and MS2 Virus Removal by MF and UF. *Journal of the American Water Works Association,* 87:9:107.

Judd, S. 2011. *The MBR Book: Principles and Applications of Membrane Bioreactors in Water and Wastewater Treatment,* 2nd ed. Oxford: Elsevier Ltd.

Mercer, T.; Bennett, J.; Fahey, R.; Moore, E.; MacNevin, D.; & Kinslow, J. 2015. Groundwater Replenishment Performance and Operations: Lessons Learned During Clearwater's One-Year Pilot. *Florida Resources Journal,* 66:8:32–40.

Schreiber, I.M. & Mitch, A.W. 2005. Influence of Order or Reagent Addition on NDMA Formation During Chloramination. *Environmental Science & Technology,* 39:3811–3818.

Snyder, S. & Wert, E. 2007. Removal of EDCs and Pharmaceuticals in Drinking and Reuse Treatment Processes. Water Research Foundation Summary of Pilot Testing of MF Pretreatment With Ozone. Project Progress Report.

USEPA (US Environmental Protection Agency). 1986. *Design Manual Municipal Wastewater Disinfection.* Cincinnati, OH: USEPA.

———. 2005. *Membrane Filtration Guidance Manual, 815-R-06-009.* Cincinnati, OH: USEPA.

Water Environment Federation (WEF). 2006. *Membrane System for Wastewater Treatment.* Alexandria, VA: WEF.

Wallis-Lange, C.; Veerapaneni, S.; Pressdee, J.; & Steichen, M. 2009. A New Advanced Water Treatment Plant in Australia Extends Potable Water Resources While Reducing Nutrient Discharges to the Environment. *Water Environment & Technology.*

Zimmermann, S., et al. 2011. Kinetic Assessment and Modeling of an Ozonation Step for Full-Scale Municipal Wastewater Treatment: Micropollutant Oxidation, By-Product Formation and Disinfection. *Water Research,* 45: 605–617.

Chapter 6

Operations

This chapter describes the operational issues related to membrane processes in water reuse applications. This chapter is not intended solely for operators of membrane systems. It is also intended for designers, planners, and managers of membrane systems so that they can effectively communicate the requirements for successful application and sustainable operation of membrane systems in water reuse.

In this chapter, the terms "low-pressure membrane," "membrane filtration," and "microfiltration–ultrafiltration (MF–UF)" membranes are used interchangeably to describe a hollow-fiber MF or UF system, configured in an inside-out flow path. Similarly, high-pressure membranes and reverse osmosis (RO) are used interchangeably to describe a multistage, polyamide RO system.

MEMBRANE SYSTEM DATA COLLECTION AND ANALYSIS

Data collection and analysis are an imperative part of operating a membrane plant in a water reuse application. When performed properly, data collection and analysis can help identify, if not prevent, problems early when issues can be resolved with less effort. The ultimate goal of data collection and analysis is to achieve plant production capacity, ensure the integrity of the membrane's contaminant removal characteristics, and maintain an efficient operations and maintenance cost.

Regular, organized data collection is essential for effective process monitoring and troubleshooting of membrane systems. It is also helpful and typically required when membrane warranty issues arise. The treatment plant operator is advised to review the specific requirements in the membrane manufacturer's warranty documents and recognize that appropriate data collection for ensuring long-term, successful operations is not limited to the warranty requirements alone.

A data collection program should begin at startup and be maintained throughout the membrane system's life. Data from startup and periods of successful operation serve as a benchmark for future performance. Meticulous data collection and record keeping are especially important for MF systems that are used in a reuse application due to the high variability of feedwater quality and operating conditions.

Without continuous data collection and analysis, membrane treatment plants can suffer a number of performance-related problems including the following:

- Irreversible loss of permeability and inability to achieve design flows (both types of membrane systems)
  - Exposure to unproven chemicals (for example, inappropriate coagulants);
  - Organic fouling due to organic material deposited onto or absorbed by the membrane surface;
  - Biological fouling resulting from biological growth within the membrane system; and
  - Inorganic fouling/scaling caused by precipitation of inorganic material in and on the element (primarily in RO systems).
- Failure to achieve contaminant removal goals (for example, broken fibers in membrane filtration, salt passage increase for reverse osmosis)
  - Introduction of contaminants that can damage the membrane system (for example, heavy solids loading in membrane filtration or sizable particulate material in RO);
  - Chemical oxidative damage to the membrane due to exposure to oxidants (RO); and
  - Salt passage increases resulting from mechanical failures such as permeate backpressure membrane damage or O-ring integrity (RO systems).

Feedwater Characteristics

Several feedwater characteristics should be monitored as part of overall data collection and monitoring for a membrane system. At a minimum, these include temperature; pH; concentrations of total organic carbon, total suspended solids (TSS), and total dissolved solids; electrical conductivity; turbidity; and chloramine residual for systems that use chloramine to prevent biological fouling.

Several treatment facilities have found that use of nontraditional analytical techniques can offer insights into the level of treatment occurring at upstream wastewater treatment processes, the membrane fouling potential of their source water, and changes in water quality that are not necessarily indicated by traditional means, such as turbidity or TSS. Two such analytical techniques are ultraviolet transmittance and fluorescence excitation and emission matrices.

Instrument Calibration

Instruments should be carefully calibrated on a regular basis. Instrumentation should be calibrated at least as often as the manufacturer's recommendations and checked when irregular unit performance is suspected.

Cleaning Logs

It is important to collect data before and after a chemical clean-in-place (CIP) to analyze cleaning effectiveness. It is also important to record the details of the cleaning process. At a minimum, this includes the following:

- CIP parameter,
- Chemical used,

- Chemical concentration,
- Solution temperature,
- Solution pH,
- Recirculation flow rate,
- Recirculation duration,
- Soak duration,
- CIP feed pressure, and
- CIP differential pressure or transmembrane pressure (TMP).

OPERATING DATA

System operating data and water quality parameters should be collected and analyzed regularly. Different parameters may require different collection frequencies based on necessity, cost, and labor to collect and review them. For example, water quality sampling data are collected on a monthly, quarterly, or yearly basis, whereas most operating data are collected daily or automatically archived in a supervisory control and data acquisition (SCADA) system.

It is imperative for data to be collected regularly. SCADA data may be recorded every 15 minutes during plant startup and daily during stable plant operation. In the absence of online instrumentation, field data must be collected at least daily at the same time each day because daily changes in feedwater quality, feed temperature, and operating conditions will otherwise create inaccuracies in data analysis.

Special attention should be given to startup data because it serves as a baseline for long-term performance. Additional data that should be collected at startup include the following:

- Membrane serial numbers,
- Installed location,
- Membrane module/element area,
- Number of installed modules/elements,
- Standard backwash characteristics (for example, interval, flow rates, aeration rates),
- Standard cleaning characteristics,
- Sensor calibration records, and
- Initial performance data.

Key design parameters for any membrane system should be monitored as part of operations, including water flux and water recovery. It is critical that the operator of a membrane system understand the design intent of the facility and be able to make knowledgeable and appropriate adjustments to the operating conditions when the need arises.

Water Flux

Water flux, also called filtrate flux or permeate flux, is a measure of the water flow per unit of membrane area over time. Typical US units are gallons of water per square foot of membrane area per day, or gfd. It is important to monitor water flux because higher flux rates result in faster deposition of contaminants or foulants on the membrane surface. This deposition can result in shorter cleaning intervals, potentially less effective cleanings, and poorer long-term performance.

Water Recovery

Water recovery is the ratio of product water flow per feedwater flow and is reported as a percentage. It is important to monitor water recovery because higher recovery rates result in higher concentrations of contaminants or foulants on the membrane surface. These concentrations can result in shorter cleaning intervals, potentially less effective cleanings, and poorer long-term performance.

Temperature or Viscosity Correction Factors

Temperature correction factors (TCFs) are used to adjust temperature-sensitive calculations, as water viscosity can significantly impact the driving pressure required to transport water through a membrane. Membrane manufacturers supply TCFs for data normalization. While membrane manufacturers and ASTM International are useful resources, TCF equations can be modified and adjusted to better suit a treatment plant's unique operating conditions.

Membrane Filtration Systems

Essential data for MF systems that should be collected on a daily basis include the following:

- Pressures and flows of feed, filtrate, and any bleed or concentrate streams,
- Feed temperature and pH,
- Feed and filtrate turbidity,
- Membrane integrity test (MIT) start pressures and results, and
- Silt density index (SDI) on the filtrate stream.

Calculated values include the following:

- TMP,
- Temperature or viscosity correction,
- Filtrate flux,
- Water recovery, and
- Specific flux, normalized transmembrane pressure, and membrane resistance.

Calculated values should be normalized to standard operating conditions. Normalized data are used to assess membrane performance, quantify fouling rates, and identify trends that might suggest the need for cleaning or impending performance problems. These values are calculated using operating data and system information such as membrane area and design flows. Normalized values indicate the performance of a system if it is operated under "standard" operating conditions. This allows for a direct comparison of data from one day to the next regardless of variations in feedwater temperature or flows.

Process trains and equipment should be evaluated individually but also be compared to other units within the system. Large deviations in the performance between individual units are often due to mechanical problems and should be investigated.

MEMBRANE INTEGRITY TEST

MITs are used to monitor the performance of individual membrane modules or racks and should be tracked throughout the life of each module. In some cases, this parameter may be a regulatory requirement, and the operator is advised to consult the treatment plant permit to determine regulatory reporting requirements. The MIT is a critical parameter for monitoring membrane performance.

It is important to monitor and record the start pressure of the integrity test along with the result. In units that have especially poor integrity, a low start pressure will result in falsely passing MIT, as illustrated in Figure 6-1. In this example, the MIT result suggests improved integrity once the start pressure for the test decreases below a critical threshold, but the true membrane integrity remains compromised.

In addition to daily integrity test data, records of module repairs, replacements and failures and their causes should be maintained. Localized integrity problems could be the cause of a specific issue, and "mapping" fiber repairs within the unit is an effective way to investigate the problem. Careful record keeping of integrity tests is also useful when dealing with potential warranty issues.

Clean Water Flux Test

During startup, most treatment facilities perform a clean water flux test (CWFT). In this process, potable water is used while the system is operated at multiple flux set points. In addition to documenting the startup performance for a new set of MF membranes, a CWFT can confirm the cleaning effectiveness of a CIP. For individual units, the resulting data can be used to confirm that a system returned to baseline performance after a cleaning procedure and often provide insight into potential optimizations for the cleaning regime. When CWFT results are compared for several units from the same MF system, this data can suggest whether a particular unit is experiencing a heavier rate of fouling (possibly due to mechanical or other issues) or whether ineffective cleanings is a systemic issue. Plotting temperature-corrected TMP against filtrate flux makes it possible to assess cleaning effectiveness and the condition of the membrane with regard to fouling.

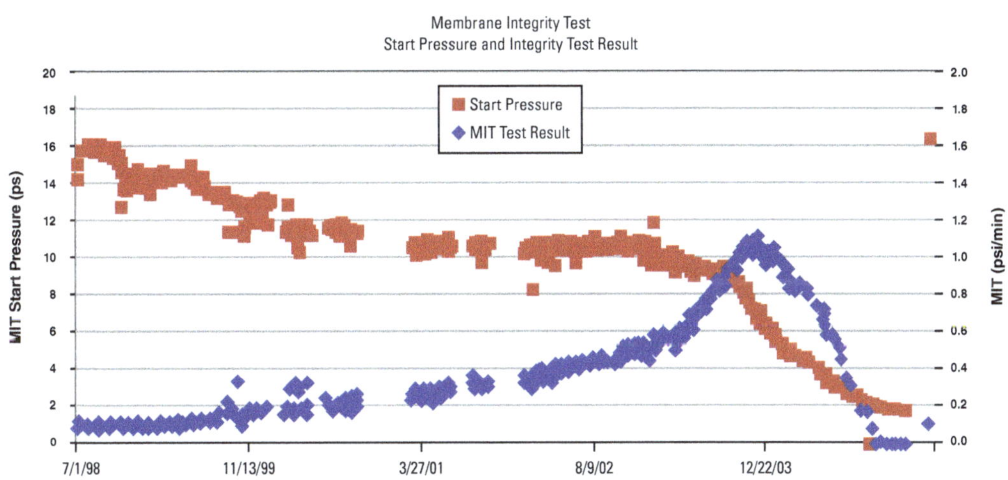

Figure 6-1 Illustration of the importance of maintaining sufficient start pressures for a membrane integrity test

Abbreviation: MIT, membrane integrity test.

RO Systems

Many modern RO treatment plants are automated to collect the appropriate data and enable operations staff to analyze real-time and historic performance trends. Table 6-1 shows the data parameters that must be continuously collected for analysis.

In order to assess the performance of an operating RO membrane and identify any trends, normalized values are calculated using the system operating data. These values are calculated using operating data and system information such as membrane area and membrane type. Normalized values indicate the performance of the system if operated under "standard" operating conditions, thereby allowing direct comparison to predicted performance from one day to the next, regardless of actual operating conditions.

Membrane Surface Fouling Indicators

Specific flux, water permeability, and normalized permeate flow are a class of parameters that measure fouling within a membrane system. It is not necessary to monitor all of these parameters; however, at least one should be calculated and continuously monitored in order to understand the fouling conditions of the membrane system. Resources for how to calculate this type of parameter are available through ASTM International and membrane manufacturers.

The specific flux ratio is the ratio of the last-stage specific flux to the total-system specific flux. This parameter is a sensitive indicator of scaling. Scale typically begins in the tail end of the RO system, causing reduced permeability relative to total system permeability. This value gradually decreases as scale formation progresses. Using this ratio, a distinction can be made between uniform permeability loss due to normal fouling and a localized loss of permeability due to scale formation. In the absence of interstage flow monitoring capability, it is likely that early tail-end permeability loss would go undetected until it develops into a significant fouling event that impacts overall system production.

Feed/Brine Flow Path Fouling Indicator

Normalized differential pressure (NdP) represents the pressure drop across the feed/brine flow channel within the RO elements. This value should monitored routinely at each stage of the RO system, as it provides an indication of a blockage or obstruction that is restricting flow through the feed/brine channel. The value should be stable for normal operation. *An increase in NdP can be an indication of particulate accumulation in the membrane ele-*ment, scaling, organic fouling, or biological growth. Resources for how to calculate NdP are available from membrane manufacturers.

Table 6-1 General reverse osmosis data collection

Raw Data Parameter	Location of Measurement
Antiscalant dose	Feed
Acid dose	Feed
Chloramine concentration	Feed
Conductivity	Feed, permeate, concentrate
Flow	Permeate, concentrate, interstage
pH	Feed, product water
Pressure	Feed, permeate, concentrate, interstage
Silt density index	Feed
Temperature	Feed

Salt Rejection or Passage

Electrical conductivity is a useful online measurement that is used as a general indicator of a membrane's salt rejection or passage. If more specific constituent rejection is required, the operator should investigate alternative online measurements or periodically sample for the constituents of interest in order to monitor the membrane's removal characteristics for that constituent.

Normalized permeate conductivity, normalized salt passage, and normalized salt rejection are a class of parameters that allow analysis of the membrane's salt removal characteristics. The normalization process adjusts for changes in operating temperature, water flux, and feed/brine concentration, allowing for direct comparison of performance over time. Any significant change in these values should be investigated, as they may indicate deterioration of membrane performance or mechanical leaks in the system. Resources for how to calculate these parameters are available through ASTM International and the membrane manufacturers.

In addition to the parameters listed in Table 6-1, water reuse plants may need to monitor additional parameters for process control or regulatory requirements.

HETEROTROPHIC PLATE COUNT

Heterotrophic plate count (HPC) represents the level of biological activity in a given sample. This measurement is typically used to confirm the presence of biological growth in a membrane system. Feed, permeate, and concentrate streams are sampled to determine the level of biological activity within the membrane system. An increase in HPC from feed to concentrate streams is expected. However, if this increase is significantly greater than that expected from the concentration factor based on system recovery, biological growth may be present. Resources for sampling and laboratory methods in determining HPC are provided by the US Environmental Protection Agency.

OPERATIONAL CONSIDERATIONS

The proper operation of a membrane system is critical to ensure equipment reliability and maintainability. This is accomplished by following general operational guidelines and specific membrane system manufacturers' recommendations. There are standard guidelines that apply to all types of membrane systems, as well as specific guidelines for low- and high-pressure membranes. The goals for the proper operation of any membrane system should be to meet the design effluent quality and quantity while ensuring the safe operation and reliability of the membrane.

Standard operating procedures (SOPs) should be developed in advance of any membrane commissioning and be updated as conditions and equipment change. These procedures provide the basic operational guidelines to safely start, operate, and shut down a membrane system. The level of detail included depends on system size and complexity. For example, a manually operated membrane system that is less than 100 gpm in size may have a one-page SOP, while a complex 100-mgd facility may have volumes of SOPs that provide operating procedures for every treatment system. SOPs are developed based on the vendors operations and maintenance manuals and are site specific.

Basic Guideline Checks

Site-specific operating guidelines are documented in the process control orders. These include operational set points as well as ranges for flows, pressures, chemical dosages, and water quality specifications. This is a living document that is periodically updated as conditions change. It details the system's operating limits to the operations staff. In addition, it should include the frequency of specific functions such as backwashing and membrane cleaning.

- Ensure influent and effluent water quality variations are within the design of the membrane system.
- Abnormal variations in operating pressures can be indicative of an equipment failure or membrane fouling. Erratic pressure fluctuations can damage the membrane and should be resolved promptly.
- Rapid flow increases can indicate a seal or membrane integrity failure.
- Ensure pretreatment chemical dosages are properly maintained and verified.
- Guidelines for short-term and long-term shutdowns.

Startup and Shutdown

Prior to starting any membrane system, it is critical to verify the proper position of all manual valves. The automated valves with feedback positioners should indicate an alarm status if they are not in their proper position. Once the valve lineup is completed, a check of the system operating parameters should be reviewed. For manually controlled systems, it is necessary to adjust valve positions to control the operating pressure and flow to achieve the desired flux rate. The programmed set points for automated systems should be verified prior to startup. This is accomplished via the human machine interface (HMI) or SCADA screens.

All rotating equipment and pretreatment systems should be checked for service. This includes chemical dosing systems. Failure of a chemical system can result in irreversible damage to the membrane or unacceptable effluent water quality. Care must be exercised when starting rotating equipment. For manually operated systems, sudden pressure increases can damage the membrane and dislodge O-ring seals and gaskets.

During any shutdown period, it is critical to minimize biological growth within the membranes. This is accomplished by daily flushes with effluent water or other clean water source when the system is shut down for short periods of time. Long-term shutdowns require not only flushing of the system but the addition of a preservative or biocide. Any chemicals introduced should be compatible with the membrane and components of the system. The membrane manufacturer should be consulted for recommendations. They will also be able to provide detailed recommendations for short- and long-term shutdown care.

Low-Pressure Membrane Systems

Proper pretreatment operation is critical to any membrane system. Typically, low-pressure membrane systems used in water reuse applications have chemical pretreatment that includes the chemicals necessary to maintain a monochloramine residual and coagulant application. Chloramine formation can be accomplished in a number of ways, including the addition of sodium hypochlorite to systems with excess ammonia in the source water or the preformation of chloramines from the combination of sodium hypochlorite and ammonium chloride. The chloramine residual controls the biological activity within the feedwater. Care must be taken to ensure that the membranes are not exposed to oxidizers, which could damage the membrane, unless the membrane is made of an oxidant-tolerant material such as polyvinylidene difluoride (PVDF). Coagulant addition aids in the filterability of particles in the feedwater. However, the membrane manufacturer should be consulted to ensure compatibility of the coagulant with their membrane chemistry. Improper dosages have the potential for irreversible fouling of the membrane surface. Therefore, all chemical pretreatment systems should be periodically inspected and calibrated.

Many low-pressure membrane system manufacturers require that the system have some type of physical prescreening. The screening system equipment can vary, but many are the automated backwash type with 100- to 500-μm screens. This process should be monitored frequently to ensure proper operation and consistent removal of large material that can damage or foul the membranes.

Many MF–UF systems are operated in a dead-end filtration mode, and periodic backwashes are necessary to remove the buildup of solids on the membrane surface. Backwash sequences typically include reverse filtration using MF–UF product water, with an aeration step on the outside of the membrane fiber to assist in scouring. The backwash cycles are dependent on the proper operation of the system valves. In the event a valve sticks or remains in an either open or closed position, the backwash will be ineffective. Valve failures can be monitored by visually inspecting them during operational cycles. In addition, rapid increases in the TMP across a membrane can be an indicator of a valve failure. For facilities with multiple membrane units in service, this can be evident when one unit experiences a rapid increase in TMP and the others do not. In this case, the first consideration should be a failure attributed to that specific unit.

Product water demand is a factor in determining how a low-pressure membrane system is operated. If there is sufficient product water storage, the operating flux rate needs to be determined. Typically, additional capacity is built into a membrane design to take into account aging of the membrane, redundancy, and diurnal flow demand. An example would be the decision to operate a multiarray membrane system with fewer arrays in service and have their operation cycled for equal run-time or to have them operate at a reduced flux rate. Each site and membrane system should have their control sequence defined in their process control orders. If this is not done, it should be addressed so that there is consistency in the operation of the membrane system.

High-Pressure Membrane Systems

In a water reuse application that uses low-pressure RO, it is critical that the incidence of membrane fouling be minimized. Since RO is the most energy-intensive membrane process in an advanced wastewater treatment facility, maintaining permeability ensures that operating pressures remain stable and production goals are met. It also represents the critical step in removing inorganics, organics, and contaminants of emerging concern. Therefore, effective pretreatment measures must be taken.

Low-pressure MF–UF treatment upstream of an RO system is generally effective in removing particulates, suspended solids, and microbial agents that are larger than the nominal pore size of the membrane. Membrane manufacturers typically publish a maximum turbidity of 1.0 ntu and a SDI_{15} value of 5 as indicators of adequate particulate removal in RO feedwater. For an RO system that operates on MF–UF pretreated municipal wastewater, it is not uncommon to see turbidity averages of <0.1 ntu and SDI_{15} values of 1.

In addition, prior to entering the RO system, feedwater may also pass through cartridge filtration to ensure that there is no foreign debris or contaminants in the system. These filters are typically constructed of a string-wound or melt-spun polymer with a nominal pore size that ranges from 10 to 20 μm. This additional filtration step prevents particulates and debris from entering the RO system and damaging the high-pressure pumps (Figure 6-2). Smaller material and debris could pass through and deposit on the lead RO elements. This would restrict feed flow and create stagnant hydraulic zones within the RO elements. This is a suitable environment for biological deposition and proliferation, that is, biological fouling; all of these conditions contribute to declining water flux and higher operating costs.

Figure 6-2 Particulate and debris collected within the cartridge filter housing (left) and on the cartridge filters (right)

As with low-pressure membrane systems, the pretreatment for high-pressure membrane systems such as RO membranes is critical for successful operation. While a cellulose acetate membrane can tolerate harsh oxidants such as free chlorine, polyamide RO membranes are not tolerant of free chlorine or harsh oxidants. More than two decades of operational experience at several facilities has shown that polyamide RO membranes are tolerant of a continuous chloramine residual within the range of 3 to 5 mg/L. However, each RO product may differ, and the operator is advised to verify their membrane's chloramine tolerance with the membrane manufacture. Any upstream chlorination that generates a free chlorine residual will need to be dechlorinated, including any carryover of free chlorine residual from the CIP processes of PVDF low-pressure membranes.

Other pretreatment chemicals used to benefit RO performance include strong acids such as sulfuric or hydrochloric for pH adjustment and scale inhibitors. Depending on the influent water quality constituents and operating recovery, pH may need to be suppressed to control the formation of scale, such as calcium carbonate or calcium phosphate scale, on the membrane surface. There are other scale-forming constituents that pH suppression is not as effective at controlling, and specialized scale inhibitors are required. These include calcium sulfates and barium sulfates, but the most common and challenging scale to control is silica. The proper dosage of these pretreatment chemicals is essential to maintaining the membrane system's performance. Often, it is not sufficient to rely on the software provided by scale-inhibitor manufacturers for scale control in the reuse environment, and only empirical performance data can reveal a product's efficacy.

As water passes through the RO system, solutes retained within the bulk stream are concentrated to a point where they may exceed their solubility limit and precipitate onto the membrane. This process is known as scaling. Scaling typically begins at the tail end of the RO system, where the concentration of solutes is the highest. A scanning electron microscopy (SEM) image of a tail-end RO membrane scaled with silicates is shown in Figure 6-3. (A clean membrane is also shown for comparison purposes.)

Scaling is primarily a function of the concentration of inorganics in the feedwater and the operating recovery of the system, which generally ranges from 75 to 85 percent, depending on the design criteria. A system that operates at a recovery of 85 percent would concentrate highly rejected constituents such as silica by a factor of nearly 6.7. A silica concentration of 50 mg/L in the feedwater would equate to nearly 335 mg/L in the concentrate stream in the tail-end of the RO system.

Figure 6-3 Scanning electron microscopy images of a clean reverse osmosis membrane surface (left) and one containing silicate scale (right)

Scaling control measures are highly dependent on the feedwater profile and generally include the addition of acid and/or scale inhibitors. The optimal strategy is determined by reviewing the water quality profile, identifying potential scalants, and targeting their control with specific scale inhibitors. Ultimately, pilot testing of any proposed chemical pretreatment scheme is recommended prior to full-scale implementation. It is important to remember that successful control at one water reuse facility with similar technologies and practices is no guarantee of successful operations at all facilities. Effectiveness is highly site specific.

The location and nature of fouling and flux decline within an RO system dictate what chemical cleaning to perform and where. Close examination of operational data generally provides the information necessary to formulate a corrective action. When reviewing operational data, the normalized data should be used. The normalization calculations compensate for changes in temperature, feed/brine flow, permeate pressure, and similar parameters, all of which can influence the operational data if not fixed to a standard set of conditions. While several operational parameters exist and can be monitored, the ones of significant importance in a water reuse application are specific flux, specific flux ratio, and normalized differential pressure. Figures 6-4 and 6-5 illustrate these parameters.

High-pressure membrane systems operate in a static mode compared to low-pressure systems. There are no backwash systems because most are a cross-flow design. An area of consideration with these systems is the rate at which the operating flow is achieved during startup. An increase that is too rapid could result in membrane damage or blown seals. Excessively high pressures can damage the membrane. Proper programming of the variable-frequency drive used to control the high-pressure feed pump can prevent this. High differential pressure can result in telescoping of spiral-wound elements. High permeate back pressure can result from any number of scenarios (where pressure on the permeate side of the membrane exceeds the pressure on the feed/brine side) and can cause a separation of the glued elements, resulting in a loss of membrane integrity.

MEMBRANE CLEANING AND FLUX RECOVERY

Membrane fouling continues to remain a weakness of both the membrane filtration and RO technologies. As a result, a host of mitigation strategies, for example, MF–UF backwashing and pretreatment chemical addition, are routinely used to minimize the incidence of membrane fouling. Despite these efforts, membranes require chemical cleaning to restore water permeability when fouling does occur.

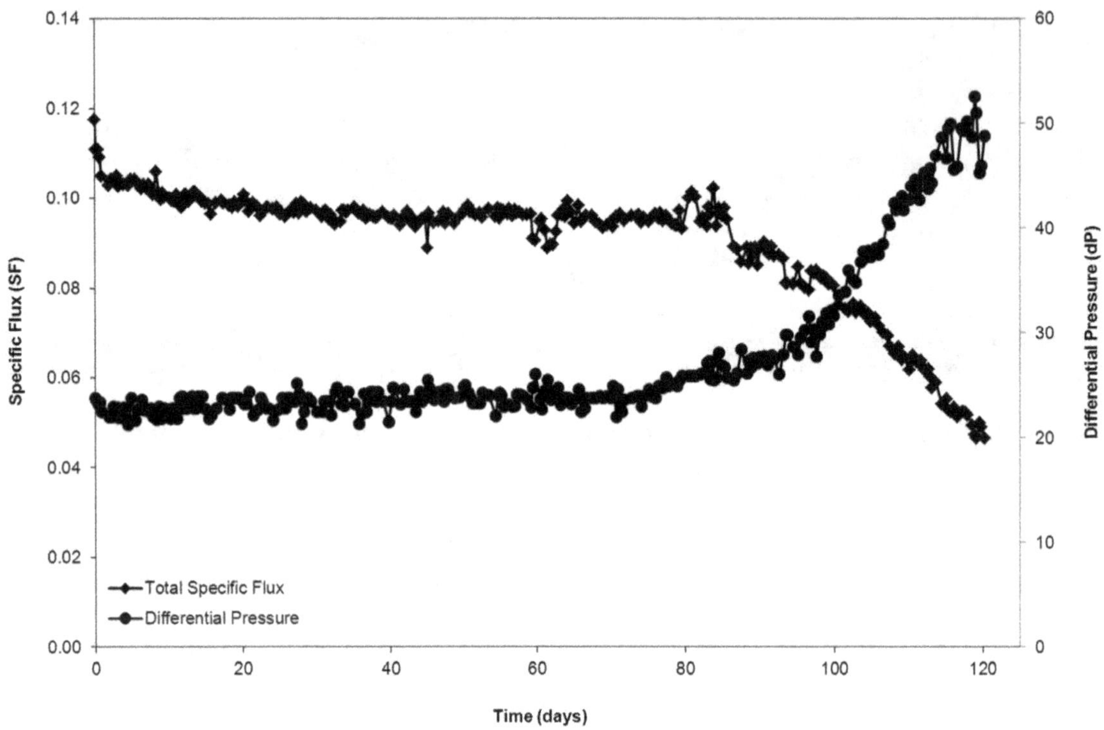

Figure 6-4 Changing reverse osmosis system performance parameters in response to a membrane fouling event

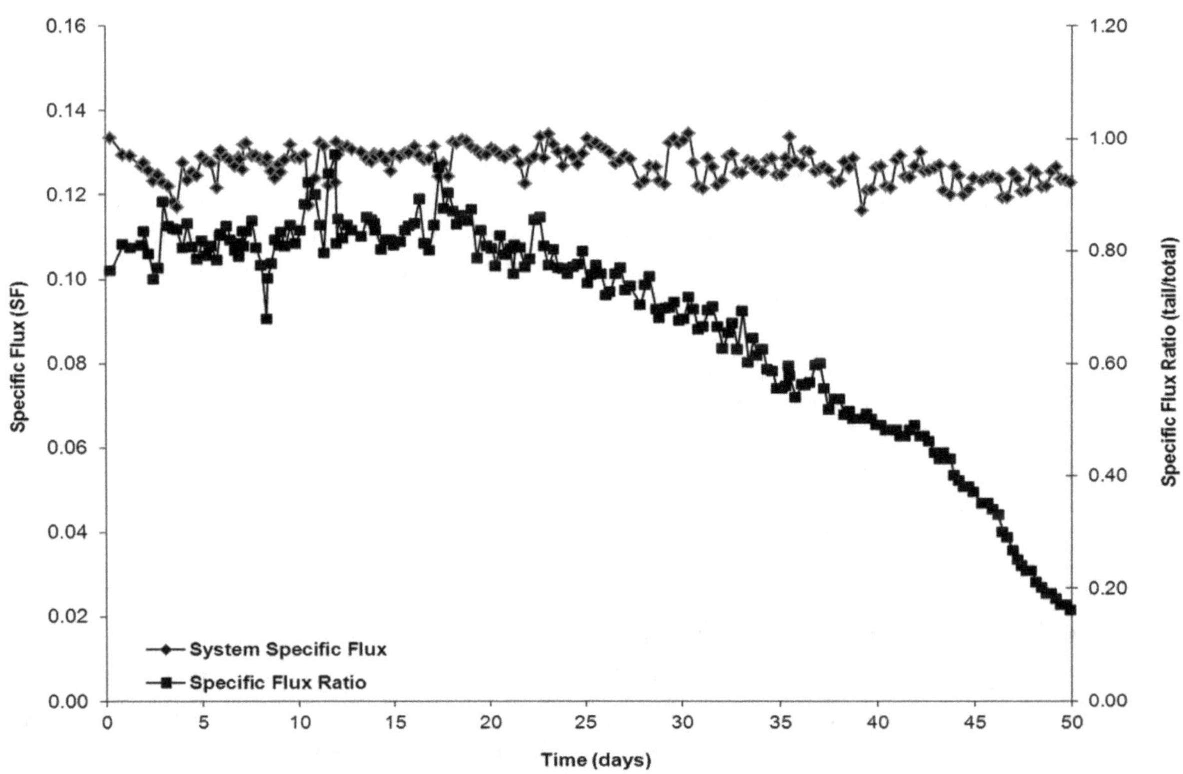

Figure 6-5 Ratio of specific flux at the tail-end stage to the total system specific flux. Note the declining ratio while the overall system specific flux remains unchanged.

Membrane Fouling

Membrane fouling is best defined as the deposition of any foreign material on the membrane surface that impedes water transport (that is, water flux) and impacts production. In the case of membrane filtration, deposition of foreign material within the membrane pores also constitutes membrane fouling. For RO membranes, foreign material can deposit within the feed/brine flow spacer, causing system pressure increases and the need for cleaning. Fouling includes the accumulation of particulates and colloidal material such as silts, clay, and suspended solids. Biological fouling consists of the deposition of viable bacteria and bacterial extracellular polymeric substances including polysaccharides, proteins, and other cellular detritus unique to municipal wastewater. Inorganic fouling typically consists of mineral scaling. Sparingly soluble salts that are concentrated beyond their theoretical solubility limit in the tail array of an RO system can precipitate and deposit onto the membrane surface. This forms a physical barrier that resists water permeability and increases the solute concentration in the permeate. Typical scales found in RO systems include calcium carbonate, calcium phosphate, calcium sulfate, barium sulfate, strontium sulfate, and silica complexes. Depending on the symptoms exhibited and the composition of the foulant, a tailored chemical cleaning is typically performed to remove the foulant and recover membrane permeability.

Low-Pressure Membranes

This discussion focuses on MF–UF membrane systems that are constructed of a hollow-fiber polymer material and operate in an outside-in configuration. In this construction, feedwater is directed from the outside of the fiber (shell) to the inside (lumen) by the pressure differential that exists across the membrane, known as the transmembrane pressure. Contaminants larger than the nominal pore size are retained on and within the membrane pores. A membrane hollow fiber and the foulant accumulated on the outside (shell side) surface are illustrated in Figure 6-6.

Accumulation of solids on the shell side of the fibers continually increases the hydraulic resistance across the membrane until a manufacturer-specified threshold maximum TMP is reached. At this point, most systems use a short-duration, low-concentration chemical wash to mitigate the fouling and extend run times between CIP. This step goes by many names, dependent on the membrane manufacturer, and is generally used with a frequency of daily to weekly. Terminology associated with this procedure includes

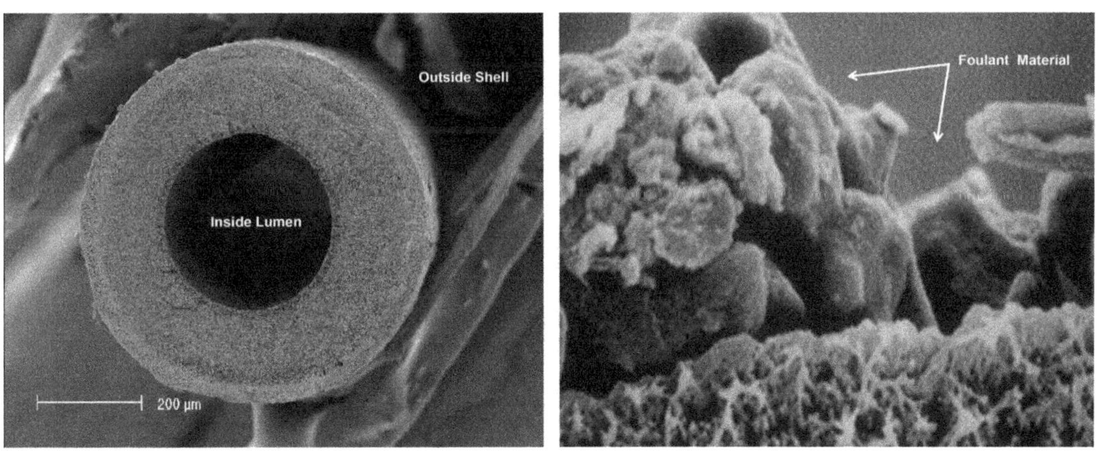

Figure 6-6 Scanning electron microscopy images of a membrane hollow fiber (left) and accumulated foulant on the outside fiber surface (right)

enhanced flux maintenance, chemically enhanced backwash, and maintenance wash. As this frequent chemical washing becomes less effective at reducing membrane fouling, a chemical CIP is generally performed to completely remove the foulant, lower the TMP, and restore permeability back to baseline.

The operational goal of all MF–UF systems is to achieve the specified time between chemical cleanings, which is determined during initial pilot and qualification testing. For a low-pressure membrane system that treats municipal wastewater, the interval is 20 to 30 days, depending on source water quality and the specific design conditions. To achieve the specified cleaning interval, many MF–UF system suppliers use an air scour and liquid backwash routine at specified intervals based on filtration time or volume of filtered water. Air is introduced into the membrane fiber bundle to dislodge material collected on the shell side of the fiber through agitation and mechanical shearing. At the same time, filtrate water is reverse flowed from the lumen side outward through the fibers to remove foulant accumulated within the pore structures.

As part of the backwash routine, some manufacturers have incorporated automated chemically enhanced backwashes. A chemical such as acid, caustic, or chlorine is added during the backwash routine in an attempt to improve its effectiveness in removing foulant and lowering the TMP. The frequency of this routine is typically less than that for the standard filtrate backwash. For example, if the standard backwash occurs every 30 min, the chemically enhanced backwash may be programmed to be daily or weekly depending on the source water's fouling propensity. Some manufacturers couple the chemically enhanced backwash with a short-term (for example, 5 min) soak in chlorine (free chlorine) or acid. This is referred to as a maintenance wash. The goal of all backwash and soak routines is to extend the duration between chemical cleanings. Over time, these practices become increasingly less effective as the foulant continues to accumulate on and within the fiber pores, increasing the hydraulic resistance until the threshold maximum TMP is reached. A chemical CIP must then be performed to remove the foulant and restore permeability.

The type of chemical CIP is generally determined by the membrane manufacturer and typically consists of a high pH alkaline or free chlorine cleaning, followed by a low pH acid cleaning. Alkaline and chlorine cleanings target particulates and biological and organic material, which are the predominant wastewater foulants that impact low-pressure membrane performance. Acid cleanings target inorganic constituents such as calcium, silica, and metal oxides.

The goal of all chemical CIPs is to restore membrane permeability to or as close to startup "baseline" permeability as possible (Figure 6-7). As the membranes age, permeability is likely to subtly decline as the membranes reach the end of their lifespan. Monitoring post-CIP TMP as filtration is initiated can assist in determining the effectiveness of the chemical cleaning. A decline in post-CIP permeability (that is, higher starting TMP) can be attributed to either an ineffective chemical cleaning or to the early stages of irreversible fouling; this is a condition whereby cleanings are ultimately unsuccessful in restoring membrane permeability.

Clean Water Flux Test

To more precisely track the effectiveness of cleanings and to monitor for the incidence of irreversible fouling, a CWFT can be conducted. For this test, the membranes are placed in filtration mode after a CIP at a specified flux and with recirulating clean water (for example, filtrate) while recording the normalized TMP. Repeatable post-CIP TMPs signify continued successful cleanings. However, a significant deviation warrants that corrective action be taken. The initial step is to review the cleaning regime to ensure that all parameters and equipment are operating correctly. Other strategies such as extended chemical

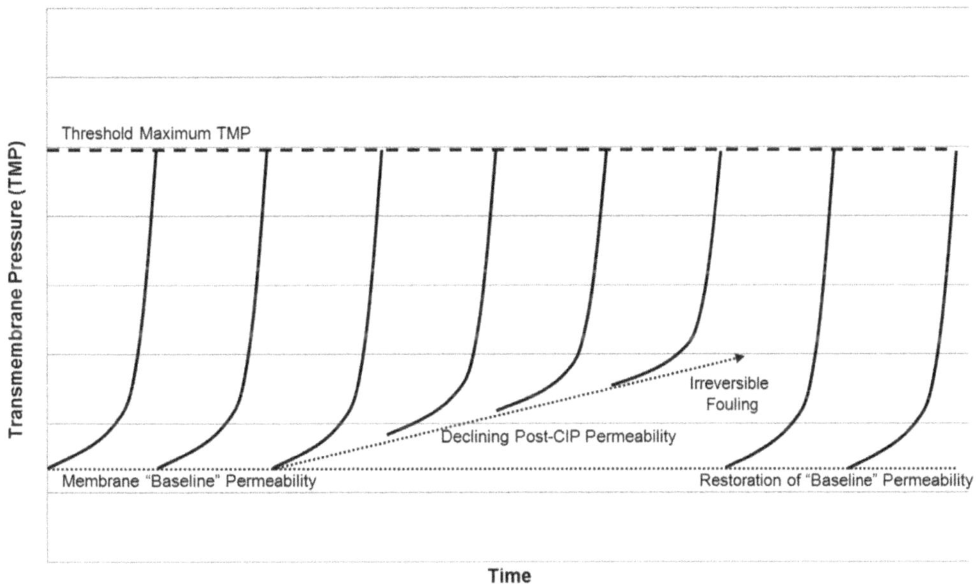

Figure 6-7 Example of transmembrane pressure increases versus time for consecutive cleaning intervals. Note the baseline, declining, and restoration of membrane permeability are reflective of potential irreversible fouling or chemical cleaning effectiveness.

Abbreviation: CIP, clean in place.

soaks and use of alternative, more aggressive chemical formulations could be considered to restore membrane permeability. Since this would be a deviation from the prescribed cleaning regime, consultation with the membrane manufacturer is recommended to ensure that warranties are not voided as a result of these actions.

High-Pressure Membranes

The goal of any RO cleaning is to restore water permeability and retain rejection without damaging the membrane's thin-film separation layer. Frequent exposure to harsh cleaning chemicals can reduce membrane life. Therefore, it is critical to select the appropriate cleaning chemicals and to target the appropriate area(s) within the RO system, for example, first stage and second stage, to restore membrane flux. Most commercially available polyamide RO membranes are susceptible to irreversible damage from exposure to strong oxidants including free chlorine and peroxide. A combined chlorine residual, for example, chloramines, has been shown to be compatible with polyamide membranes and can help alleviate the incidence of biological fouling.

Chemical cleanings are generally conducted at elevated temperatures and at either an acidic or alkaline pH depending on the foulant. This assists in removal of the foulant from the membrane surface. The membrane manufacturer should be consulted to determine the chemical compatibility and membrane limitations before conducting a CIP or using any pretreatment chemical.

In water reuse, RO has become increasingly engineered and precise as the technology is refined. However, membrane fouling and mitigation through chemical cleanings are more subjective and based on such things as operator experience, practices, and historical optimization efforts. Numerous generic, proprietary, and specialized chemical formulations and practices have been developed to address most membrane foulants. Their effectiveness is site specific and dependent on a host of parameters including the CIP system, staff's ability to operate the system, and the magnitude and composition of the foulant. Heavily fouled membranes limit the penetration of cleaning chemicals throughout the foulant material and therefore may require multiple cleanings to achieve effective removal. Alternatively, extended recirculation and/or soaks can be an effective strategy. Only through trial and error will the optimal cleaning strategy be developed. Table 6-2 includes a list of general cleaning solutions. While there are exceptions, acidic cleanings are generally targeted to dissolve inorganics, while alkaline cleanings are targeted to organic and biological material.

A successful chemical cleaning is determined by its effectiveness in recovering water flux to initial stabilized baseline permeability; this is similar to the post-CIP permeability after a low-pressure membrane cleaning (see Figure 6-6). In a water reuse application, it is not uncommon to see initial startup membrane permeability drop 10 to 15 percent within the first 15 to 30 days of operation. Once stabilized and under normal operating conditions, water flux subtly declines over time as material deposits on the membrane, reducing permeability and increasing operating pressures. As the membrane ages, it is common to see the inherent permeability begin to decline as well.

If the cleaning was successful, permeability should be restored reasonably close to the initial, stabilized permeability demonstrated during the first few weeks after startup (Figure 6-8). A review of the post-CIP permeability is an adequate means of assessing the effectiveness of the cleaning, as well as monitoring for the incidence of irreversible membrane fouling.

Table 6-2 General reverse osmosis cleaning chemical guidelines

Foulant	Chemical, wt %*	pH†	Temperature†
Inorganic scale:			
Carbonates (e.g., $CaCO_3$) and phosphates (e.g., $CaPO_4$)	2% citric acid, 0.5% hydrochloric acid (HCl – [muriatic acid])	Low	Elevated
Sulfates (e.g., $CaSO_4$, $BaSO_4$)	0.5% hydrochloric acid, 0.1% sodium hydroxide + 1% sodium salt of ethylene diamine tetraacetic acid	Low High	Elevated Elevated
Metal oxides/Hydroxides (e.g., Fe, Mn, Zn)	2% citric acid, 1% sodium hydrosulfate	Low Low	Elevated Elevated
Organic/Biological/Biofilms	2% sodium tripolyphosphate + 0.2% sodium salt of dodecylbenzene sulfate	High	Elevated
Inorganic colloids (e.g., silt)	0.1% sodium hydroxide + 0.05% sodium dodecylsulfate, 2% citric acid	High Low	Elevated Elevated
Silica/Silicates	0.25–0.50% ammonium bifluoride (effective at removing early-stage and well-established scale)	Low	Elevated

* All chemical concentrations are in percent by weight and are typically prepared in reverse osmosis permeate.

† Consult the membrane manufacturer regarding maximum temperature and pH limits.

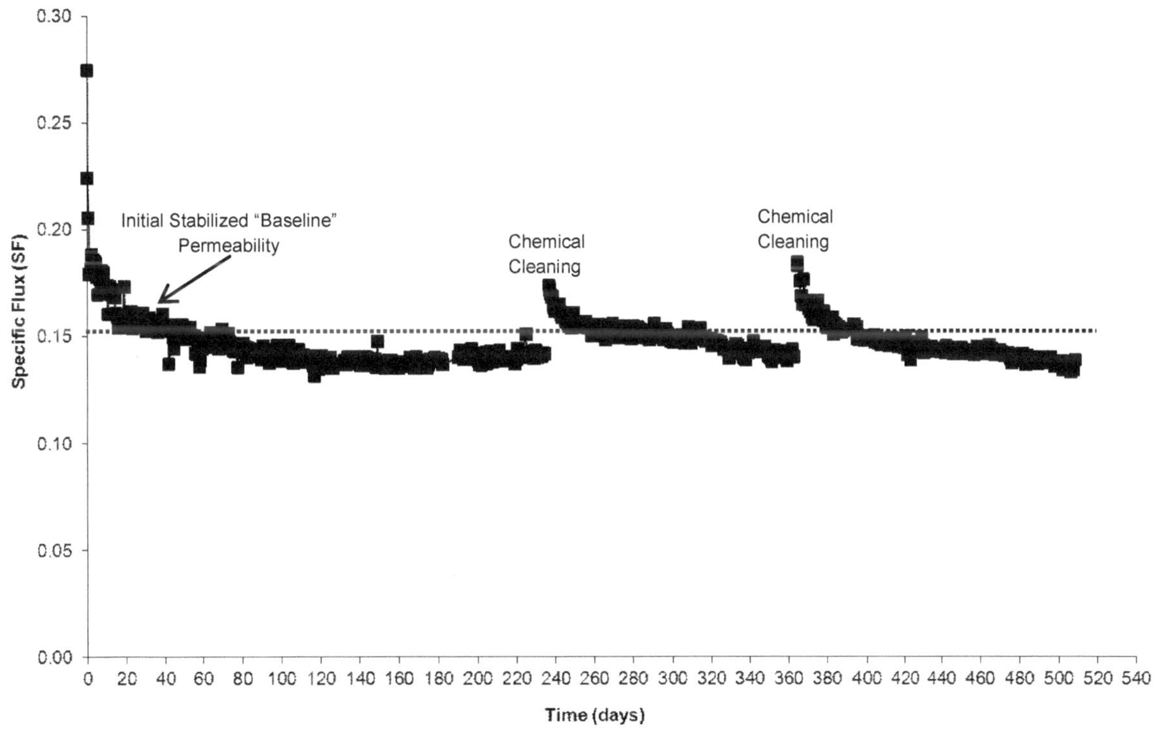

Figure 6-8 Restoration of reverse osmosis membrane flux to initial baseline permeability after chemical cleaning

INTEGRITY TESTING

The effectiveness of any membrane system is dependent on its ability to separate the contaminants from its source water. To ensure that this is being accomplished, the integrity of the membrane system must be monitored and tested. Due to the nature of the separation that occurs between low-pressure and high-pressure membranes, each has its own unique requirements for monitoring and testing membrane integrity. Both types of membrane systems can be tested using direct methods or indirect methods by analyzing the effluent water quality.

Low-Pressure Membranes

There are several ways to determine the membrane integrity of low-pressure membrane systems. These include testing of the fibers to retain air on one side of the membrane or an evaluation of the filtrate water quality. The results of testing the membrane integrity can be influenced by equipment failures within the system, such as those caused by valve seats or actuators. These must be considered when reviewing test results and diagnosing the failure.

The most common MIT method involves the application of air to the membrane. This is done as a directly applied pressure decay test (PDT) or a vacuum decay test (VDT). In both methods, the capillary force that holds the water within the membrane pores is used to evaluate the membrane's integrity. This force is referred to as the "bubble point." Air pressure or a vacuum is applied to one side of the membrane. The applied pressure must be kept below the bubble point. If it is not, the water will be forced out of the pores and the air will pass through the fiber, resulting in a false integrity failure. These methods are described in detail in the ASTM D6908-06, Standard Practice for Integrity Testing for Water Filtration Membrane Systems.

The PDT method involves the application of air pressure to one side of the membrane. This can be applied to either side of the membrane. Typical test pressures range from 10 to 30 psi depending on the membrane type and manufacturer. For packaged membrane systems, the PDT is programmed as a sequence in the operation. The frequency is typically daily but, depending on the application, it can be set more frequently per regulatory requirements. The test involves draining the side of the membrane to which the air pressure will be applied. Once the air is applied and allowed to stabilize, it will be shut off and the air side isolated. The other side of the membrane is vented to the atmosphere. An initial pressure is recorded. After an elapsed time, typically several minutes, a final pressure is recorded. The difference in pressure over the elapsed time is the pressure decay rate. This value is greater than zero due to the diffusion rate of air into water. Pressure decay rates that are less than 1 psi/min are indicative of adequate membrane integrity. Several membrane manufacturers have conducted tests to correlate PDT results with log-reduction rates.

The VDT method is similar to the PDT method except that a vacuum is applied to one side of the membrane. In addition, both sides of the membrane are drained of any liquid. While the vacuum is applied, the other side is vented to the atmosphere. The applied vacuum is measured at the beginning of the test. After an elapsed time, the vacuum is measured again. The difference in values over the elapsed time is the vacuum decay rate. This test method is less common than the PDT method but is used in some vacuum-driven membrane applications.

Membrane integrity can also be monitored based on filtered effluent water quality. This has been implemented in continuous process monitoring and individual sample testing. Continuous monitoring methods include the use of process instrumentation such as inline turbidity meters and particle counters. This has even been applied as a regulatory requirement for certain water reuse applications. The water quality is dependent on the type of filter being used. There are limitations with either type of instrumentation, which are dependent on the types of particles that are being filtered and the characteristics of the particles. There may be significant integrity breaches that these instruments do not detect. It is critical the instrumentation's suitability be evaluated before using it to validate the integrity of any membrane system.

There are other, less common test methods that can be applied. These include an air flow test, which is similar to the PDT and VDT methods. With this test, the air flow rate is monitored to identify any breaches within the membrane. Another method is the filtered water quality test, which uses marked particles that pass through the membrane, or SDI testing to monitor the effluent water quality.

High-Pressure Membranes

The integrity of high-pressure membranes such as an RO membrane is continuously monitored using conductivity measurement. Gross membrane integrity failure can be quickly identified by an increase in permeate conductivity. If the failure is severe, it would also be observed by an increase in the specific flux of the system. Elevated permeate conductivity could be the result of membrane scaling or changes in feedwater quality. The investigation into a rise in permeate conductivity should include a review of all system data and performance. A mechanical integrity failure of a high-pressure membrane involves a systematic investigation to identify the specific element or seal that is the cause of the failure.

If there is an increase in system permeate conductivity that is not attributable to feedwater changes or scaling, the cause of the failure must be determined. The first step is to perform a conductivity profile test of all vessels; this is done using a hand-held conductivity meter. Typically, this test is performed periodically as a check of membrane performance. The goal is to identify the specific vessel or vessels that have elevated salt passage. Once the vessel with the integrity breach is identified, a vessel probing is performed. This involves insertion of small-diameter plastic tubing through the entire length of the vessel's permeate tube. The conductivity sample is then measured from the end of the tubing, and the tubing is drawn out of the tube. A significant change in conductivity indicates the location of the integrity breach. The location will be an interconnection between two elements. Staff can then remove the elements at the breach and inspect the O-rings on the interconnector. If the O-rings appear to be in good condition, the adjoining elements should be inspected and tested as appropriate.

The use of continuous online analyzers is becoming more common; some regulatory agencies in California require their use. The most common analyzers are used to measure total organic carbon and ammonia. These analyzers provide a real-time indication of a loss of integrity at low levels and can detect the presence of chemicals in the feedwater that are not well removed by RO.

MEMBRANE STORAGE

Most manufacturers package their membrane elements in oxygen-barrier bags that are filled with a preservative solution and vacuum sealed. This solution must be flushed out during normal operation after the membrane is installed. If not installed immediately, new membranes must be properly stored to maximize their performance and useful life. Most manufacturers recommend that new membranes be stored in a cool, dry place and out of direct sunlight. However, the membrane manufacturer should be contacted for specific storage recommendations, as storage requirements differ by manufacturer.

Membranes that are in service can be stored in place for short periods if a rack or train is taken out of service. Low-pressure membranes should be kept wet either with product water or feedwater, as recommended by the manufacturer. A backwash cycle can be initiated periodically to ensure the membranes are wet. High-pressure membranes should be flushed with permeate periodically, as recommended by the manufacturer. If membranes need to be stored in place for longer than one or two weeks, the manufacturer may recommend storing the membranes in a preservative such as sodium bisulfite.

MEMBRANE REPLACEMENT

Useful Membrane Life

Most membranes have a useful life of 5 to 10 years and need to be periodically replaced. Feedwater quality has a major impact on a membrane's life. For low-pressure membranes, poor feedwater quality can result in more broken fibers and frequent backwashes and chemical cleanings, which shorten membrane life. For high-pressure membranes, poor feedwater quality can result in more frequent chemical cleanings for organic fouling or mineral scaling. These can also shorten membrane life.

Membrane Replacement

There are two main reasons to replace membranes. The first is high TMP in a low-pressure membrane or high feed pressure in a high-pressure membrane even after chemical cleaning. A higher initial or "new membrane" TMP could result in shorter intervals between chemical cleans and ultimately force a decrease in flows to maintain a reasonable chemical cleaning interval. Shorter cleaning intervals and reduced throughput increase unit production costs and affect the ability to meet water demands. Higher feed pressures increase unit production costs, could damage the membrane if it is caused by a high differential pressure, or could eventually exceed pumping equipment capacity.

The second reason to replace membranes is a degradation in filtrate or permeate quality such that it no longer meets permitted, contractual, or process requirements. For low-pressure membranes, this could result in an unacceptable MIT result, filtrate with SDI values consistently greater than 3, or filtrate turbidity values greater than 0.1 ntu because of continuing integrity issues. For high-pressure membranes, as the membranes wear over time, levels of certain key constituents in the permeate could exceed permitted or contractual requirements.

NEW MEMBRANE PROCUREMENT

Low-Pressure Membranes

Currently, there is no standard configuration for low-pressure membranes. Membrane manufacturers have developed many types of membranes with numerous inlet and outlet configurations and made from various materials. Many low-pressure membrane systems are proprietary, meaning replacement membranes can be purchased only from the original membrane supplier. This limits a membrane system owner's flexibility in looking for the most cost-effective replacement. To get the best pricing for replacement membranes, the owner should negotiate the price of replacement membranes when procuring the initial low-pressure membrane system. By doing this, the total life-cycle cost of a low-pressure membrane system can be evaluated, including initial capital cost and ongoing operating, maintenance, and membrane replacement costs.

The industry seems to be moving away from proprietary systems. New low-pressure membrane systems are being developed that will make it possible to purchase replacement membranes from multiple manufacturers, which is the most cost-effective replacement solution. However, the owner may not receive the long-term performance warranty that comes with proprietary systems.

High-Pressure Membranes

Most high-pressure membranes have an 8-in. (20-cm) diameter, 40-in. (101-cm) long membrane element or a 16-in. (41-cm) diameter, 40-in. (101-cm) long element. This allows owners to replace high-pressure membranes with membranes from different manufacturers. However, not all membranes are equal. Some have better rejection; some have a lower operating pressure; and some better withstand the rigors of cleaning.

Historically, replacement membranes were selected based on the lowest purchase price. This is still a very valid approach; however, it does not capture the differences in ongoing operations and maintenance costs. Some owners use a different approach to maximize the cost and performance differences between membranes. A procurement process that includes more than just pricing when evaluating replacement membranes is as follows:

- Manufacturers are required to prequalify their membranes in a long-term demonstration test before they are allowed to supply replacement membranes. In order to prequalify, a membrane is tested for a minimum of 5,000 hours with the feedwater on which they will be operating. The results are compared with the results for a baseline membrane over the same operating period. The same operating period is used to account for any changes in feedwater quality over time. Membranes that meet specific criteria for pressure, specific flux decline, water quality, and other criteria that may be important to a particular application pass and are qualified to be used as replacement membranes.

- Based on the qualification testing data, the power required to operate the membrane is calculated. Higher operating pressures during the test result in higher calculated power costs. Differences in permeate water quality can be taken into account by assigning point or dollar values to different levels of key constituents in the permeate.
- Prequalified manufacturers are asked to submit bids for replacement membranes. The cost of the membranes is added to the power cost and water quality "cost" to determine which membrane has the lowest total "cost" to the owner.

Even if power use and water quality of the permeate are not used to determine the lowest cost membrane, membranes should still be prequalified to give the owner more confidence that they will perform well and meet water quality requirements.

MAINTENANCE

Membranes must be maintained to maximize useful life, preserve production capacity, and ensure good water quality. With low-pressure membrane operation, the backwash and chemical cleaning intervals must be set so that TMPs do not reach excessive levels. Low-pressure membranes also require regular integrity testing. This is usually accomplished by performing a pressure decay test on the system. If the system cannot hold the pressure as required by the manufacturer, there is a leak in the system that must be repaired. Leaks are located by visual inspection on some systems. On others, listening devices are used. Leaks caused by fiber breaks must be repaired by removing the affected module and pinning the individual fibers. When the fiber breaks are too numerous, the module should be replaced. Defective or worn valve seats and O-rings can also cause leaks; these should be replaced if they are leaking.

High-pressure membranes should be monitored for indications of scaling, fouling, significant leakage around membranes, and degradation of the membrane itself. Items to monitor include specific flux, differential pressure, and permeate constituent levels. In addition, system conductivity profiles and pressure vessel probing are good ways to find major leaks around brine seals and O-rings and to find major membrane failures. If a membrane is fouled or scaled, it should be chemically cleaned with an appropriate cleaner. If the chemical clean does not restore the specific flux or water quality to acceptable levels, the membrane should be replaced.

TROUBLESHOOTING

Regardless of the type of membrane system, there are two aspects of membrane performance that must be considered and monitored thoroughly: integrity and permeability. Most troubleshooting activities are performed to improve or correct problems related to one of these two areas. Various tools can be used in membrane system to troubleshoot and optimize performance. Some tools are similar between MF–UF and NF–RO systems, while others are specific to one technology or the other. For instance, membrane autopsies are often the most effective way to identify a problem within the system, this applies to both MF–UF and NF–RO systems. However, pressure hold tests are specific to MF–UF systems, and conductivity profiling is only relevant for NF–RO membrane systems.

Membrane Integrity

Membrane integrity refers to the membrane's ability to reject constituents of concern for a particular system. Physical or chemical damage to the membrane can result in a breach of membrane integrity, which results in poorer product water quality.

When membrane integrity issues are observed or suspected for an MF system, direct and indirect integrity test methods can be used to evaluate the system. The direct integrity tests include pressure hold tests, sonic or visual integrity tests, and bubble point tests. Indirect integrity tests include turbidity, particle counts, and SDI testing. Troubleshooting tools for RO membrane integrity include conductivity profiles, vessel probing and flow tests, CIP effectiveness evaluation, and normalized performance data evaluation.

The most important tool for evaluating MF membrane issues is the pressure hold test. It is a timed sequence of applying low-pressure air to one side of a membrane and monitoring the pressure decay over time. It allows for testing of all modules within a membrane unit and allows for sensitivity on the order of a single fiber break. It is considered the standard for MF systems, is automated, and has widespread acceptance in the industry. The test allows for testing of both the membrane and piping for leaks and is a precursor to repairing broken fibers or correcting system leaks.

Another test that is important for evaluating MF system integrity is the SDI test. The test is an RO fouling indicator test that uses a 0.45-μm cellulosic membrane filter to track MF filtrate quality as it relates to the impact on RO fouling potential. SDI data are very useful in integrated membrane plants and can be used to link the performance of the two membrane systems. The test indicates when the MF system is having integrity problems even before turbidity indicates there are problems with the system. It also reveals changes in RO feedwater quality that could result in higher RO fouling rates.

In addition to permeate water quality analyses, permeate conductivity or total organic carbon is an online indication of the NF–RO membrane system performance and can identify when there are problems with the membranes salt rejection. These problems are typically the result of membrane age, chemical damage, physical damage, O-ring disconnects, or O-ring deterioration. Conductivity profiles and vessel probing are useful tools in assessing the integrity of RO systems and in determining whether a problem is localized within a particular stage or pressure vessel.

Ultimately, periodic collection of product water quality samples for constituents of concern and noting any changes over time is the best method for confirming a membrane integrity problem. Membrane autopsies can be performed to identify the mechanism of damage, with the intent of identifying the cause of the integrity loss. While fiber repair can be performed on MF–UF membranes, membrane cleaning or replacement is often the only resolution to a loss of membrane integrity for NF–RO systems.

Membrane Permeability

Membrane permeability is a good indication of membrane performance and can indicate when actual membrane issues are occurring, such as tightening of the membrane, fouling, scaling, or loss of flow. For MF systems that use dead-end filtration, TMP is typically used to directly monitor the membrane's fouling rate. As membrane fouling occurs, the TMP will increase and permeability will decrease.

Clean water flux test data are an important tool for documenting the startup performance for a new set of MF membranes, as well as for confirming the cleaning effectiveness of MF systems.

As discussed previously, a variety of issues contribute to membrane fouling and the loss of permeability in RO systems. Biological, colloidal, organic, or inorganic fouling is common and can result in an increase in feed pressure or an increase in differential pressure across the stages of an NF–RO system. As feed or differential pressures increase in an

RO system, membrane permeability can decrease. Locating the problem is key to identifying its cause and resolution. For instance, a loss of permeability in the tail end of a membrane train may be an indication of inorganic scaling, while loss of permeability in the lead end of an RO train is potentially related to biofouling or particulate fouling.

In addition to data evaluation for identifying problems with the membrane systems, there are evaluations of the CIP data that can identify effectiveness of the cleaning, including lower feed pressure, lower differential pressure, and the "temporary" effect on product water quality. Ideally, cleaning will return the membrane to its startup permeability value. However, in many cases it is more practical to expect an effective CIP to return the system to a consistent post-clean permeability, without a negative impact on membrane integrity.

When irrecoverable loss of permeability occurs, membrane autopsies and foulant analyses are beneficial in determining an appropriate resolution. For instance, foulant analysis can indicate whether the cleaning chemicals being applied are appropriate for the particular foulant. Similarly, membrane autopsies can reveal hydraulic or mechanical issues that may be contributing to a loss of permeability, such as hollow-fiber tin-canning or RO membrane compaction.

While membrane integrity is critical to ensure product water goals are achieved, acceptable membrane permeability is critical to ensure product capacity goals are reliably achieved. Membrane permeability is inversely proportional to the operating pressures and energy costs associated with operating a membrane system. As membrane permeability decreases, the energy cost associated with maintaining design capacity increases.

This page intentionally blank.

Chapter 7

Residuals Management

This chapter describes management of concentrated waste streams (residuals) from both low-pressure (microfiltration–ultrafiltration [MF/UF]) and high-pressure (nanofiltration–reverse osmosis [NF/RO]) membranes. The quantity and quality of residuals produced by low- and high-pressure membranes vary and can limit disposal methods. Issues relevant to disposal of residuals from high-pressure membranes (also known as concentrate) as well as their disposal options are described. This chapter also presents regulatory issues that must be addressed when planning and designing membrane facilities and the impacts on the water reclamation processes that must be addressed to avoid water quality violations.

Membrane treatment systems generate significant volumes of waste streams that must be managed properly to ensure successful operation. Low-pressure membranes generate backwash streams that are up to 10 percent of the incoming flow. These streams contain solids removed by the membranes, including suspended solids, pathogens, and any coagulants added upstream of the membranes. As low-pressure membranes do not remove soluble constituents (unless precipitated first), these backwash wastes are typically returned ahead of the membrane treatment process. These wastes can be returned upstream of the membrane system for suspended solids removal or to intermediate treatment to remove solids. A periodic clean-in-place (CIP) membrane cleaning procedure that uses acids, chlorine, and caustic generates a waste stream that differs from the normal backwash residual stream and may require a different handling and disposal approach.

High-pressure membranes generate volumes of concentrate waste streams that are larger than those generated by low-pressure membranes; volumes range from 10 to 30 percent of the incoming flow. These systems remove dissolved constituents, and the residual streams contain the concentrated salts that were removed from the influent or feed along with removed organics and a small amount of suspended solids. Waste from membrane CIP procedures are also generated; these include cleaning chemicals and typically lower levels of salts compared to the concentrate streams. Depending on the reclaimed water source, concentrate may contain high concentrations of salinity, heavy metals, and other contaminants that may be toxic. Consequently, concentrate streams must be disposed of in compliance with the Clean Water Act and the Safe Drinking Water Act. In many cases, local regulations restrict the discharging of concentrate solutions to publicly owned treatment works.

RESIDUALS MANAGEMENT FOR LOW-PRESSURE MEMBRANES

This subsection presents the residuals generated by the various types of low-pressure membranes, residual water quality, disposal issues, and residual management techniques.

Types of Low-Pressure Membrane Residuals

Low-pressure membranes that are used for reuse applications include the following types of systems:

- Submerged MF or UF modules mounted in basins and
- Pressure vessel–enclosed membrane modules.

Submerged MF or UF modules mounted in basins. Submerged MF or UF modules are mounted in basins and used to treat secondary effluent. Submerged membranes are backwashed regularly, typically every 10 to 30 min for 1.5 to 2.5 min (WEF 2006). The membrane supplier generally provides the backwashing frequency and duration. Periodically, the entire basin is dumped to flush out the solids removed from the membranes. This flow is a spike of several thousand gallons per minute and is normally sent to an equalization basin. The volume of backflush water ranges from 5 to 10 percent of the feed flow. The equalized backwash waste can be sent to the head of the wastewater treatment plant (WWTP) (if possible) or to a separate treatment process for removal of the solids and organics prior to recycle to the membrane treatment system. Treatment methods are described under management techniques.

Maintenance cleans and recovery cleans are performed for submerged membranes. Similar chemicals and concentrations are used in the CIP recovery process. The frequency of maintenance cleans ranges from daily to weekly and recovery cleans range from 30 to 90 days. The frequency of maintenance and recovery cleans is highly dependent on the quality of the feedwater.

Pressure vessel–enclosed membrane modules. Pressure vessel membranes consist of hollow-fiber modules mounted in a pressure vessel. Membrane backflushing occurs at a rate of 1.5 to 2 times the normal flux during operation. Backflushing is normally performed every 10 to 30 min and lasts for 1 to 2 min (WEF 2006). The membrane supplier should be consulted for the frequency and duration of backflushing, which are based on the equipment and the quality of the water being treated.

Since the pressurized membrane system backflushes one unit (rack) at a time, the peak flow rate of the backflush water is less than that in a submerged system and therefore requires a smaller equalization basin. For systems larger than 10 mgd in capacity, the backflush flow is nearly continuous. The chemicals used for maintenance and recovery cleans are the same as those used for submerged membranes. The volume of spent solution is typically less because of smaller dead volume in the pressure vessels. The membrane supplier should be consulted for the actual CIP waste volumes.

RESIDUALS MANAGEMENT TECHNIQUES

Methods for handling different types of residuals from low-pressure membrane systems are discussed in this section.

Backwash Waste

Discharge to sewer. Discharge of backwash waste to a sewer system is successfully practiced at many locations. A reuse facility that is located at the WWTP sends the residual stream to the front of the plant or ahead of secondary treatment. For satellite treatment facilities, the residuals are discharged to the closest wastewater collection system. Discharge to a sewer is simple and convenient. However, with both approaches, there is an increase in hydraulic and solids loading of total suspended solids (TSS) and biological oxygen demand (BOD) to the local WWTP. The peak and average loadings should be determined to ensure that these loads do not impact the WWTP. Connection fees and user charges should also be calculated to determine the cost of this residuals management method.

Treatment and recycle. Treatment of backflush waste and recycling it to the head of the membrane system is often a cost-effective approach. The turbidity of the backflush wastewater can range from 30 to 150 ntu. The treatment system should use coagulation, flocculation, and some form of sedimentation to remove TSS, colloids, dissolved organic carbon, and bacteria and viruses. The residuals from the coagulation–flocculation–sedimentation (CFS) process can be sent directly to the solids handling system at the WWTP or dewatered separately with drying beds or mechanical dewatering. Examples of CFS systems used to treat membrane backflush waste include the following:

- Gravity thickener,
- Plate/tube settlers,
- Dissolved air flotation, and
- Ballasted flocculation.

Table 7-1 summarizes design considerations for each CFS process. Typical coagulants used to treat waste backflush water include alum, ferric chloride, and aluminum chlorohydrate and polymer. The plate settler and dissolved air flotation processes work best with flocculation, with at least 15 min of contact time. The ballasted flocculation systems typically have a built-in reaction/flocculation system, for example, the ACTIFLO system.

Another approach for handling waste backflush water is to use a second stage of membranes to concentrate the waste and produce a filtrate that can be combined with the treated flow from the first set of membranes. Secondary membranes must be operated at a flux that is lower than the flux for primary membranes and may require more frequent maintenance and recovery cleans. The residuals from secondary membranes can be handled with the same techniques used for primary membranes.

Table 7-1 Summary of coagulation–flocculation–sedimentation processes

System	Design Criteria	Effluent Quality	Solids Characteristics, %TSS
Gravity thickener	Hydraulic loading, 0.2 gpm/ft^2; TSS loading, 2.2 lb/ft^2/d	<2.0 ntu with coagulant and polymer	0.75–1.5
Plate settler	Hydraulic loading, 0.35 gpm/ft^2 on plate area	<2.0 ntu with coagulant and polymer	0.5–1.5
Dissolved air flotation	Hydraulic loading, 2–4 gpm/ft^2; TSS loading, 15–20 lb/ft^2/d	<2.0 ntu with coagulant and polymer	1.0–2.0
Ballasted flocculation	Hydraulic loading, 20 gpm/ft^2 on plate area	<1.5 ntu; requires coagulant and polymer to operate	0.05–0.1

Abbreviation: TSS, total suspended solids.

Maintenance Wash Waste and CIP Wastes

It is important to note that maintenance wash waste contains high concentrations of chlorine (up to 1,000 mg/L) and consequently needs to be neutralized with sodium bisulfite. In addition, the acid solution needs to be brought back to neutral pH prior to returning it to the upstream processes or disposing it to the sewer. The spent organic acids (citric or oxalic) have a very high BOD (5,000 to 10,000 mg/L) that should be considered in the WWTP design. Batch discharges of high-BOD maintenance wash waste can upset the biological process. Adding an equalization tank to the system and continuously discharging high-BOD wastes back to the WWTP at a low flow rate is a better approach for disposal. Equalization of the high-strength waste should be considered even for discharge into the WWTP collection system.

CIP wastes are more difficult to handle because they contain chlorine, metals, high concentrations of spent organic acids, ethylenediaminetetraacetic acid (EDTA), and organics removed from the membranes. After pH neutralization and dechlorination with sodium bisulfite or calcium thiosulfate, these wastes, if not discharging to the sewer, can be recycled along with the backflush waste or be treated separately to remove the acids. It is difficult to recycle the CIP waste because citric acid and EDTA are chelating agents that interfere with coagulation. Testing at the West Basin Water Recycling Plant, Carson, California showed that citric acid could be recycled only if the spent acid solution was separately collected and bled back into the recycled stream at a concentration of less than 10 mg/L. The best way to deal with citric acid and EDTA is biological treatment. If recycle or biological treatment is not possible, precipitation of citric acid with calcium chloride can be considered. The precipitated solids can be removed and handled with the other solids streams.

Membrane preservation and storage. Many membranes come shipped in special solutions such as glycerin to keep them wet and free from bacteria. The solutions are disposed of into a sanitary sewer; the BOD of the solution should be considered to prevent shock loading to the plant. An equalization tank for managing high-BOD waste streams will reduce the adverse impact on the WWTP biological process.

CONCENTRATE MANAGEMENT FOR HIGH-PRESSURE MEMBRANES

Membrane separation processes such as RO and NF are used in water reclamation to remove dissolved contaminants from treated wastewater. In addition to product water, these processes generate a by-product stream, referred to as concentrate, that contains all of the contaminants removed by the membrane process. The concentrate stream is typically 10 to 30 percent of the flow treated, and it has total dissolved solids (TDS) concentrations 3 to 10 times greater than those in treated wastewater. Managing the concentrate stream is often one of the greatest challenges at water reclamation facilities.

In pressure-driven membrane processes, the product water passes through the membrane, and contaminants are separated from treated water by rejection on the feed side of the membrane. The RO and NF membranes remove dissolved ions, organic contaminants, and pathogenic microorganisms that are collected in the concentrate at concentrations several times greater than those in the feedwater. Within the pressure-driven membrane category, RO membranes have higher rejection rates than NF membranes and therefore generate concentrate that contains higher levels of rejected contaminants.

Concentrate management is complicated by the volume of the concentrate stream, health and environmental issues, cost and energy issues, and the often complex interaction of federal, state, and local regulations that govern concentrate disposal.

ISSUES RELATED TO CONCENTRATE DISPOSAL

Concentrate Volume

The volume of the concentrate stream is determined by the recovery achieved by the membrane process. For reclaimed water, product water recovery ranges from 70 to 90 percent; therefore, the concentrate stream is 10 to 30 percent of the feed flow. The options for disposing of large concentrate streams are limited due to water quality, cost, and space requirements issues.

Blending concentrate with a WWTP influent or effluent is challenging because the ratio of concentrate to influent or effluent increases. When dilution is inadequate, the WWTP National Pretreatment Program may not allow the discharge of concentrate to the WWTP. The salinity is normally not toxic to microbial organisms in the WWTP until the chloride concentration is greater than 5,000 mg/L. Batch discharges of concentrate cause a sudden increase in TDS at the WWTP; however, the microbes can adjust to sudden increases in TDS with little process disruption. A sudden cessation of concentrate can cause a sudden drop in the TDS of the wastewater that can be fatal to the microbes. A sudden decrease in the wastewater TDS causes a large osmotic pressure differential across the microbial cell wall. This pressure differential causes the microbes to swell and, if high enough, will rupture the cell wall, killing the microbe. This sudden decrease in concentrate discharge may be unique to reclaim systems because reclaim water production could be stopped under certain conditions.

Direct discharge of concentrate to surface waters and too much concentrate discharged to the WWTP are both affected by effluent toxicity. Discharges to surface waters must meet acute or acute and chronic aquatic toxicity testing criteria. The total effluent flow to surface water or a receiving stream must meet the aquatic toxicity testing protocol. A TDS concentration greater than 1,500 mg/L exceeds a concentration threshold where aquatic toxicity can occur. As the effluent TDS concentration increases, the probability of effluent toxicity increases. While there is no absolute limit for toxicity, effluent with a TDS concentration of 2,500 mg/L is likely to fail the aquatic toxicity test.

One trigger for aquatic toxicity is the effluent's ionic composition. If the effluent has a high sodium or high chloride concentration (in relative terms, a higher proportion/molar concentration of sodium or chloride compared to the other ions present), toxicity is more likely to occur. The onset of toxicity is usually lessened (higher concentrations of TDS can be tolerated) as the effluent hardness or alkalinity is increased.

The ability to discharge to a specific receiving stream is based on the stream's volume of flow. If the summer low-flow periods produce a flow that is equal to or less than the proposed concentrate flow, it is likely that the discharge will no be allowed. The acceptance of a concentrate stream for discharge is also impacted by state rules pertaining to the aquatic toxicity test. Whole effluent toxicity (WET) testing is the most restrictive aquatic toxicity test. WET testing does not allow blending of the effluent with stream water. If allowed by a state, serial dilution is a better choice for surface discharges. For serial dilutions, the flow in the receiving stream dictates the amount of dilution allowed. Contact your state National Pollution Discharge Elimination System (NPDES) permitting authority for more information.

Many states have general permits for cooling tower blowdown discharge to surface waters. While cooling tower and membrane concentrates are not produced via the same mechanism, their characteristics are similar. To determine how a state might evaluate concentrate discharge to surface water, check the regulations regarding cooling tower discharges to see if surface water discharge is allowed.

Large concentrate volumes can make some treatment options too expensive or logistically impractical. For example, solar evaporation can be an effective option in arid regions; however, even in arid climates, the land area required for evaporation ponds can make this option impractical. A reclamation plant that treats a 10-mgd flow and achieves 70 percent RO recovery might require 1,000 to 1,500 acres of pond area to evaporate the concentrate in an arid climate (with loading rates between 1.4 gpm/acre and 2.0 gpm/acre). Land requirements of this magnitude can preclude pond construction in many situations.

HEALTH AND ENVIRONMENTAL ISSUES

State and federal regulations that relate to water reuse protect the public from pathogenic microorganisms. The water reclamation treatment process includes RO membranes and disinfection to provide barriers to these pathogens. However, concentrate is not treated by RO and is not disinfected; therefore, the public health protection barriers applied to reclaimed water do not exist for concentrate. Furthermore, all of the contaminants removed by RO and collected in the concentrate are at much higher concentrations than in reclaimed water. Consequently, standards for water reuse can restrict options for concentrate disposal where human exposure is likely.

Discharge of concentrate can also have detrimental environmental effects. TDS concentrations in reclamation plant concentrate typically range from 3,000 to 6,000 mg/L, levels that are toxic to many plants and animals. The high salinity levels in concentrate limit its disposal by surface water discharge or land application.

In addition to elevated salinity levels, concentrate can contain a variety of individual contaminants that adversely affect plants, fish, and wildlife. Domestic and industrial water uses add contaminants to wastewater that were not present in the natural water source. Examples include heavy metals, synthetic organic chemicals, pharmaceuticals and personal care products (PPCPs), and endocrine-disrupting compounds (EDCs).

EDCs are a large group of chemicals shown to interfere with the natural action of hormones in organisms and to induce adverse reproductive and developmental effects in wildlife. Examples of EDCs include natural and synthetic hormones such as estradiol and ethinylestradiol, degradation products of phenols and plasticizers, and pharmaceutical compounds such as hormones, lipid regulators, antibiotics, and analgesics. EDCs are not completely removed in wastewater treatment and have been found in wastewater plant effluents. Because many of these compounds are well rejected by NF and RO membranes, they end up in NF and RO concentrate at many times the concentrations found in wastewater effluent. When present in concentrate, environmental issues may arise if the concentrate directly or indirectly reaches surface water or groundwater.

COST AND ENERGY ISSUES

High concentrations of inorganic and organic compounds in concentrate make concentrate treatment more expensive (typically 10 times more expensive) and more energy intensive than treatment of water or wastewater. An example of this is treatment of concentrate with thermal desalination devices. Concentrate is treated and recovered as product water with a brine concentrator followed by a crystallizer. However, the capital cost and energy requirements are high. Seawater desalination with RO membranes is considered an expensive and energy-intensive process for generating drinking water. The cost and energy needed to treat concentrate with thermal technologies can be compared with RO seawater desalination. On a gallon-per-gallon basis, life-cycle costs for treatment with a brine concentrator is typically 4 times as expensive and consumes 8 times as much energy compared to an RO process. Crystallizer treatment is even more expensive and energy intensive. Compared to RO seawater desalination, the capital cost of using a crystallizer is approximately 125 times greater and energy consumption is 350 times greater.

CONCENTRATE DISPOSAL OPTIONS

The following methods have been used to manage concentrate:

- Discharge to surface water,
- Discharge to sewer,
- Inject underground,
- Use evaporation ponds,
- Use treatment, and
- Apply to land.

Discharge to Surface Water

Historically, discharge of concentrate to surface water has been the most widely used management method (Mickley 2006). Relative to other options, surface water discharge typically is low cost. However, there are high standards for environmental protection. Discharge of concentrate to surface water is regulated under the federal Clean Water Act (CWA) through the NPDES, which was created to establish minimum standards for surface water discharge. Any direct discharge to waters of the United States requires an NPDES permit. The permit specifies effluent limits, pretreatment requirements, and standards for inspection, monitoring, and reporting.

The effect of concentrate discharge on receiving waters depends on the water's dilution and mixing capacity. The amount of dilution and mixing required increases in proportion to the concentration of contaminants in the concentrate relative to the contaminant standard. Environmentally safe surface water discharge becomes more challenging as the volume of the concentrate stream increases and as the salinity and concentration of contaminants of concern increase.

Density of the concentrate relative to the receiving water also has an effect. If the concentrate has higher salinity than the receiving water, it will be denser and tend to sink. Conversely, if the concentrate has lower salinity, it will rise. Either condition could give rise to localized salinity gradients that are harmful to local plants and animals.

Surface water discharge systems include a pump, pipeline from the plant to the shore, pipeline from the shore to the outfall, and the outfall structure. Outfall structures frequently include a diffuser to promote dilution of the discharge. Diffuser design is complex and often conducted using software programs designed for this purpose. Design variables include the following:

- Discharge flow rate,
- Shape and hydraulics of the receiving body,
- Temperature of the effluent and receiving body,
- Density of the effluent relative to the receiving body, and
- Receiving stream flow rate, especially stream low flow.

While surface water discharge is often the least expensive concentrate management option, many factors affect its cost and feasibility. These include uses and quality of the receiving water, distance to the receiving water, environmental impacts, and outfall design.

Discharge to Sewer

Historically, discharge of concentrate to a sewer has been the second most widely used management method (Mickley 2006). An NPDES permit is not required for discharge to a sanitary sewer system. However, the publicly owned treatment works (POTW) that receives and treats the wastewater does have an NPDES discharge permit, and the ability of the receiving POTW to comply with its permit can be diminished if the concentrate is of extremely poor quality.

Concentrate discharge to a sewer system can have the following adverse effects:

- Corrosion of conveyance systems,
- Scaling of conveyance systems,
- Degradation of activated sludge settling,
- Salt inhibition of the activated sludge process,
- Toxicity in the WWTP effluent, and
- Contamination in the concentrate, for example, microconstituents, PPCPs, and EDCs.

Sulfate in concentrate can increase concrete corrosion within the collection system, including piping, manholes, and pumping stations. Sulfate can induce concrete corrosion directly (Lea 1998) or indirectly (Morton et al. 1991). In direct attack, sulfate reacts with calcium hydroxide in the cement to form calcium sulfate and with hydrated calcium aluminates to form calcium sulfoaluminate. In indirect attack, sulfate is converted to sulfides under the anaerobic conditions within the collection system; this leads to corrosion of concrete pipe and metal surfaces.

Concentrate can adversely affect the settling characteristics of biological floc in the secondary clarifier (Higgins & Novak 1996; Murthy et al. 1998; Bott & Anan 1999), and inhibition of methanization of volatile fatty acids in anaerobic digestion (Feijoo et al. 1995).

Furthermore, inorganic constituents such as chloride in the concentrate are not removed in biological treatment. As a result, discharge of concentrate to the sewer may cause the POTW to exceed its permit limits for inorganic contaminants. Contaminants in concentrate that are removed by biological treatment end up in the waste sludge and can affect its suitability for land application. POTWs must be willing to accept the concentrate, and they typically impose local limits on concentrate water quality to protect treatment efficiency and ensure compliance with NPDES permit limits.

The CWA also contains national water quality provisions for discharges to sanitary sewers under the National Pretreatment Program. These regulations are intended to protect POTWs from high concentrations of industrial pollutants. The standards are organized into two categories: categorical pretreatment standards and specific prohibitions. Categorical standards are applied to specific wastewater discharges of particular industrial categories. POTWs are not included in any industrial category and therefore are covered under the specific prohibitions. Specific prohibitions are intended to protect the receiving POTW from contaminants that could create a hazard, adversely affect treatment, cause corrosive damage, interfere with flow, or pose health risks to plant personnel. Authority to administer the Industrial Pretreatment Program is typically delegated to the POTW; however, several states have retained the authority to administer the program. In most states, any POTW with a design capacity that is greater than 5 mgd is required to develop and maintain an approved pretreatment program. All POTWs are required to develop a sewer use ordinance that gives them the legal authority to manage and regulate discharges into the collection system.

The requirement to control TDS in surface waters in the United States has been part of the CWA since its inception. However, in recent years states have begun to adopt water quality goals and are beginning to regulate TDS in general and to regulate specific ions. Many states have incorporated the secondary maximum contaminant levels (MCLs) for TDS, sulfate, and chloride into their water quality standards. Unlike secondary MCLs, water quality standards are not voluntary. Rather they are hard limits and, when incorporated into discharge permits, are limits that must be enforced and compliance determined.

Discharge to a sewer typically is not a sustainable option for reclaimed water because, in most cases, the sewer discharges back to the WWTP where the concentrate originated. Recycling the concentrate back to the WWTP would create a TDS sink at the WWTP, leading to increased TDS concentrations in the plant influent and effluent. If a large portion of the POTW-treated effluent is harvested as reclaimed water, it is likely to increase the TDS at the POTW to unacceptably high concentrations. This would limit the POTW ability to discharge to surface waters and possibly limit or increase the cost of producing reclaim water of acceptable quality.

Inject Underground

Underground injection is a disposal option in which waste streams are injected into porous subsurface formations for storage and confinement. Underground injection has been one of the most frequently used methods for disposal of hazardous wastes in the United States. Injection wells are also being widely used for concentrate disposal, particularly in Florida.

Underground injection is regulated by the Underground Injection Control (UIC) standards authorized under the Safe Drinking Water Act (SDWA). The UIC regulates injection of fluids into wells that are defined as follows:

- Can be bored, drilled, or driven and with a shaft that is not deeper than the largest surface dimension,
- A dug hole whose depth is greater that the largest surface dimension,
- An improved sinkhole, or
- A subsurface fluid distribution system.

The intent of the regulations is to protect underground sources of drinking water (USDW). The US Environmental Protection Agency (USEPA) established a classification system for injection wells based on location and the type of waste injected. Class I, nonhazardous, is the well classification pertinent to concentrate injection. Class I wells are used for the injection of nonhazardous industrial or municipal waste beneath the lowermost formation of any USDW within one-quarter mile of the well bore. A USDW is defined as having TDS less than 10,000 mg/L. Underground injection wells must be cased and cemented to protect USDW. UIC permits also include operating, monitoring, and reporting requirements.

There are several site requirements for class I nonhazardous underground injection. The injection zone must be below any USDW and separated from the USDW by a hydrologically impermeable formation that prevents migration of the injected fluid into the USDW. Any USDW must not be within one-quarter mile of the disposal well. The injection zone must meet water quality and hydrologic criteria. The water quality of the receiving formation must be physically, chemically, and bacteriologically compatible with the injected concentrate. The receiving zone must have adequate permeability and storage volume so that injection does not increase pressure in the receiving formation.

Underground injection is an isolation method for disposing of concentrate. The intent is that the injected waste remain immobilized in the isolation zone indefinitely, making site selection critical. Geologic and hydrologic evaluations are required to make certain the receiving formation has the capacity to hold and confine the waste volume. Porosity and permeability are two important hydrogeological characteristics that determine the suitability of subsurface formations for injection of wastes.

Porosity is the percentage of pore volume in the formation. The pores may be connected or discrete. Effective porosity is the percentage of interconnected pore volume. To be suitable for injection, a subsurface formation must have sufficient effective porosity to store large volumes of liquid. Sandstone is an example of a highly porous material; sandstone stratum generally is a good formation for underground injection.

Permeability describes the ability of water to move through the formation. Permeability is related to porosity by the interconnectedness of pores in the formation. A high degree of interconnectivity facilitates water movement through the formation and increases permeability. In addition to being porous enough to store concentrate, a suitable subsurface formation must be bound by impermeable strata that prevent the migration of injected concentrate. Shale is an example of an impermeable material.

UIC regulations specify construction requirements for class I injection wells. These wells must be cased and cemented to prevent leakage. A class I injection well is made up of casing, tubing, and packers. Casing is the outer pipe that is cemented in place. The casing is used to maintain the structural integrity of the bore hole and seal off intermediate aquifers from the injected concentrate. There may be several stages of casing comprising successively smaller pipe diameters with depth. The space between casing stages is filled with cement. Tubing is the innermost pipe that conveys the injected fluid to the injection zone. Packers are used to seal the annular space between the innermost casing and the tubing. The annulus between the casing and tubing is filled with a noncorrosive liquid that is maintained under positive pressure. Pressure in the annulus fluid is monitored to detect leaks.

Concentrate solutions are typically supersaturated with respect to sparingly soluble salts. Precipitation of salts after the concentrate is injected can reduce subsurface permeability and requires higher pumping pressures to maintain the design flow rate. Consequently, treatment with an antiscalant to inhibit precipitation may be necessary prior to injection.

Underground injection is an environmentally safe disposal method as long as there is no leakage of concentrate during injection or subsurface movement of concentrate after it reaches the injection zone. Strycker and Collins (1987) noted the following five ways in which an injected waste may contaminant adjacent underground water supplies:

- Leakage due to insufficient casing or corrosion of the casing;
- Escape vertically outside of the well casing;
- Escape vertically from the injection zone through confining strata that are inadequate due to high permeability, solution channels, joints, cracks, or induced fractures;
- Escape vertically from the injection zone through nearby wells that are improperly cased; and
- Lateral migration of concentrate to a region of freshwater.

Some regulatory agencies require monitoring wells for the detection of contaminants in nearby aquifers from the injection well. Samples are collected from the monitoring wells periodically and analyzed to ensure there has been no migration of concentrate into adjacent aquifers.

Use of Evaporation Ponds

Evaporation ponds have been used for decades to manage liquid waste from desalination, industrial treatment, and power plant facilities (Westbrook & Wirth 1977; Ahmed et al. 2000; Salzman et al. 2001), and they have been used for centuries to recover salt from seawater. Evaporation is the process by which water is transformed from the liquid phase to the vapor phase by diffusion. Diffusive mass transfer is driven by a concentration gradient that, for evaporation, is the difference between the actual vapor pressure of air at the air–water interface and its vapor pressure at saturation. Saturation vapor pressure, which is an indication of the capacity of air to hold moisture, increases as temperature increases. Therefore, evaporation ponds are most effective in dry, warm climates. However, even in arid regions, evaporation requires large surface areas. Consequently, practical use of evaporation ponds for concentrate disposal is usually limited to small concentrate flows and locations where land is available and not prohibitively expensive.

Evaporation ponds are sized based on the local evaporation rate. The rate of evaporation from a pond depends on site-specific climatological conditions including temperature, solar radiation, wind speed, and humidity (Burman & Pochop 1994). Evaporation pans are often used to simulate pond evaporation to develop site-specific design evaporation rates for pond sizing. The standard evaporation pan in the United States is the class A pan of the US Weather Bureau. This pan is a cylindrical container with a 4-ft diameter and 10-in. depth. Water is maintained between 2 and 3 in. from the rim; water level, wind speed, and water temperature are measured continuously.

Due to differences in scale and sidewall effects, evaporation from a pond differs from that in a pan. A pan coefficient, defined as the ratio of pond evaporation to pan evaporation, is frequently used to correlate the two. Kohler (1954) suggested a pan coefficient of 0.7. Linacre (1994) surveyed pan coefficients reported in 18 articles and found a wide scatter around an average value of 0.77.

The effect of salinity must also be accounted for in sizing evaporation ponds for concentrate management. Water vapor pressure decreases as salinity increases; therefore, the rate of water evaporation also decreases. At the saturation concentration for sodium chloride (NaCl; 26.4 percent), the evaporation rate is approximately 70 percent the rate for freshwater (USDI 1971), and the evaporation rates measured for lesser NaCl concentrations were 98 percent for 2 percent NaCl, 97 percent for 5 percent NaCl, 93 percent for 10 percent NaCl, and 78 percent for 20 percent NaCl (Bureau of Reclamation 1969). Mickley (2006) noted that an evaporation pond is expected to reach saturation during its life cycle. In the absence of site-specific data, a 70 percent reduction in the evaporation rate is reasonable to account for salinity effects.

Evaporation ponds are an effective concentrate management strategy for small concentrate volumes in arid climates. They are generally easy to construct and, because no mechanical equipment is required, operational attention and costs are minimal. Most states require a leak-proof lining and monitoring systems to reduce the risk of groundwater contamination; these features substantially increase construction costs. Even with state requirements to line the pond, evaporation ponds can still be a cost-effective concentrate management approach.

A primary disadvantage of evaporation ponds is the large land area required. The Bureau of Reclamation (1971) evaluated evaporation ponds for desalination of a brackish groundwater in central Arizona. A production capacity of 2 mgd and a concentrate flow of 0.5 mgd were projected. Based on an evaporation rate of 84 in. per year, it was estimated that 80 acres of evaporation ponds would be required for the 2-mgd plant. This calculation did not include reductions for pan coefficient and salinity effects. With a pan coefficient of 0.7 and 70 percent reduction for salinity effects, the calculation would have yielded 163 acres of evaporation ponds. Even in central Arizona, a region with one of highest evaporation rates in the United States, large land areas are required to handle significant concentrate flows.

Enhanced evaporation systems can be used to reduce the area required for evaporation ponds. This is done by using mechanical energy to increase evaporation rates relative to conventional evaporation ponds. The enhanced evaporation systems that have been tested can be categorized into the following general modes of operation:

- Spray system that increases the surface area available for evaporation by projecting concentrate into the air as droplets or mist and
- Film evaporation system that increases surface area available for evaporation by cascading concentrate down strips of material in a film.

The purported advantage of spray systems is that they are inexpensive to build; however, they can pose environmental risk. The salt precipitate that remains after the liquid portion of the concentrate evaporates can be carried by the wind to contaminate the surrounding area with a salt residue.

Wind-aided intensified evaporation (WAIV; Gilron et al. 2003) is the best known film evaporation system. WAIV uses vertically mounted and continuously wetted evaporation surfaces to increase surface area and enhance evaporation rates. Concentrate is pumped to the top of the strips where it falls as a film down the length of the strips. The strips increase the water surface exposed to air.

Water that is not evaporated falls into a holding pond under the WAIV strips and is recirculated. Researchers have reported an evaporation rate that is 10 times that of an evaporation pond with an equivalent footprint (Gilron et al. 2003; Katzir et al. 2009). Salts that precipitate on the strips fall into the holding pond and must be cleaned out periodically.

The keys to the design of WAIV are the strip material and the spacing of the strips. The material must be hydrophilic enough to allow spreading of the concentrate over the strip area but not so hydrophilic as to reduce vapor pressure. The strips must be spaced to allow sufficient air flow between them while minimizing the footprint required. Currently the WAIV technology is owned and sold by Lesico CleanTech of Holon, Israel. WAIV has been pilot tested but currently there are no full-scale installations.

Advantages of WAIV are that it takes advantage of a natural renewable energy source and that the only mechanical equipment required is a pump to lift concentrate from the holding pond to the top of the strips. Because WAIV is an emerging technology and has only been demonstrated in pilot tests, there is no record of long-term operation from which to assess treatment performance and cost. As is the case with evaporation ponds, WAIV is effective in arid climates.

Apply to Land

An NPDES permit is not required for spray irrigation unless there is a potential for the runoff to reach a receiving water. Mickley (2006) noted that in many cases it is less expensive to obtain a permit than it is to prove runoff will never reach a receiving water. According to USEPA guidelines (EPA/625/R-95/008), spray irrigation may be viable under the following conditions:

- TDS, chloride, or salinity does not exceed a level that will damage crops.
- There is a requirement for irrigation in the vicinity.
- A backup disposal or storage system is available for use during periods of sustained rainfall.
- The water quality of the surficial groundwater is protected from degradation.

- Local, state, and federal regulations are met.
- A system of monitoring wells is in place. (Each state may have its own policy on groundwater monitoring, so state regulations should be checked.)

The following are important criteria in the evaluation of spray irrigation:

- Hydraulic loading rate (HLR);
- Land requirement;
- Effect of salinity, sodium, sodium adsorption rate (SAR), boron, and carbonate on soil permeability; and
- Vegetation selection.

The HLR is used to determine the land area required for spray irrigation. Consequently, an accurate determination of the HLR is the most important step in designing a spray irrigation system. The HLR depends on soil permeability and precipitation and evapotranspiration rates.

The water balance for spray irrigation can be represented with the following equation:

$$HLR = E + T - P \qquad \text{(Eq 7-1)}$$

Where:

HLR = hydraulic loading rate
E = evapotranspiration
P = precipitation
T = percolation

Percolation describes the water infiltration into the ground. The amount of percolation depends primarily on soil permeability, but other factors such as depth to water table and duration of the drying period between spray applications must be accounted for in the calculation. Evapotranspiration is the sum of water evaporated from the plant surfaces and plant transpiration. Transpiration refers to the water that is taken into the plant through its roots and released to the atmosphere through stomata in the plant leaves. Rest periods should be built into the design calculations to allow the soil to dry and get air into the pore spaces.

Percolation, evapotranspiration, and precipitation all vary seasonally. Evapotranspiration also varies with plant species because transpiration rates vary among plant species. Consequently, mass balance calculations should be performed monthly and annually to ensure adequate irrigation area throughout the year. The land requirement should include space for a buffer zone around the irrigation field, storage capacity, and possibly land for capacity expansion or contingencies. An irrigation field should be well drained, have loamy soils, have a minimum of 5 ft to the water table, and have a ground slope of less than 20 percent.

Water quality characteristics may rule out spray irrigation of concentrate for many applications. Sodium can reduce the permeability of soil. The US Department of Agriculture developed the following equation for evaluating the effects of sodium on soil permeability, where sodium, calcium, and magnesium are normalities in milliequivalents per liter.

$$SAR = Na/[(Ca + Mg)/2^{1/2} \qquad \text{(Eq 7-2)}$$

Concentrate with an SAR value greater than 9 can impair soil permeability and may be toxic to plants.

High concentrations of calcium and carbonate can lead to precipitation of calcium carbonate in soil pores, causing blockage of flow paths and reduction in soil permeability. The high salinity levels in concentrate can be toxic to plants and can damage the soil for future agricultural use due to accumulation of salt. Accumulation of heavy metals in the soil is another concern with spray irrigation.

Irrigation water that is not taken up by the plants percolates down to the water table. Therefore, the impact of concentrate on the aquifer below the irrigation field must be considered, particularly if the water table is shallow. Drainage systems may be required to capture percolated water before it reaches the water table.

Good farming practices on the land application sites are critical. As the SAR is depleted, periodic lime addition may be needed to reestablish the SAR. TDS and specific ions accumulate in the soil and can inhibit or become toxic to plants. Periodic field flooding is a standard farming practice performed to flush accumulated salts out of the plant root zone. The local soil conservation service should be contacted to learn more about the typical practices in the area around the land application site.

Depending on the state rules that govern the control of runoff from land application and land disposal sites, each site may need an equalization or storage basin to manage the liquid volume (this cannot apply when the soil is saturated with water). Tailing ponds to catch site runoff may also be required. Sizing of these systems is regulated by the state.

One drawback of land application is conservation of mass. The salts do not disappear and will run off the site if concentrate is applied during wet weather or the periodic saturation of the site by natural rainfall will transport TDS into the groundwater.

Treatment

Treatment involves treating concentrate to reduce its volume or to eliminate discharge of concentrate from the plant. In general, treatment is more expensive than disposal. However, more stringent standards for concentrate discharge and the need to conserve water have led to increased attention on concentrate treatment options.

Historically, concentrate has been treated using thermal desalination processes. These processes have been used for decades to desalinate water, particularly in the Middle East where energy has been relatively inexpensive. In the United States, thermal desalination primarily has been used to treat industrial water and cooling water from power plants.

Thermal desalination is accomplished by evaporating the water, condensing the resulting steam, and recovering the condensate as high-quality product water. It is a very energy-intensive process because water has a high heat of vaporization (2260 kJ/kg). Thermal desalination produces a high-quality distillate, typically with TDS less than 10 mg/L.

A typical treatment system for concentrate includes a brine concentrator followed by a crystallizer or spray dryer. A brine concentrator (Figure 7-1) is a falling film evaporator with a sump at the bottom of its column and condensation tubes that extend vertically above the sump. Feedwater is heated and discharged to the sump. Vapor above the sump is drawn off and compressed in a vapor compressor. Heated water in the bottom of the sump is pumped to the top of the condensate tubes and allowed to fall down the inside of the tubes. Compressed vapor is discharged to the outside of the condensate tubes. The temperature of the compressed vapor is greater than that of the water inside the tubes, and heat exchange between the vapor and the water inside the tubes causes the vapor to condense on the outside of the tubes. This condensate is captured as product water.

Recovery of RO concentrate in a brine concentrator used to treat RO concentrate from a reclamation facility typically ranges from 95 to 98 percent. The remaining water is discharged from the sump as a highly saline concentrate. Recovery in the brine concentrator is limited by the scaling potential of the salts; concentrations of TDS in the sump can reach 250,000 mg/L. Corrosion-resistant and, in some cases, highly specialized materials are required for construction of the brine concentrator to protect it from corrosion over its life span. Consequently, capital costs for brine concentrators are high.

If further reduction of the concentrate volume is required, the concentrate discharged from the brine concentrator can be treated with a crystallizer (Figure 7-2) or spray dryer. The crystallizer is a falling film evaporator, similar to the brine concentrator, where the more soluble salts (sodium sulfate, sodium chloride, calcium chloride) are precipitated. The brine fed to the crystallizer is distilled so that the only products from the crystallizer are distilled water and salts that are approximately 80 percent dry. Recovery of concentrate as product water using a brine concentrator and crystallizer approaches 100 percent.

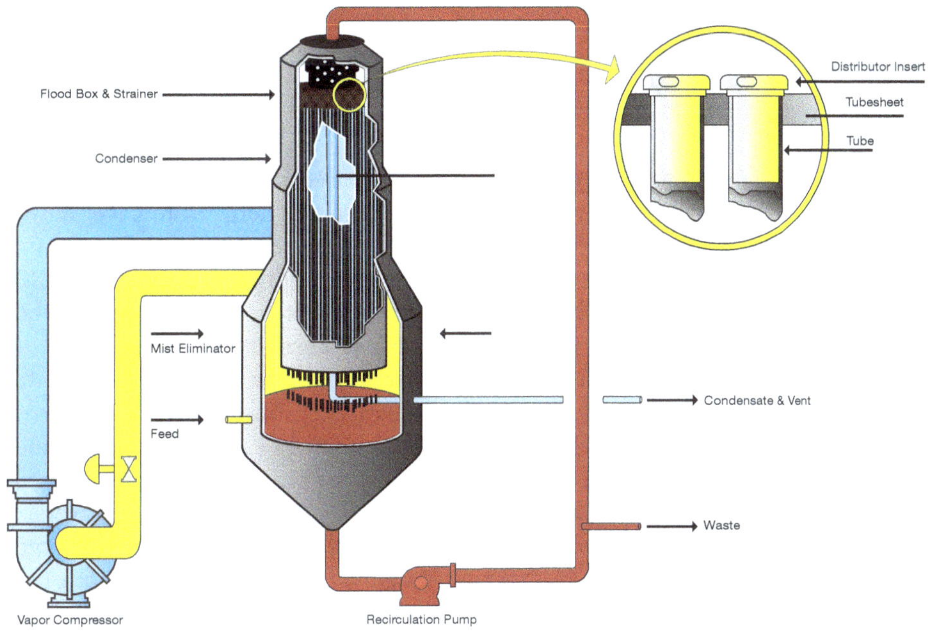

Courtesy of GE.

Figure 7-1 Mechanical vapor compression brine concentrator

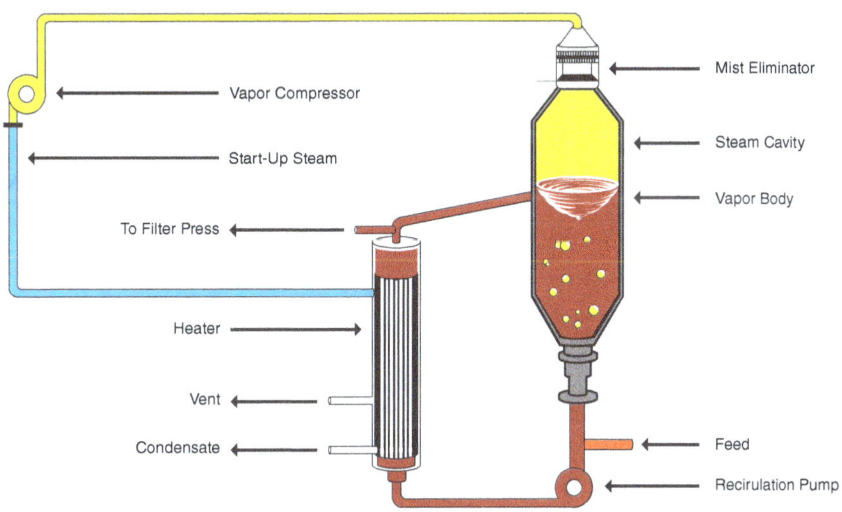

Courtesy of GE.

Figure 7-2 Mechanical vapor compression crystallizer

Spray dryers are an alternative to crystallizers for small flows. Crystallizers are typically more cost effective than spray dryers for flows greater than 10 gpm (Mickley 2006). For treatment with a spray dryer, concentrate is atomized and fed into a chamber along with hot air. The water is vaporized, and solids are captured in a bag filter. Water is discharged to the atmosphere in the heated air rather than recovered.

Thermal technologies are energy intensive. Typical energy consumptions for thermal treatment of concentrate are as follows:

- Brine concentrator, 80 kWh/kgal,
- Crystallizer, 250 kWh/kgal, and
- Spray dryer, 740 kWh/kgal.

CURRENT RESEARCH

Considerable research has been focused on reducing the cost of concentrate treatment. Because thermal desalination is expensive and energy intensive, new approaches that reduce dependence on thermal desalination and evaporation ponds by increasing the percentage of water recovered by membrane separation processes such as RO or electrodialysis (ED) are being investigated. There are variations of this approach, but the theoretical basis is fundamentally the same as described here.

As recovery in desalination membrane processes increases, concentrations of organic and inorganic species in concentrate increase. If concentrations increase to the point that they cause solids to precipitate, those precipitates can adhere to the membrane surface, causing membrane fouling and reducing membrane permeability. Consequently, recovery of product water during desalination of brackish or reclaimed water is limited by the fouling potential of inorganic and organic compounds. Sustainable recovery limits are reached at concentrate TDS levels well below those where the osmotic pressure difference becomes the limiting factor, as is the case with seawater desalination.

One way to reduce the costs for concentrate treatment is to treat concentrate to reduce its membrane fouling potential and then desalinate the treated concentrate in a second membrane application. Treatment processes that are used to reduce the membrane fouling potential of concentrate include lime softening, fluidized bed crystallization, use of activated alumina, coagulation with metal salts, ion exchange, and ED. The most highly developed options for the second membrane desalination step are RO and ED.

An example of this approach is the use of conventional lime softening or lime softening in a fluidized bed crystallizer to remove calcium from concentrate. Some total organic carbon (TOC) is removed in lime softening. For concentrate with sufficient magnesium, lime softening also removes silica by adsorption to magnesium hydroxide precipitates. The process train is as follows:

Primary RO → Lime Softening → Filtration → Secondary RO or
Electrodialysis Reversal (EDR) → Thermal Device or Evaporation Pond

Supernatant from the softening process is filtered by granular media filters or membranes and then desalinated in a second device, either RO membranes or EDR. Recovery in the second desalination device depends on the level of concentrate treatment applied. However, it is reasonable to expect 70 to 80 percent recovery, which represents a significant increase in product water recovery and a significant reduction of concentrate volume. For example, if concentrate from an RO system that achieves 75 percent recovery is treated with lime softening and desalinated in a second RO system with 75 percent recovery, overall product water recovery is increased from 75 percent to 96 percent and the concentrate volume is reduced by a factor of 4. The remaining concentrate, 4 percent of the feed flow, can then be treated more economically with thermal desalination or discharged to a more reasonably sized evaporation pond. This approach has been estimated to reduce capital and operating costs by more than 50 percent compared with thermal desalination without concentrate treatment (Bond & Veerapaneni 2007).

Another technology being investigated for concentrate treatment appears to offer a unique advantage by using ion-selective membranes to circumvent precipitation of inorganic and organic compounds, which is the limiting factor for recovery in RO, NF, EDR, and ED. Electrodialysis metathesis (EDM) is an technology that contains repeating units comprising four compartments and four ion-selective membranes (Alheritiere, Ernst, & Davis 1998). The repeating unit contains one diluate, one sodium chloride, and two concentrate compartments.

The driving force for ED membrane systems is the electrochemical potential difference between a positively charged anode and a negatively charged cathode. A conventional ED or EDR stack has hundreds of repeating cell pairs between a cathode and anode, each containing a cation- and an anion-exchange membrane, a diluate compartment, and a concentrate compartment. As water is pumped through the diluate compartment, anions and cations are transported through ion-exchange membranes into the concentrate compartment and held there by electrostatic repulsion from ion-exchange membranes with the same charge. The diluate stream is demineralized, and all ions removed from the diluate are contained in the concentrate stream. Recovery in ED is limited by the precipitation potential of the same sparing soluble salts that limit RO recovery.

Rather than a cell pair, the EDM repeating unit includes four compartments and four membranes. This unique configuration causes EDM concentrate to be separated into two streams of highly soluble salts, one containing sodium with anions and the other chloride with cations. This characteristic provides significant advantage when treating concentrated streams because ion pairs that form low-solubility salts such as calcium sulfate and calcium carbonate are not present in the same concentrate stream. Furthermore, silica and TOC are not concentrated in EDM concentrate because they carry insufficient electrical charge to be transported through the ion-exchange membranes.

The process train for this approach is as follows:

Primary RO → EDM → Thermal Device or Evaporation Pond

Primary RO concentrate is desalinated in EDM without prior treatment with a process such as softening to reduce precipitation potential. This is possible because the EDM process does not concentrate ion pairs of sparingly soluble salts, silica, or natural organic matter quantified as TOC. This approach was tested on water sources with high TOC concentrations similar to reclaimed water (Bond et al. 2011). The treatment cost using any form of ED is dependent on TDS concentrations. In the range of TDS concentrations found in reclaimed water, this approach was estimated to reduce treatment costs and energy consumption by more than 50 percent compared with thermal desalination without concentrate treatment.

REGULATORY ISSUES

Several federal regulations are applicable to concentrate, including portions of the SDWA and the CWA. Some of these regulations are mandatory, while others are intended as guidelines. Most states have delegated authority from USEPA to develop and administer state regulations that ensure federal regulations are met. States can also adopt regulations that are more stringent than federal regulations. In addition to federal and state regulations, concentrate disposal can be further governed by local regulations. Consequently, permitting concentrate disposal can be a challenge and an important consideration in evaluating concentrate management options.

As suggested from Table 7-2, the CWA and the SDWA can impact concentrate, backwash, and waste chemical cleaning disposal practices. The CWA regulates all discharges to surface waters, activities in wetlands, disposal of sludge, and dischargers to wastewater collection systems through the NPDES permit program. Although discharge of concentrate to a POTW does not require an NPDES permit, the concentrate stream must meet local pretreatment requirements established under the USEPA National Pretreatment Program (40 CFR 403). Solid residuals, including soluble materials that adsorb onto biosolid surfaces in the WWTP, are likely to be regulated under the 40 CFR part 503 sewage sludge disposal regulations because the WWTP is required to protect the biosolids quality for beneficial reuse.

The 1977 amendments to the CWA required a POTW to set pretreatment standards for each significant local source. Concentrate, backwash, and waste chemical cleaning streams are not included among the 35 industrial sectors for which USEPA set the uniform categorical standards for concentrations of specific pollutants that are technology based and applicable nationwide. Therefore, the POTW must develop local limits to regulate the discharge of these materials to the WWTP.

The USEPA Local Ordinance Development document (2004), which is the latest revision to the Pretreatment Guidance standards, is intended to complement categorical standards. In 1984 the original Pretreatment Guidance Manual was revised to require that each POTW develop "technically based local limits"; this requirement was retained in the 2004 revisions. Each POTW must set local limits based on site-specific conditions that include the efficiency of the treatment process, history of compliance with NPDES permit limits, the water quality standards that are applied to the receiving water, and biosolids handling.

USEPA developed guidelines for establishing local ordinances that begin with the maximum allowable headworks loading (MAHL) and the maximum allowable industrial loading (MAIL) concept. The MAIL is that portion of the MAHL that is contributed by industrial users (IUs) and is thus considered readily controllable. A reclaimed water treatment plant that discharges concentrate, backwash, and/or waste chemical cleaning streams into the wastewater collection system falls under the category of an IU. MAILs can be uniform concentrations that apply across all IUs, individual waste load allocations, or a combination of the two.

The first step in establishing a MAHL is to determine the pollutants of concern (POC). The general criteria to define a POC are deleterious effects in the collection system, interference with treatment such violation of NPDES limits, interference with biosolids handling, jeopardy to workers, and operational problems at the WWTP, specifically biological process inhibition and even physical maintenance issues that result from a given waste (corrosion would be considered an interference if the corrosion rate were accelerated by TDS and specific ion concentrations). At a minimum, USEPA recommends that each WWTP screen for the presence of the following pollutants: ammonia, arsenic, BOD5, cadmium, chromium, cyanide, lead, mercury, molybdenum, nickel, selenium, silver, TSS, and zinc.

In addition, USEPA recommends examination of POCs that can limit biosolids disposal to the land. These include arsenic, cadmium, copper, lead, mercury, molybdenum, nickel, selenium, and zinc.

Table 7-2 Microfiltration–ultrafiltration residuals and applicable regulations

Residual	Typical Contaminant	Typical Concentration	Regulation	Representative Limitation
Backwash bleed-off discharges	Raw water, precipitated solids, algae, pathogens, possible chemical residues	TSS: 5–200 mg/L Cysts: 7–50 × raw conc Acids and bases: pH <6 or pH >9 Chlorine: >5 mg/L	CWA	pH: 6–9 TSS: 30 mg/L + raw TSS Chlorine: <0.2 mg/L BOD_5: 30 mg/L
			Safe Drinking Water Act*	If TSS and cyst removal = 90 percent recycled at WTP and recycle flow <10 percent of WTP flow
Solids/sludge	Raw water, precipitated solids, algae, pathogens	TSS: 500–2,000 mg/L for settled backwash solids Cysts: 15–200 × raw conc.	CWA	pH: 6–9 TSS: 30 mg/L + raw TSS BOD_5: 30 mg/L
			Resource Conservation and Recovery Act	No free liquid (>20 percent solids after dewatering) Pass toxicity characteristic leaching procedure Possible use as landfill cover or for sod farming
Spent cleaning solutions	Residual cleaning chemicals, TOC, BOD_5,dissolved solids	TSS: <500 mg/L Acids and bases: pH <6 or pH >9 Chlorine: >5 mg/L TOC: <100 mg/L BOD_5: 5,000–10,000 mg/L (if citric acid is used) Surfactants: low concentrations	CWA, US Environmental Protection Agency, and local industrial pretreatment and wastewater collection system use rules	pH: 6–9 TSS: <400 or 500 mg/L BOD_5: <400 or 500 mg/L Chlorine: <10 mg/L Nothing that will harm or interfere with wastewater collection systems, wastewater treatment plant. or its operation

* The Safe Drinking Water Act rules apply indirectly because they are a basis used in the development of local limits under the pretreatment program. The publicly owned treatment works must control the discharge of pollutants to the receiving stream to protect the uses of the stream; potable water production is one of the stream uses. The methodology for developing local limits can be found USEPA (1987).

Abbreviations: BOD_5, biological oxygen demand 5-day test; CWA, Clean Water Act; TOC, total organic carbon; TSS, total suspended solids; WTP, water treatment plant.

Experience with regulating concentrates, backwash, and waste chemical cleaning streams from membranes is very limited. The permit requirements for discharge of concentrate from membrane operations at reclaimed water treatment plants to the wastewater collection system vary widely (Zander 2000). For instance, local limits may be established that go beyond the criteria set forth by USEPA. In particular, limits may address the corrosive nature of RO brines, pH above or below neutral, and unacceptably high concentrations of TDS and heavy metals. States including California, Arizona, New Mexico, Texas, Oklahoma, Missouri, Illinois, and Florida have developed instream water quality standards that address the control of inorganic constituents in concentrate streams by various parameters (TDS, chloride, sulfate, salinity, and conductivity) in the WWTP effluent. These types of limitations will get passed upstream to specific IUs to limit the source quality and quantity of the high TDS and specific ion concentrations.

A disposal fee is established based on the volume and chemical characteristics of the concentrate, backwash, or waste chemical cleaning stream (Zander 2000). The cost of disposal to the wastewater collection system is often far less than for other options. However, the total environmental burden was found in one study to be greater for discharge to the wastewater collection system than for other options (Zander 2000). Disposal fees are highly variable, depending on the impact the concentrate has on use of treatment capacity and hydraulic capacity and the potential for concentrate to damage treatment plant components.

If the WWTP that receives the membrane brine reuses effluent, TDS and specific ions may be regulated based on the requirements of the end user of the reclaimed water, including reuse guidelines for water quality for specific uses such as crop irrigation. If any reclaimed WWTP effluent is returned to the WWTP, evaporation can cause the TDS and other ion concentrations to increase in concentration. Brine discharges to the WWTP will not be beneficial in this case and may be strictly regulated and possibly even prohibited.

An IU is defined as any user that discharges more than 25,000 gpd or enough mass of any pollutant that can cause pass-through or cause interference at the WWTP. Pass-through and interference are wastewater terms defined by USEPA in 40 CFR part 403. Interference is defined as any negative impact on the WWTP, the biological processes, or WWTP equipment including the collection system. Because the collection system is considered to be an extension of the WWTP, brine corrosion and erosion are considered to be interference. Pass-through applies to any material that passes through a WWTP and is present in the effluent. This general definition for pass-through allows the pretreatment authority to regulate TDS and specific ions and even metals and nonbiodegradable organics present in the WWTP if they enter the system through industrial discharges. The pretreatment coordinator has the legal authority to include technically based local limits for any pollutants defined as exhibiting "pass-through" or "interference" and can develop system-wide limits or specific limits for a single discharger.

If the WWTP is owned and operated by the city and the reclaimed water plant is also owned by the city, does the City have to permit the reclaimed water plant (conventional or membrane) for discharges to the sewer system? The city can issue a pretreatment permit to one of its facilities or it can internally regulate the IU discharge. However, the city, by way of the industrial pretreatment authority, is still required to protect the WWTP and ensure NPDES permit compliance. The city has the latitude to decide where to invest its funds to comply with the NPDES permit. However, ultimately the City is responsible for taking the appropriate action to meet its NPDES discharge permit, even if it means restricting discharges from the water treatment plant.

If the reclaimed water plant and WWTP are owned by different agencies, the WWTP, through the industrial pretreatment program, must regulate discharges from the reclaimed water plant if, in the judgment of the pretreatment coordinator, the discharge from the reclaimed water plant has the potential for pass-through or interference. Violating any NPDES numerical effluent limit is considered to be interference and the WWTP, through the pretreatment program, must control the discharges that caused the violation. In this case, the reclaimed water plant would be considered another IU on the system.

There may also be aquatic toxicity testing criteria in the WWTP's NPDES permit. The NPDES permit program has two components. One component, the numerical limits portion of the permit, defines the specific concentration limits that must be met to maintain stream water quality and protect biosolids quality. The other component is not as easily defined because it is based on the logic of what may have been missed in setting numerical standards that could affect stream quality. Thus, there is the need for an aquatic testing program.

Aquatic toxicity testing is very similar to setting human health standards for drinking water. The impact of specific or even general contaminants on the target aquatic species is evaluated. Aquatic toxicity testing is used to set the numerical water quality standards. There are many chemicals present in the WWTP effluent than have not been rigorously tested to set numerical limits. Consequently, aquatic testing is used to determine if anything is present that poses a danger to aquatic life.

Aquatic toxicity testing is divided into two criteria: acute and chronic. Acute testing is used to determine whether the WWTP effluent will kill the target species on contact. Different states have variations of this rule as it relates to testing blends of water representative of what is actually occurring in the stream versus WET testing. WET testing is the most extreme because the test is performed using 100 percent WWTP effluent and is used by regulators to determine the survival rate of aquatic life.

If the permitting authority does not require WET testing, it is likely that serial dilution aquatic toxicity testing will be required. In this test method, the effluent is allowed to be blended into different amounts of receiving stream water to identify the necessary dilution needed for no observable effects concentration (NOEC). An in-stream river diffuser may be required to attain the blending needed to attain NOEC in the receiving stream.

The framework already exists within the NPDES permitting process to address any new POCs that are added in the future, specifically EDCs. Once a toxicity threshold for a specific EDC, or for any other chemical, is determined it can be written into an NPDES permit or the utility can use that data to develop a local limit to protect effluent quality. When embarking on a reuse project, always contact the regulatory authorities to identify the authority that will regulate discharge from the reuse facility, whether it is the State permitting authority or the local city industrial pretreatment program administrator. Start the permitting discussion early to avoid costly surprise complications later in the project.

REFERENCES

Ahmed, M.; Shayya, W.H.; Hoey, D.; Mahendran, A.; Morris, R.; & Al-Handlaly, j. 2000. Use of Evaporation Ponds for Brine Disposal in Desalination Plants. *Desalination*, 130:155–168.

Alheritiere, C.; Ernst, W.R.; & Davis, T.A. 1998. Metathesis of Sodium Salt Systems by Electrodialysis. *Desalination*, 115:189–198.

AWWA. 2005. M53, *Microfiltration and Ultrafiltration Membrane for Drinking Water.* Denver, CO: AWWA.

Bond, R.G. & Veerapaneni, S. 2007. *Zero Liquid Discharge for Inland Desalination*. Denver, CO: AwwaRF and AWWA.

Bond, R.G.; Batchelor, B.; Davis, T.A.; Klayman, B.; Spencer, C.; & Veerapaneni, S. 2011. *Zero Liquid Discharge Desalination*. Denver, CO: Water Research Foundation.

Bott, C.B. & Love, N.G. 2002. Investigating a Mechanistic Cause for Activated-Sludge Deflocculation in Response to Shock Loads of Toxic Electrophilic Chemicals. *Water Environment Research*, 74:306–315.

Burman, R. & Pochop, L.O. 1994. *Developments in Atmospheric Science; Evaporation, Evapotranspiration, and Climatic Data*. New York: Elsevier Science Publishers B.V.

Campos, J.L.; Mosquera-Corral, M.; Sanchez, R.M.; & Lema, J.M. 2002. Nitrification in Saline Wastewater With High Ammonia Concentration in an Activated Sludge Unit. *Water Research*, 36:2555–2560.

Feijoo, G.; Soto, M.; Mendez, R.; & Lema, J.M. 1995. Sodium Inhibition in the Anaerobic Digesting Process: Antagonism and Adaptation Phenomena. *Enzyme and Microbial Technology*, 17:180–188.

Gilron, J.; Folkman, Y.; Savliev, R.; Waisman, M.; & Kedem, O. 2003. WAIV–Wind Aided Intensified Evaporation for Reduction of Desalination Brine Volume. *Desalination*, 158:205–214.

Higgins, M.J. & Novak, J.T. 1996. Settling and Dewatering of Activated Sludge: The Case for Using Cation Analysis. *Water Environment Research*, 69:225–232.

Katzir, L.; Volkmann, Y.; Daltrophe, N.; Korngold, E.; Mesalem, R.; Oren, Y.; & Gilron, J. 2009. WAIV–Wind Aided Intensified Evaporation for Brine Volume Reduction and Generating Mineral Byproducts. *Desalination and Water Treatment*, 13:63–73.

Kohler, M.A. 1954. Lake and Pan Evaporation. Technical Report, Paper 269, Geological Survey, Washington, DC: US Department of the Interior; pp. 127–148.

Linacre, E.T. 1994. Estimating U.S. Class-A Pan Evaporation from Few Climate Data. *Water International*, 19:5–14.

Lea, F.M. 1998. *Chemistry of Cement and Concrete*. London, UK: Elsevier Butterworth-Heinemann Publishers.

Mickley, M.C. 2001. Membrane Concentrate Disposal: Practices and Regulation. Denver, CO: US Department of Interior, Bureau of Reclamation, Technical Service Center.

———. 2006. Membrane Concentrate Disposal: Practices and Regulation (Second Edition). U.S. Department of Interior, Bureau of Reclamation. Desalination and Water Purification Research and Development Program Report No. 123.

Miller, A. & Stratton, R. Planning and Design Considerations for Membrane Plant Residuals Handling. California–Nevada AWWA 2009 Conference.

Morton, R.L.; Yanko, W.A.; Graham, D.W.; & Arnold, R.G. 1991. Relationships Between Metal Concentrations and Crown Corrosion in Los Angeles County Wastewater Collection Systems. *Journal Water Pollution Control Federation*, 63:789–798.

Panswad, T. & Anan, C. 1999. Impact of High Chloride Wastewater on an Anaerobic/Anoxic/ Aerobic Process With and Without Inoculation of Chloride Acclimated Seeds. *Water Research*, 33:1165–1172.

Salzman, S.A.; Allinson, G.; Stagnitti, F.; Coates, M.; & Hill, J.R. 2001. Performance of Constructed Evaporation Ponds for Disposal do Smelter Waste Water: A Case Study at Portland Aluminum, Victoria, Australia. *Water Research*, 35:9:2121–2128.

Strycker, A. & Collins, A.G. 1987. State of the Art Report: Injection of Hazardous Wastes into Deep Wells. Ada, OK: Robert S. Kerr Environmental Research Laboratory.

USEPA (US Environmental Protection Agency). 1987. *Guidance Manual on the Development and Implementation of Local Discharge Limitations Under the Pretreatment Program*. Washington, DC: USEPA.

———. 2004. *Local Limits Development Guidance*. Office of Wastewater Management 4203. EPA 833-R-04-002A. Washington, DC: USEPA.

———. *Technology Transfer Handbook: Management of Water Treatment Plant Residuals*. EPA/625/ R-95/008. Washington DC: USEPA.

USDI (US Department of the Interior). 1971. Disposal of Brine by Solar Evaporation: Field Experiment. Washington, DC: Office of saline Water, Government Printing Office.

Westbrook, G. & Wirth L. Jr. 1977. Water Management at Power Plants. *Environmental Science & Technology*, 11:2:140–143.

Yoon, S.H. 2007. *Five Year Experience With Membrane Performance Enhancers for Membrane Bioreactors: From Mechanistic Studies to Full-Scale Implementation*. WEFTEC 2007

Zander, A.K. 2000. *Current Management of Membrane Plant Concentrate*. Denver, CO: American Water Works Research Foundation.

Chapter 8

Cost of Treatment

During the initial stages of a water reuse project, decision makers need accurate cost estimates that can be used to forecast complete project costs. This information is used to obtain project funding or financing and to evaluate various water supply alternatives. In developing these cost estimates, it is important to consider that both the capital and operations and maintenance (O&M) costs are functions not only of the plant itself but also of the facilities associated with plant implementation, including various upstream and downstream facilities. The intended use of the product water is also a major factor, as potable reuse facilities are generally more costly than facilities that produce water for nonpotable uses.

While it is always difficult to generalize the factors that influence facility costs, this chapter provides a framework for developing preliminary costs by considering that actual costs for facilities must account for site-specific differences that cannot be generalized. Costs included here are broken into two major categories: the cost of potable reuse facilities and the cost of membrane-based nonpotable reuse facilities. Costs include the treatment facility only and do not incorporate costs of transmission, distribution, or injection into aquifers or reservoirs, which are beyond the scope of this manual. Assumptions about the types of treatment processes are identified, while it is understood that not all facilities use identical process trains. The framework for cost considerations presented here is a useful tool for development of site-specific capital and O&M costs for membrane-based water reuse facilities.

SUMMARIZING PROJECT COSTS

Project costs generally fall into one of two cost categories:

- Capital costs include the cost of equipment, materials, and labor to construct the treatment facility as well as costs to plan, develop, permit, design, build, and finance the project. Public outreach is often a major component of potable reuse facility development. As these costs vary drastically from project to project, they are not included in the capital costs identified here.

- O&M costs include the cost of labor, power, chemicals, repairs, replacement parts, and other consumables (for example, membrane replacement, cartridge filters) necessary for O&M of the plant. Other costs include those associated with source water purchase, residuals disposal, project administration, water quality monitoring, and permit compliance.

Typically both capital and O&M costs are evaluated to determine a total cost of water or to develop life-cycle costs for a project. These costs are discussed in more detail later in this section.

Capital Costs

Capital costs include two cost components:

- *Construction costs:* the cost of equipment, buildings, pipelines, and other physical facilities that make up the project, including the contractor's general and administrative costs, costs of bonding and insurance, profit, and financing costs. These costs are often referred to as "hard capital costs" and typically represent 60 to 80 percent of the total capital costs. For membrane-based reuse projects, the membranes and associated facilities are typically the major portion of the capital costs.
- *Owner's indirect costs:* the cost of engineering, environmental and siting issues, legal expenses, administration, and financial services needed to plan, develop, permit, and design the project. These costs are often referred to as "soft costs" and typically represent 20 to 40 percent of the total capital costs.

The construction cost of a membrane-based reuse project depends on many factors including the following:

- Source water quality (upstream treatment processes, organic content, total dissolved solids);
- Intended finished water use;
- State and local treatment requirements;
- Size (capacity) of the project;
- Need for source water equalization;
- Means of residuals disposal (membrane backwash waste, reverse osmosis [RO] concentrate, cleaning waste);
- Site development issues such as land availability, land cost, and site access;
- Geotechnical conditions;
- Architectural constraints;
- Environmental and permitting requirements; and
- Other community concerns.

This section includes cost curves that present typical capital costs from existing reuse facilities. To determine how these costs are affected by site-specific factors, reference designs for typical potable reuse and nonpotable reuse facilities are included in this section. Cost changes can be determined based on these reference designs.

O&M Costs

A plant's O&M costs includes the cost of operating and maintaining the facility to produce water for the intended use. O&M costs include cover labor, power, chemicals, repair, replacement parts, and other consumables (for example, membrane replacement, cartridge filters) necessary for plant O&M, as well as the costs associated with source water purchase, residuals disposal, project administration, water quality monitoring, and permit compliance.

O&M costs may be further broken down into two categories: fixed and variable costs. Fixed costs are independent of the amount of water treated and may range from 10 to 40 percent of the total O&M costs (excluding debt service, which is sometimes considered a fixed O&M cost). Fixed O&M costs include labor costs (including wages/salaries and fringe benefits) and administration costs.

Variable O&M costs are dependent on and vary with the amount of water that is treated at a given time. Power and chemicals are examples of variable costs. The power costs for reuse facilities vary greatly depending on the salinity of the source water, the type of plant, the unit power cost, the efficiency of plant operations, and the flow variability.

DEVELOPMENT OF CONSTRUCTION COST MODEL

Figures 8-1 and 8-2 present typical construction costs for membrane-based nonpotable and potable reuse plants in US dollars per gallon per day installed plant capacity. Capacity is defined here as the maximum daily production for which the plant is permitted. All costs are presented in 2013 dollars represented by an *Engineering News Record* construction cost index value of 9668. The costs for potable reuse projects include facilities incorporating membrane filtration (microfiltration [MF] or ultrafiltration [UF]) and RO) with the majority of the facilities also including ultraviolet light (UV) disinfection and/or advanced oxidation. To simplify the cost summaries, nonpotable reuse projects include projects with tertiary membrane filtration only and do not include membrane bioreactors or facilities that provide partial or full demineralization with RO, electrodialysis reversal, or other desalination technologies. Plants that use full demineralization would be expected to have comparable costs to potable reuse facilities. The costs for membrane bioreactor facilities are discussed in Water Environment Federation Manual 36 (WEF 2011).

The costs presented in these figures are "on-site construction" costs only. They do not include indirect costs, for example, engineering, legal, financial, permitting, and construction management, or contingencies. To the extent possible, they do not include "off-site" costs for conveying water to the site, transporting product water to offsite uses, bringing power to the site, and similar processes. The costs do not include wastewater treatment facilities such as primary and secondary treatment or nonmembrane-based tertiary treatment. Because these cost curves are based on actual costs from existing facilities, it must be considered that the unique features of each facility impact the curves' accuracy for the planning future facilities. This is particularly true at the higher capacities on the curve where the number of existing facilities is limited.

CONTRIBUTING FACTORS TO CAPITAL COSTS

The following section briefly discuss the factors that affect capital costs.

Source Water Quality

Source water quality in a reuse project is impacted by both the types of wastewater dischargers in the collection system and by the level of wastewater treatment upstream of the reuse facility. Collection systems with high contributions from industrial dischargers can have higher variability in water quality, which can be more challenging to treat. Source control programs are therefore an important feature in reducing the costs of advanced wastewater treatment facilities. The following water quality factors will have a significant impact on the cost of treatment:

- Full nitrification and denitrification in the wastewater treatment facility will reduce the fouling potential on MF–UF and RO membranes.
- Waters with higher total organic carbon (TOC) concentrations (>10 mg/L) will have higher fouling potential.
- Chemicals used in wastewater treatment, such as alum and organic polymers, can increase fouling and disinfection by-product formation, resulting in higher treatment costs for potable reuse facilities.
- Waters with higher total dissolved solids (TDS) concentrations will require more energy to treat with RO.

The temperature of the source water is also important because higher temperatures generally increase membrane permeability, reducing the feed pressure required to generate a given quantity of water. However, warmer water can also contribute to increased biological fouling on the membranes, reducing the potential energy savings from the higher permeabilities. In general, based on source water quality, the MF–UF and RO fluxes are approximately 20 to 30 percent lower in advanced wastewater treatment than in typical drinking water applications, adding significantly to the capital costs.

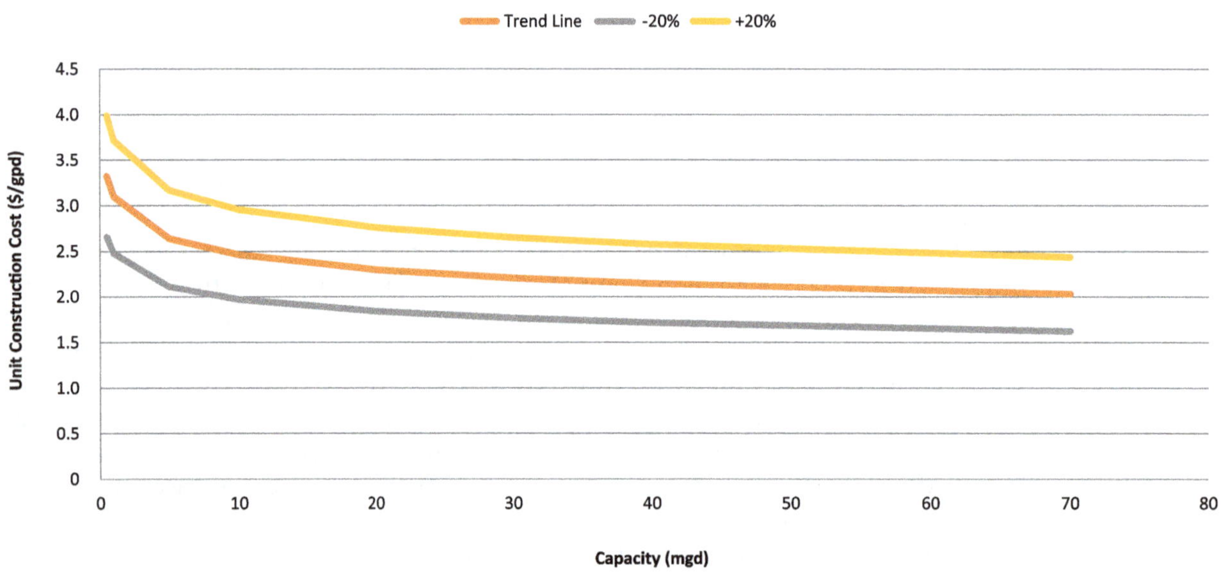

Figure 8-1 Nonpotable reuse treatment plant construction cost for a microfiltration-based facility

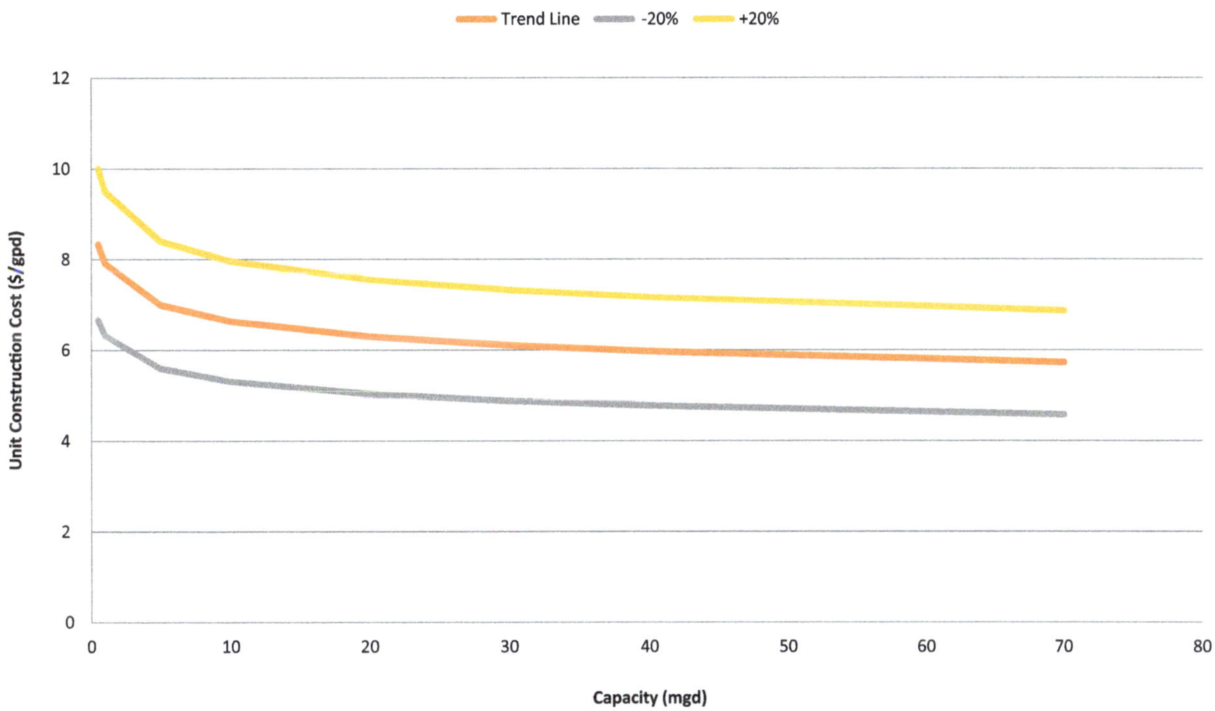

Figure 8-2 Potable reuse treatment plant construction cost for a microfiltration–reverse osmosis–based facility

Intended Finished Water Use

In general, more stringent product water quality goals increase the cost of treated water. Water treated primarily for agricultural and irrigation reuse have less stringent water treatment requirements than water treated for potable reuse or some forms of industrial reuse. For example, production of boiler makeup water at a membrane plant typically involves a two-pass RO system, where product water from the primary RO system is pumped through a second pass system, producing a water with virtually no dissolved salts. Two-pass systems may have lower overall system recovery, increased capital costs, and higher operating costs due to increased power, chemical, and membrane replacement costs compared to single-pass RO systems, which are used most commonly in potable reuse. Potable reuse facilities often have some level of treatment after RO, such as disinfection, advanced oxidation, product water stabilization, nitrogen polishing, or some combination of these post-treatment processes. The higher the level of treatment, the higher the costs that should be anticipated.

State and Local Treatment Requirements

Reuse regulations are rapidly changing throughout the country, particularly indirect and direct potable reuse regulations, which are still in developmental stages in most states. The US Environmental Protection Agency has issued regular updates of their *Guidelines for Water Reuse* (USEPA 2012), which include summaries of federal and state regulations for water reuse. A number of states, including Washington, California, Florida, and Arizona, prescribe different levels of treatment depending on the end use. This recognition, referred to as "fit for purpose," provides a framework for cost-effective treatment that is sufficient to achieve the quality appropriate for the intended use.

In the case of indirect potable reuse (IPR), the minimum product water quality goals must meet drinking water regulations as well as state-specific criteria for potable reuse, which are often more stringent than the drinking water regulations. IPR facilities often require monitoring of constituents of emerging concern (CECs), with additional treatment steps required to remove compounds not currently regulated for drinking water facilities, such as *N*-nitrosodimethylamine, 1,4-dioxane, and other CECs.

These treatment steps and the associated monitoring add considerably to treatment costs as well as the costs of preliminary project development. An example is the San Diego (Calif.) demonstration project that was used to aid in regulatory approval and public outreach for a future 15- to 20-mgd advanced water purification facility for reservoir augmentation. The project is reported to have cost $11.8 million over 3 years (Bailey 2013), or approximately 8 percent of the projected cost for the future treatment facility. While there were unique factors that contributed to the costs of the San Diego project, regulatory drivers can add substantially to capital costs at any reuse facility.

Source Water Equalization

Because of the high diurnal and seasonal variability of wastewater flows and the need to maintain relatively uniform flows in RO systems, some recently constructed indirect potable reuse facilities have incorporated flow equalization basins for the supply water. These basins hold either secondary or tertiary effluent and make it possible for the plant to maintain minimum operating flows throughout the evening, when wastewater flows often decrease. Without this flow equalization, some wastewater RO facilities experience membrane fouling or scale formation related to insufficient concentrate flows in the second or third membrane stages.

Source water flow equalization has also been used at nonpotable reuse facilities that do not include RO. In these cases, the primary purpose of equalization is often to maximize plant production with a smaller membrane capacity. This allows treatment of the average diurnal flows, without the need to install additional membrane capacity to treat higher flows during the day. Because the use of source water flow equalization is not widespread, the costs described in this chapter do not include such facilities.

Ways to Dispose of Residuals

Reuse facilities that do not include RO typically do not have residual disposal costs that differ from those for conventional wastewater treatment facilities. MF–UF backwash waste can often be recycled to the wastewater treatment plant that supplied the flow or be discharged to the nearest sewer without major impact. However, RO facility waste flows can be high in TDS and often require special considerations for disposal. Disposal costs for RO facilities are discussed in detail in AWWA Manual M46 (2007). For reuse facilities, it is important that RO waste streams not be returned to the same wastewater treatment plant that supplied the flow unless the RO plant represents only a small portion of the overall plant flow. This will prevent concentration of salts within a closed loop between the two plants. The most common methods for wastewater RO concentrate disposal are ocean discharges combined with existing wastewater outfalls, sewer disposal with flows directed to a downstream wastewater treatment facility, and surface water discharge and deep well injection, where ocean discharges are not available.

Where waste disposal options are limited, costs are typically higher and additional treatment may be required. As an example, the 2013 expansion of the Leo J. Vander Lans facility in Long Beach, Calif., involved a production increase from 3.0 mgd to 8.0 mgd, which had to be accomplished without increasing the waste flows due to sewer capacity limitations. To achieve this, the facility incorporated dissolved air flotation, a secondary MF system, and a secondary RO system, resulting in a plant with an overall hydraulic recovery of 92 percent compared to a 78 percent hydraulic recovery at the original facility. It was projected that this increased recovery requirement added approximately 15 percent to the cost of the expansion (Chalmers et al. 2013).

Site Development Issues and Geotechnical Conditions

Some site development issues are straightforward. For example, it can readily be determined whether the parcel of land on which a reuse facility is to be built is of adequate size. The availability and cost of property are easily determined. Reuse plant compatibility with surrounding land uses is also relatively easy to determine. However, it is more difficult to determine the other aspects of construction costs associated with a particular site, such as geotechnical conditions that may increase the cost of construction and community concerns such as traffic and noise.

Architectural Constraints

Architectural constraints can impose cost increases compared to an industrial building design, as reuse facilities may be located near or within residential areas, requiring significant costs for architectural features. In addition, many facilities, particularly those used for potable reuse, include public education features to maintain community support, adding to the cost of architectural features compared to a more traditional wastewater treatment facility.

Other Community Concerns

Public outreach can often be a major component of reuse facility costs, especially where recycled water is not already used. Customer and public education can be critical components in building support for the project and in making its implementation a success. Various misconceptions about reuse may need to be dispelled and replaced with accurate information about the safety, quality, and uses of the water. Costs for such public education can be substantial. The San Diego demonstration project included a reported $1.7 million for public outreach and education before design of the full-scale facility even began (Bailey 2013). Public outreach features can be incorporated into reuse facility designs, with education facilities, large meeting rooms, and safety features needed to accommodate tours; all of these things add to the construction cost of the facility.

Indirect Costs

Indirect costs typically include engineering, permitting, construction management, and legal, financial, and administrative services. Each project is unique; therefore, the magnitude of the indirect costs as a percentage of the construction cost can vary widely depending on the project's size and complexity.

A percentage of the estimated construction cost is usually assumed for indirect costs when initially planning a reuse project. Typically, indirect costs used in the planning stages of a project are 30 to 40 percent of the total project construction cost.

Contingency Allowance

Project contingency is generally a function of the level of engineering design that has been completed and the level of detail of the cost estimate. In the initial planning stages of a reuse project, there will be a number of unknowns, which are elements that have not yet been identified. Even for elements that are known, such as the plant, unexpected costs are likely to arise as the project becomes more defined. Therefore, it is necessary to include a contingency allowance in the capital cost budget for the project. The contingency allowance is larger at the beginning of the project and decreases as planning and design proceed. However, some of the contingency allowance originally included in the preliminary project budget may be reallocated to construction costs or indirect expenses as the project elements and associated costs are further defined.

The contingency allowance is typically expressed as a percentage of the estimated construction cost. In the early stages of planning, the contingency allowance may be 30 percent or more of the estimated construction cost. After the plans and specifications have been finalized, the contingency allowance may be approximately 10 percent. Even after construction bids are received for the project, a contingency allowance is necessary to cover the unexpected costs that arise on essentially all construction projects. The contingency allowance must be individually assessed for each project, taking into account project complexity and level of design.

DEVELOPMENT OF AN O&M COST MODEL

With respect to O&M costs, both fixed and variable costs must be planned for during project development. Labor is typically considered a fixed O&M cost, and while wage levels change with time, labor costs are not directly impacted by the amount of water that is treated. Another fixed O&M cost is membrane replacement. In general, membranes do not age more rapidly when they are in operation than when they are in standby mode. For example, if a membrane plant is operated only 50 percent of the time, it is not assumed that the membrane life would be twice as long as it would be if the plant was operated 100 percent of the time. In addition, some allowance should be included for repairs and replacement parts as a fixed O&M cost for planning purposes; an allowance of 1 to 2 percent of the plant construction cost is typically included to cover this item. There may be other fixed O&M costs specific to a given project.

Variable O&M costs are those costs that directly depend on the amount of water that is treated. Variable costs include power, chemicals, UV lamp replacement, cartridge filter replacement, concentrate disposal, and other miscellaneous costs. The total annual O&M cost is the sum of the fixed and variable O&M costs in dollars per year. Breakdowns for typical O&M costs at nonpotable and potable reuse facilities are shown in Figures 8-3 and 8-4. These costs are based on a hypothetical 10-mgd facility and show a relative distribution of costs between different O&M components.

Fixed O&M Costs

Fixed O&M costs include labor, membrane replacement, and typical repairs and replacement.

Labor. The cost for labor is dependent on a number of factors, including the following:

- *Size (capacity) of the plant.* A small plant may have only one or two permanent staff, with unattended operation during some periods of the day. In contrast, larger plants typically require operators to be in attendance at all times to keep up with routine maintenance, record keeping, and similar tasks.
- *Complexity of the plant.* If the treatment process is complicated, it may be necessary to have a larger operating staff to cover all the skills needed for plant operation. Because of the treatment complexity of potable reuse facilities, the staffing requirements are often quite high, averaging between one and three full-time equivalents per million gallons per day of production capacity.
- *Regulatory agency requirements.* Local or state regulatory agencies may have minimum staffing requirements for a potable reuse plant. Requirements for nonpotable reuse are often less stringent.

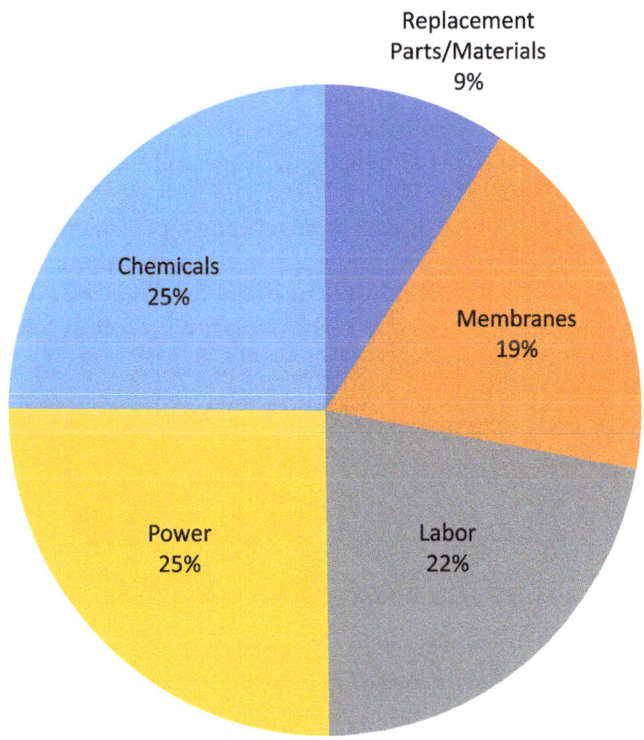

Figure 8-3 Typical nonpotable reuse treatment plant operations and maintenance cost breakdown for a microfiltration-based facility

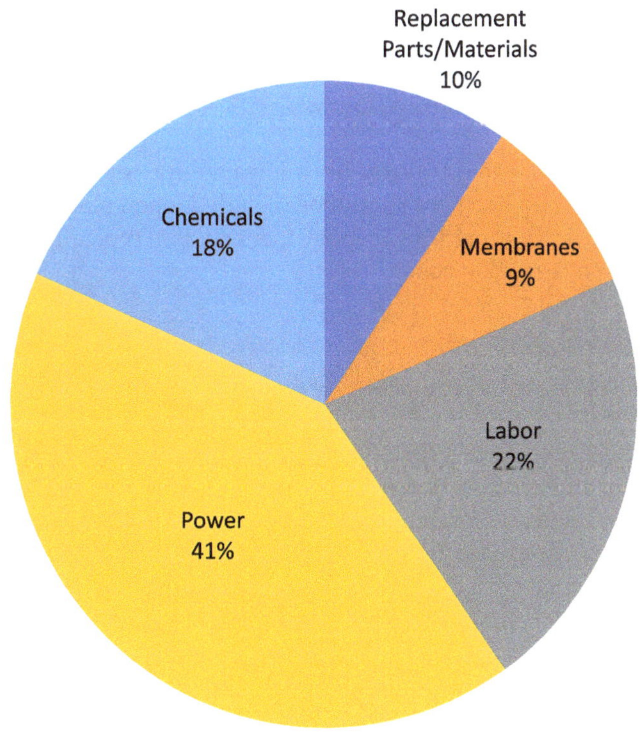

Figure 8-4 Typical potable reuse treatment plant operations and maintenance cost breakdown for a microfiltration–reverse osmosis–based facility

- *Owner's policy.* Some plant owners may want to have at least one operator at the plant around the clock. Another owner may want to have as small an operating staff as possible, with the plant operating unattended at times.
- *Local wages and benefits.* Some areas, particularly in or near densely populated metropolitan areas, may have higher wage and benefit costs than more rural areas.

Membrane replacement. The cost of membrane replacement is usually regarded as a fixed O&M cost because once a membrane is placed in operation, it is assumed that membrane degradation begins. A membrane replacement fund may be established to cover the cost of membrane replacement, as needed. The two primary factors to consider when establishing a membrane replacement fund are:

- Total cost of the membranes and
- Expected average life of the membranes.

Determining the initial cost of the membranes depends on several factors, including:

- Capacity of the membrane process,
- Membrane design flux, and
- Membrane system recovery.

Estimating the annual membrane replacement cost is a function of the number of membranes installed, the projected membrane replacement cost, and the anticipated useful life of the membranes. The useful life primarily depends on the feedwater quality and the aggressiveness of the design operating conditions. Early reuse facilities often assumed a membrane life of 3 to 5 years; however, most facilities have been operating 7 to 10 years between membrane replacements. Operating costs tend to increase as the membranes age, requiring more frequent cleaning and higher transmembrane pressures. This may drive a more frequent replacement schedule for some plants; however, a projected membrane life of at least seven years is typical today.

Repairs and replacement parts. An allowance for repairs and replacement parts other than membrane replacement should be included when planning a membrane-based reuse project. The cost for repairs and replacement parts is typically assumed to be 1 to 2 percent of the plant construction cost.

Variable O&M costs

Variable O&M costs include power, chemicals, and other consumables.

Power. Power costs vary substantially based on the types of processes used and the level of treatment achieved. In nonpotable reuse facilities without RO, the primary power uses are for membrane feed or permeate pumps and for aeration during membrane backwash, if used. For potable reuse facilities, the RO feed pumps typically account for roughly half of the overall power use, with membrane filter pumps and UV systems each accounting for 5 to 10 percent of the total power. Higher salinity water requires more power to treat with RO, and water with high fouling potential (such as a nonnitrified source water) results in higher power costs for all of the membrane processes. Finished water pumping may also be a substantial cost depending on the distance and elevation that the water must be pumped to.

It is important to obtain as accurate estimate of power costs since power can represent 25 to 50 percent of the overall O&M costs. Most of the power cost is a variable cost;

however, a small percentage is a fixed cost because even when the system is not operating or is operating at reduced capacity, there will be some power demand for building loads and site lighting.

Chemicals. The types and amounts of chemicals required in a reuse plant vary significantly. Chemicals are required for the following purposes:

- *Preliminary disinfection.* It is common to provide a continuous chloramine residual through the membrane systems to prevent biological growth on the MF–UF and/or RO membranes. Depending on ammonia concentration in the source water, this is done by injection of ammonium hydroxide and sodium hypochlorite or by injection of only sodium hypochlorite if there is sufficient ammonia in the source water. The typical chloramine residual concentration is between 2 and 4 mg/L before the RO membranes. Some facilities also use sodium bisulfite to quench residual chlorine upstream of the RO membranes in the event that concentrations exceed allowable limits from the membrane vendors.
- *RO feedwater.* Historically, it was common to add a scale inhibitor and an acid ahead of the RO membranes to inhibit scale formation and reduce the buildup of silica, calcium phosphate, and other scalants on the membranes. Some wastewater RO facilities have been operating successfully without acid or with limited acid doses. The need for acid depends on the water quality characteristics for a given plant.
- *Advanced oxidation.* Potable reuse facilities typically require advanced oxidation for removal of trace organic compounds. This is commonly accomplished by adding 2 to 3 mg/L of hydrogen peroxide upstream of high-dosage UV reactors.
- *Product water stabilization.* Wastewater RO plants typical reduce hardness and alkalinity to very low levels, which results in a relatively high corrosion potential of RO permeate. For these plants, it is common to add hardness and alkalinity to stabilize the RO permeate. Lime (calcium hydroxide) or calcium chloride is commonly added to increase the hardness of the finished water. Caustic soda is commonly used for pH adjustment when lime addition is insufficient for meeting the target pH.
- *Membrane cleaning.* MF–UF and RO membranes require periodic cleaning. The frequency of cleaning and the type of chemicals required vary for each plant. Chlorine-tolerant MF and UF membranes are commonly cleaned using a high pH sodium hypochlorite solution and a low pH citric acid solution. RO membranes and MF membranes that are not chlorine tolerant are cleaned using high pH and low pH solutions, typically caustic soda and either citric or sulfuric acid. In addition, various types of detergents and proprietary cleaning chemicals may be used from time to time with any type of membrane.

The cost for chemicals depends on the quality and quantity of the water being treated and the product water quality goals. Chemical consumption data from similar plants can be used to estimate chemical treatment needs and costs for a new facility. However, product water stabilization requirements are very site specific and should be estimated using water quality modeling for the specific source water. Water quality projections for RO permeate can be obtained using membrane vendor projection software.

Bench-scale chlorine demand tests can be conducted relatively inexpensively. However, to obtain accurate site-specific chemical demand data, it may be necessary to perform pilot testing to evaluate the fouling potential, projected cleaning frequency, and recommending cleaning chemicals for a membrane-based reuse plant.

Other consumables. Costs to replace cartridge filters and UV lamps and ballasts are usually considered variable costs, as their replacement frequency depends on the frequency of use and on the water treated. UV lamp and ballast replacement frequency varies by manufacturer and the type of UV unit used. Cartridge filter replacement is more variable and depends on the water's fouling potential, which can range from biweekly to twice a year. Cartridge filter replacement downstream of MF–UF treatment tends to be infrequent, since the majority of the potential foulants have been removed by the upstream membrane process. More frequent cartridge filter replacement is common at facilities that struggle with biofouling (Fuchs, Tirado, & Owens 2013).

COST OF WATER

The cost of water is calculated as follows: annualized capital cost plus annual O&M costs divided by the volume of water produced during the year. It is usually expressed as dollars per unit volume, for example, dollars per thousand gallons, dollars per cubic meter, or dollars per acre-foot.

Total treatment costs are dependent on the site and application; cost of source water; treatment technologies used; plant facilities constructed; power, labor, and chemical costs; annual water production; and many other factors. Figures 8-5 and 8-6 present typical costs for treated water from nonpotable and indirect potable reuse facilities (2013 dollars). Treated water costs for a 5-mgd membrane-based tertiary filtration facility range from $0.9/kgal to $1.3/kgal. Recent cost estimates for large plants (40 mgd and larger) indicate that costs range from $0.8/kgal to $1.2/kgal. In contrast, costs for a 5-mgd indirect potable reuse facility range from $2.8/kgal to $4.1/kgal, while costs for a large plant (40 mgd and larger) range from $2.6 to $3.9/kgal.

Caution should be used when extrapolating the cost data presented here to a specific current or future project. This is particularly true for larger-scale projects, as the number of operational facilities used to develop these cost curves is quite limited. These costs are presented to give the reader an approximate idea of costs and detailed estimates should be made as part of any membrane-based reuse plant feasibility analysis.

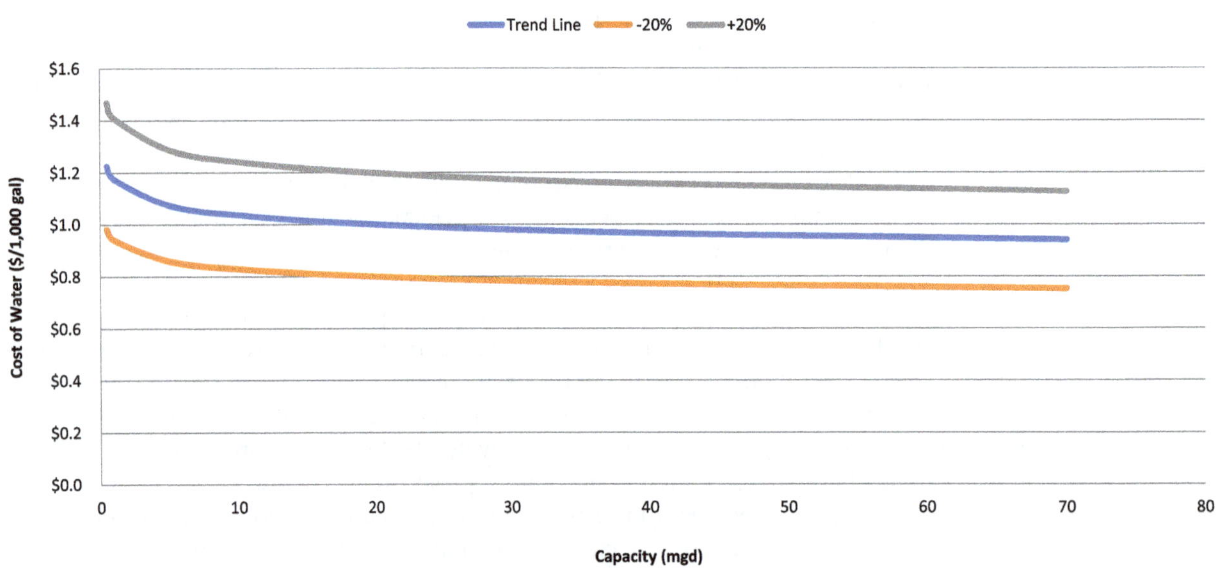

Figure 8-5 Nonpotable reuse treatment plant cost of water for a microfiltration-based facility

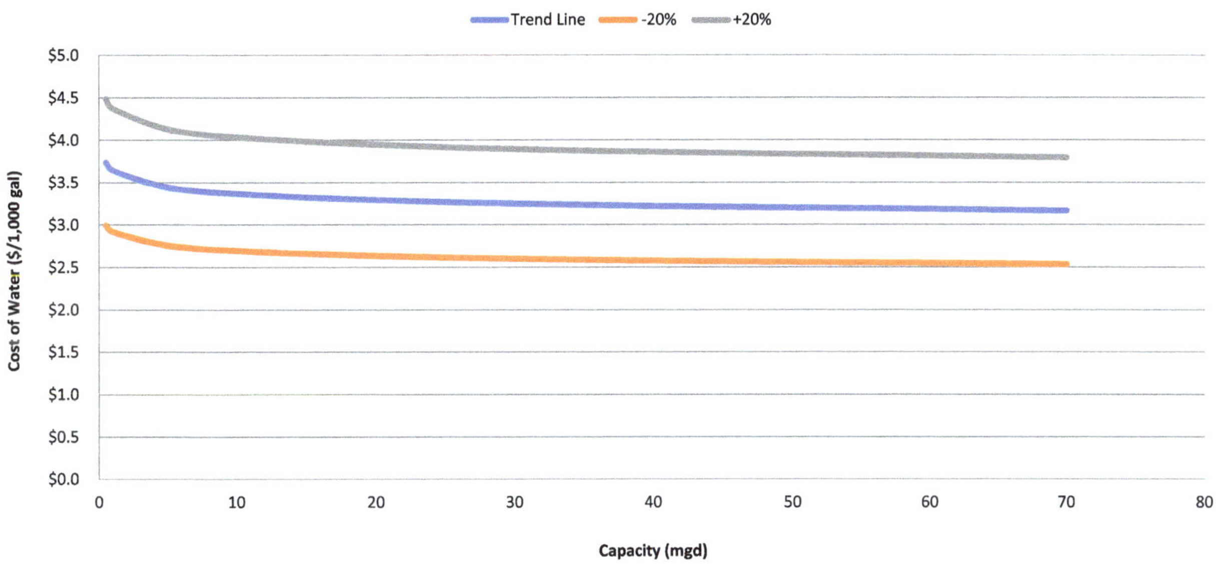

Figure 8-6 Potable reuse treatment plant cost of water for a microfiltration–reverse osmosis–based facility

EXAMPLE COST ESTIMATES

Tables 8-1, 8-2, and 8-3 present how a preliminary cost estimate for a potable reuse and nonpotable reuse project might be presented. The examples are based on the following:

- Membrane filtration and chlorine disinfection for nonpotable reuse facilities;
- Membrane filtration, RO, advanced oxidation with UV, and post-stabilization with calcium chloride and caustic soda for indirect potable reuse facilities;
- Product water design capacity = 10 mgd;
- Annual product water production rate based on:
  - 100% of rated plant capacity = 3,650 mgy
  - Online factor of 90% = 3,285 mgy;
- Membrane filtration recovery = 95%;
- RO recovery = 85%;
- Secondary effluent source water with TDS between 800 and 1,500 mg/L, TOC between 7 and 12 mg/L, and temperature between 20 and 25°C; and
- Average electrical costs of $0.10/kWh.

Table 8-2 illustrates how O&M costs for these plants might be presented. Two online factors are included to show the impact on the cost of water of spreading the fixed costs of O&M over a greater volume of water treated. As shown in the table, the variable O&M costs were adjusted to reflect the fact that if more product water is produced, the annual costs for power and chemicals will increase correspondingly. Annual fixed costs do not change. The result is that the unit O&M cost per thousand gallons of water produced at the annual production rate of 3,650 mgy is lower than the unit O&M cost for producing only 2,920 mgy from this facility.

Table 8-3 compares the cost of water for the same annual production rates. The annual capital repayment cost is constant as are the fixed O&M costs. Only the variable O&M costs change with the change in annual production rate. As demonstrated with the cost of water values in Table 8-3, producing more water from the reuse facility reduces the cost of water.

The annual capital cost was calculated using an interest rate of 5 percent and a 20-year amortization period.

Table 8-1 Membrane-based reuse treatment plant capital cost examples, with plant capacity of 10 mgd

Parameter	Potable	Nonpotable
Direct capital (construction) costs		
1. Site civil and yard piping	$4,000,000	$1,600,000
2. Influent pumping	$1,200,000	$1,000,000
3. Membrane filtration system	$14,500,000	$12,300,000
4. Break tank and transfer pumps	$1,200,000	N/A
5. RO system equipment	$13,000,000	N/A
6. UV advanced oxidation	$5,500,000	N/A
7. Chemical storage and feed	$2,600,000	$800,000
8. Finished water storage and pumping	$2,000,000	$2,000,000
9. Administration, operations, and laboratory building	$2,500,000	N/A
10. Electrical and instrumentation	$8,800,000	$3,200,000
Subtotal direct capital (construction) costs	$55,300,000	$20,900,000
Construction contingency (30%)	$16,600,000	$6,300,000
Total Direct Costs	$71,900,000	$27,200,000
Indirect capital costs		
1. Engineering and construction management	$14,400,000	$5,400,000
2. Administration, contracting and legal	$3,600,000	$1,400,000
3. Environmental permitting	$7,200,000	$2,700,000
Total Indirect Capital Costs	$25,200,000	$9,500,000
Total Capital Costs	$97,100,000	$36,700,000
Unit Cost, $/gpd	$9.71	$3.67

Abbreviations: N/A, not applicable; RO, reverse osmosis; UV, ultraviolet.

Table 8-2 Membrane-based reuse treatment plant operations and maintenance cost examples, with plant capacity of 10 mgd

Parameter	Microfiltration–Reverse Osmosis	Microfiltration–Ultrafiltration
Fixed costs		
1. Operation and maintenance labor	$1,500,000	$375,000
2. Replacement parts and materials	$663,000	$246,000
3. Replacement membranes	$638,000	$319,000
Variable costs		
4. Power	$2,873,000	$433,000
5. Chemicals	$1,277,000	$426,000
Annual production at rated capacity, mgy	3,650	3,650
Annual O&M cost at rated capacity	$6,951,000	$1,799,000
Unit cost at rated capacity, $/kgal	$1.90	$0.49
Average day demand/rated capacity	80%	80%
Annual production at average day demand, mgy	2,920	2,920
Annual O&M cost at average day demand	$6,121,000	$1,627,000
Unit cost at average day demand, $/kgal	$2.10	$0.56

Table 8-3 Membrane-based reuse treatment plant cost of water, with plant capacity of 10 mgd

Parameter	Potable	Nonpotable
Production costs at rated capacity		
1. Equivalent annual capital cost	$7,790,000	$2,940,000
2. Annual O&M cost–fixed	$2,801,000	$940,000
3. Annual O&M cost–variable	$4,150,000	$859,000
Annual production at rated capacity, mgy	3,650	3,650
Total annual cost at rated capacity	$14,741,000	$4,739,000
Annual production cost at rated capacity, $/kgal	$4.04	$1.30
Average day capacity/rated capacity	80%	80%
Annual production at average day capacity, mgy	2,920	2,920
Total annual cost at average day capacity	$13,911,000	$4,567,000
Cost of water at average day capacity, $/kgal	$4.76	$1.56

Abbreviation: O&M, operations and maintenance.

REFERENCES

AWWA. 2007. M46, *Reverse Osmosis and Nanofiltration*. Denver, CO: AWWA.

Bailey, R. 2013. The City of San Diego Report to the City Council, March 11, 2013. Report No. 13-27.

Chalmers, R.B.; Fu, P.; & Cutler, D. 2012. Expanding the Leo J. Vander Lans Water Treatment Facility From 3 MGD to 8 MGD With No Additional Waste Flows. WateReuse Association Annual Conference, Hollywood, FL.

Chalmers, R.B.; Wetterau, G.; You, E.; & Alexander, K. 2013. Increasing the Overall Recovery Rate at the Leo J. Vander Lans Water Treatment Facility to Greater than 92%. AMTA/AWWA Membrane Technology Conference.

Fuchs, F.; Tirado, K.; & Owens, E. 2013. When Replacing Modules Just Isn't Enough: West Basin's Microfiltration System Replacement Evaluation. AMTA/AWWA Membrane Technology Conference.

Gerringer, F. 2011 Ozone Pretreatment of Non-nitrified Secondary Effluent Before Microfiltration. California Water Environment Association Annual Conference, April.

Patel, M. 2010. RO CIP and Energy Optimization for the GWRS Project. WEF Membrane Applications Conference.

USEPA (US Environmental Protection Agency). 2012. *2012 EPA Guidelines for Water Reuse.* EPA/600/R-12/618. Cincinnati, OH: USEPA.

Water Environment Federation. 2011. *Membrane Bioreactors WEF Manual of Practice No. 36.* New York, NY: McGraw Hill.

Won, W.; Walker, T.; Patel, M; & Owens, E. 2010. Comparing Membrane Operations at Three of the World's Largest Advanced Water Treatment Plants. *IDA Journal*, 2:3:10.

This page intentionally blank.

Chapter 9

Case Studies

The case studies presented in this chapter represent a broad sampling of water reuse applications in which low- and high-pressure membrane processes play an important and integral part of the overall treatment system. In all cases, the raw water (reuse facility feedwater) is secondary effluent from a municipal wastewater treatment plant. With one exception (CAPCO Industrial Wastewater Reclamation Facility), the treatement plants described herein can be classified as potable reuse facilities in that they are capable of producing a treated water that, when blended with naturally occurring source waters (either groundwater or surface water), can or does serve as feedwater to drinking water plants.

The case studies are organized by the intended end use of the treated water, grouped under the following categories: indirect potable reuse, including groundwater replenishment and surface water augmentation; direct potable reuse (supplementation of drinking water treatment plant raw water supply); multiuse, where the intended end use includes indirect potable reuse and industrial process water; and industrial process water exclusively.

Given the number of case studies included and the extent of information presented on each, tables that contain information on feedwater and treated water quality as well design criteria for major treatment processes are presented in the appendices at the end of this chapter in order to manage the length of this chapter and improve readability.

CASE STUDY: THE GROUNDWATER REPLENISHMENT SYSTEM

By Mehul Patel, P.E., Orange County Water District

The Orange County area of southern California is subject to a semi-arid climate and has depended on imported water from northern California and the Colorado River for much of its drinking water needs. With ongoing droughts, environmental constraints, and increasing costs of imported water, new sources of water are needed to support a growing population. It is estimated that the population could grow by 7 million in southern California and by 300,000 in Orange County by 2020. In the search for new water supplies, the

Orange County Water District (OCWD) has sought water recycling as a key component to diversify its water supply and meet future needs.

The OCWD is a groundwater management agency created by the State of California in 1933 to manage the use of and protection for the Orange County groundwater basin. This basin provides the majority of drinking water needs for the central and northern portions of Orange County. In the mid-1970s OCWD began using recycled water to protect the groundwater basin through a facility known as Water Factory 21. The facility was built by OCWD and operated from 1976 to 2004. This facility took secondary effluent wastewater and purified it through a conventional drinking water treatment process along with a reverse osmosis (RO) membrane system. Water Factory 21 was capable of producing 15 mgd of purified water for injection into wells along the coast to form a seawater intrusion barrier. However, in the late 1990s, the facility was near the end of its useful life. The state-of-the-art technology developed in the mid-1970s had been replaced by newer, more cost-efficient technologies. OCWD looked to replace Water Factory 21 with a larger facility that could not only meet the needs of existing seawater intrusion barrier demands but could also provide a means of replenishment to the groundwater basin.

In the early 1990s, OCWD and the Orange County Sanitation District (OCSD) joined efforts to develop the Groundwater Replenishment System (GWRS). The GWRS is a joint project developed to help supply a new source of water for OCWD to replenish and protect the Orange County groundwater basin. The GWRS, which went into operation in January 2008, consists of three major components: the advanced water purification facility (AWPF), the GWRS pipeline, and expansion of the existing seawater intrusion barrier. The locations of these three components are shown in Figure 9-1. The AWPF is the main component of the GWRS. The AWPF process train was developed by OCWD using the best available technologies to deal with a large suite of organic and inorganic compounds. This is considered an indirect potable reuse project since the purified water is introduced to the groundwater basin to become a component of the drinking water supply. The AWPF treats secondary effluent wastewater from the OCSD through a three-step process that consists of microfiltration (MF) followed by RO followed by ultraviolet (UV) light with hydrogen peroxide to form an advanced oxidation process (AOP). A detailed process flow diagram is shown in Figure 9-2.

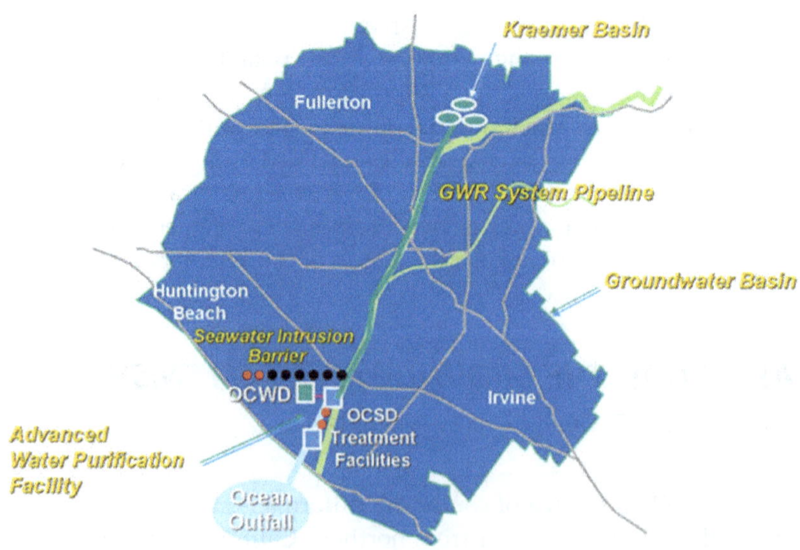

Figure 9-1 Groundwater Replenishment System components

Abbreviations: GWR, groundwater replenishment; OCSD, Orange County Sanitation District; OCWD, Orange County Water District.

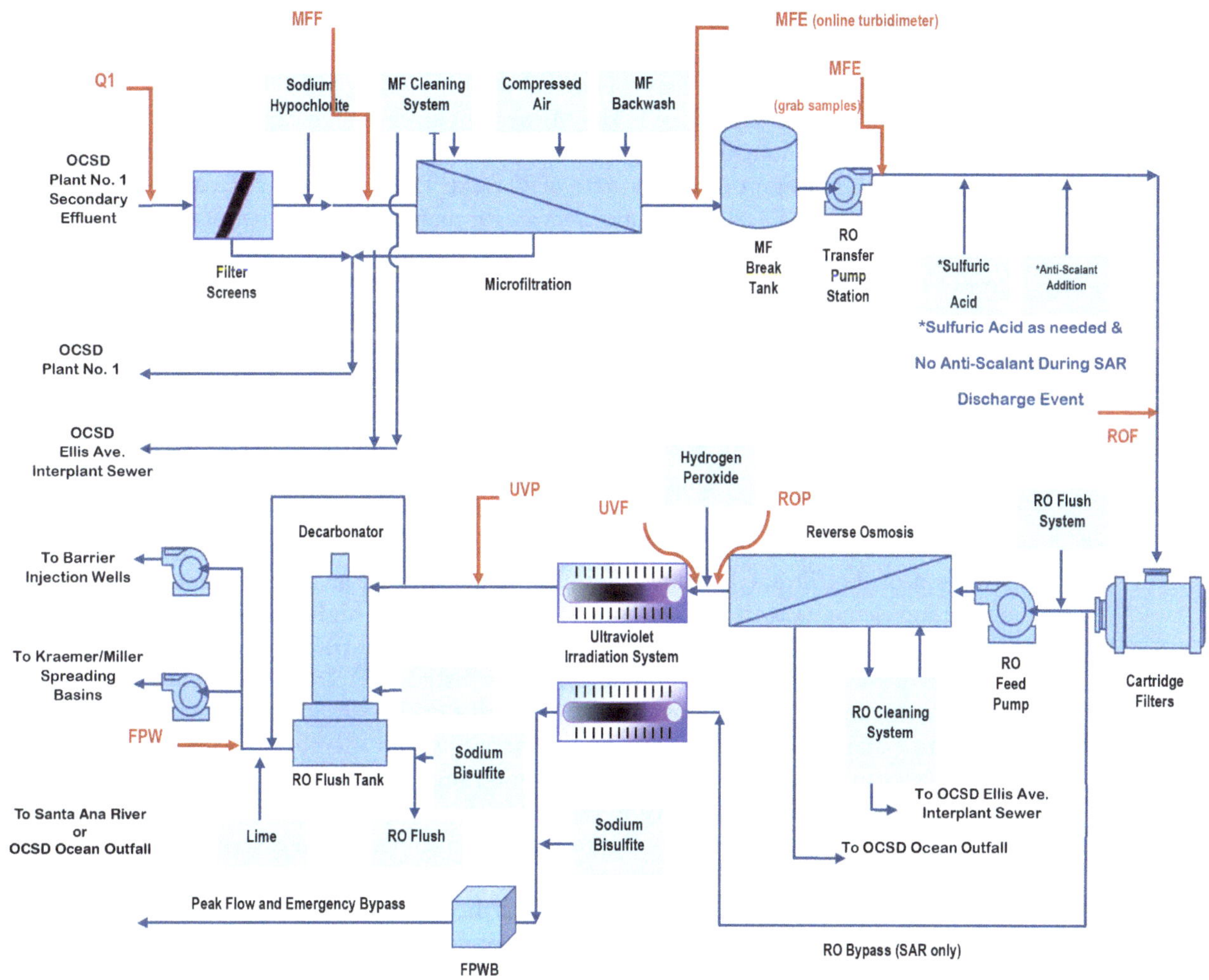

Figure 9-2 Groundwater Replenishment System process flow diagram and sampling points

Three major treatment systems make up the AWPF: microfiltration (MF), RO, and AOP. However, the first treatment processes at the AWPF are the pretreatment systems that include enhanced OCSD source control and clarified secondary treatment, fine screening, and chloramination. Clarified secondary effluent from OCSD plant 1 travels by gravity to the screening facility and to the MF system. The screening facility has four rotating 2-mm gravity screens to remove larger particles from the secondary effluent ahead of the MF process. Upstream screening is vital to MF system performance and is required as part of the MF membrane manufacturer warranty conditions. During normal operation, the MF system requires upstream chloramination, which helps to extend the interval between full MF chemical cleanings and decreases the rate of biofouling in the first stage of RO. Chlorine in the form of sodium hypochlorite is mixed into the screened secondary effluent, forming chloramines from the abundant ammonia in the secondary effluent. Chloramines are necessary to minimize biological fouling of the MF membranes without oxidizing the membranes, which are made of polypropylene and are susceptible to oxidation by free chlorine. The MF system treats the screened secondary effluent to a filtrate water quality that has no significant particulate materials as indicated by low turbidity (<0.2 ntu) and a 15-minute silt density index below 3 at a suitable cleaning interval.

Key design criteria for the MF system are listed in Appendix A, Table A-1. The four units comprise three trains of eight cells each and a fourth train of two cells. Pore size of the hollow fiber membranes is 0.2 μm. The backwash from the MF system is returned to the primary clarifiers at OCSD plant 1. The MF system has a design capacity of 100 mgd when operated at an instantaneous design flux. Approximately 82 mgd is required to meet the downstream RO design production rate of 70 mgd. The additional MF capacity is needed for occasions when MF cells are out of service for maintenance or chemical cleaning.

The second component of the AWPF process is the RO system. This system includes the RO transfer pump station, RO pretreatment chemical addition, cartridge filtration, high-pressure membrane feed pumps, RO treatment trains, flushing systems, and clean-in-place systems. The RO transfer pump station pumps MF effluent from the MF filtrate tank through the cartridge filters to the RO high-pressure membrane feed pumps. The pretreatment chemical addition includes facilities for the addition of sulfuric acid and a proprietary threshold inhibitor to the MF filtrate for pH adjustment and for prevention of sparingly soluble salt precipitation on the membranes. Cartridge filters with a nominal 10-μm rating are located upstream of the RO membranes to ensure removal of suspended solids that may be introduced through pipeline residuals or impurities in the chemical treatment of the filtrate, thereby providing an additional level of protection for the RO membrane elements. After the cartridge filters, the in-line high-pressure pumps boost the pressure of the pretreated MF effluent to the RO treatment trains.

Key design criteria for the RO system are listed in Appendix A, Table A-2. The 15 trains are arranged in 5 groups of 3 with 14 units on line at any one time. The 15th unit serves as a standby. Each vessel in a unit houses seven membrane elements. Each unit has a rated capacity of 5 mgd, operating over a range of flows from 4 to 5 mgd. Each unit operates at a recovery of 85 percent, with all concentrate piping sized to accommodate a recovery of as low as 80 percent. The reject stream (concentrate) is sent to OCSD where it is combined with secondary effluent and disposed of in the OCSD ocean outfall. The RO permeate is sent to the UV–AOP system for further treatment. Tables A-1 and A-2 provide a summary of key design criteria for both membrane processes.

The final components of the AWPF are the post-treatment systems, which include advanced oxidation (UV treatment with hydrogen peroxide), decarbonation, lime stabilization, and product water pumping. Direct photolysis (UV alone) provides disinfection and is also effective in reducing levels of *N*-nitrosodimethylamine (NDMA), which is a low-molecular-weight organic (LMWO) that is not fully removed by RO. Since direct photolysis is not effective for all LMWOs, approximately 3 mg/L of hydrogen peroxide is added upstream of the UV system for hydroxyl radical generation. These hydroxyl radicals are more effective for reduction of contaminants such as 1,4-dioxane. The AWPF UV facility consists of nine trains; each train is designed to treat 8.75 mgd. A Trojan UVPhox system is used for the UV facility. The UV system is designed to provide a 4-log reduction of viruses and a 1.2-log reduction of NDMA. The UV system uses low-pressure/high-output lamp technology to treat the RO permeate. After treatment by the RO and UV systems, the product water still has a high carbon dioxide content created by the pH reduction ahead of the RO system. The AWPF uses a decarbonation system that consists of blowers and pack towers to partially strip the carbon dioxide from the water to an optimal level. After partial decarbonation, the UV product is stabilized with hydrated lime. The stabilized product water is then pumped to the seawater intrusion barrier and to groundwater recharge basins located 14 miles away from the plant site to supplement other sources.

The GWRS is governed by an operating permit issued by the California Regional Water Quality Control Board. The parameters in the permit are determined by the California Department of Public Health (CDPH). The GWRS falls under the Draft Recharge Criteria regulations set by CDPH. The key regulatory criteria7that sets the performance goals for the AWPF is a finished product water limit for total organic carbon of 0.5 mg/L and a total nitrogen limit of 5 mg/L. Appendix A, Table A-3, lists typical water quality achieved by the GWRS for each major treatment step. The table also shows the permit requirements that are based on the finished product water location.

The GWRS includes multiple components, as mentioned previously. Among those components are a distribution pipeline and extension of an existing seawater intrusion barrier. The existing seawater intrusion barrier was expanded from 28 to 36 injection well sites. A breakdown of the total capital cost of the GWRS is shown in Table 1. A portion of the capital cost was offset by the receipt of approximately $100 million in state and federal grants. In addition, the net capital cost that is left after subtracting the grant money was split equally between OCWD and OCSD. OCSD agreed to pay the net capital costs to ensure the GWRS would always take a minimum amount of secondary effluent produced by their wastewater treatment plant. The OCSD ocean outfall currently in place is undersized in high-flow conditions. Construction of the GWRS relieved the capacity issues associated with the existing ocean outfall pipeline. OCSD believed that investing the nearly $200 they originally set aside for an additional ocean outfall capital improvement project into the GWRS project made fiscal sense. It also provided OCSD with a more environmentally responsible solution to their disposal capacity issues.

The GWRS AWPF operating costs are tracked closely and reported on an annual basis. The key operating costs are for power, labor, chemicals, maintenance, replacement and refurbishment, and capital debt payback. Table 9-2 summarizes the yearly operating costs for the most recent fiscal year of 2011–2012. The MF membranes have a warranty period of seven years from the manufacturer, and the UV lamps have a warranty period of 12,000 hours. Replacement of the MF membranes, RO membranes, and UV lamps make up the majority of replacement and refurbishment costs shown in Table 9-2. OCWD invests yearly into a replacement and refurbishment fund to maintain enough money for large replacement expenditures associated with the membrane and UV processes. Also, the GWRS receives a partial operating subsidy each year from the Metropolitan Water District and an additional cost credit for participation in an electrical load shedding demand response program.

Table 9-1 Groundwater Replenishment System capital cost breakdown

Facility	Cost
Microfiltration and ultraviolet treatment vendor design services	$860,000
Advanced water purification facility plant	$298,330,000
Groundwater Replenishment System pipelines	$63,010,000
Barrier facilities (8 well sites)	$16,380,000
Miscellaneous projects*	$20,490,000
Other	$14,950,000
Engineering, legal, and administration	$66,890,000
Total project cost (2008)	$480,900,000

*Projects include a 5-mgd temporary treatment plant to continue operation of water facility-21 during advanced water purification facility construction, Southern California Edison substation, and project management trailer complex.

Table 9-2 Groundwater Replenishment System operating costs

Cost Category	Cost 2011–2012	Cost per Acre-Foot*	Cost per 1,000 Gallons
Electricity	$7,815,069	$109	$0.33
Chemicals	$4,246,328	$59	$0.18
Labor (32 percent benefits)	$7,897,139	$110	$0.34
Maintenance	$3,197,130	$45	$0.14
Resource and Recovery fund contribution	$5,151,972	$72	$0.22
Debt service	$11,499,996	$160	$0.49
Metropolitan Water District Local resources program subsidy	($8,084,408)	($113)	($0.35)
Demand response credit	($1,273,050)	($18)	($0.06)
Total	$30,449,988	$478	$2.11

* Based on production of 66,084 acre-feet (21,543 mil gal) of which 66,815 (21,782 mil gal) is eligible for the Metropolitan Water District subsidy.

The GWRS project is an example of a successful partnership between two large public agencies to address a common issue of water availability in Orange County, Calif. The project was successful due, in part, to a large-scale continuous public outreach campaign to avoid any misperceptions on the nature and source of the water produced. In addition, the plant design and operational costs have made the water produced by GWRS cost competitive with other available sources such as imported water. This has, in turn, allowed the project to be a model for implementation of indirect potable reuse schemes on a large scale.

PROJECT PHOTOS

Figure 9-3 Aerial view of the Groundwater Replenishment System and adjacent Orange County Sanitation District facility

Figure 9-4 Groundwater Replenishment System facility entrance

Figure 9-5 Microfiltration system

Figure 9-6 Reverse osmosis treatment trains

Figure 9-7 Ultraviolet light–advanced oxidation process treatment facility

CASE STUDY—BROAD RUN WATER RECLAMATION FACILITY

By Tom A. Broderick, Loudoun Water

The Loudoun County Sanitation Authority (doing business as Loudoun Water) provides water and sewer services to a rapidly growing population in the northern Virginia suburbs of Washington, D.C. Sewage flows from the service area have historically been conveyed through an interceptor system to Washington, D.C., for treatment at a large regional facility, the Blue Plains Advanced Wastewater Treatment Plant. Loudoun Water is limited by agreement to an average daily allocation of 13.8 mgd of treatment capacity at this regional facility. When growth projections for the service area indicated that additional treatment capacity would be needed by 2008, Loudoun Water began planning for the design and construction of its own treatment facility.

Loudoun Water subsequently constructed the Broad Run Water Reclamation Facility (WRF) to treat flows in excess of its allocation at the regional facility. The initial capacity of the facility is 11.0 mgd, and the ultimate capacity of the facility may exceed 30 mgd. The completed $190 million facility began operations and surface water discharge in 2008.

Virginia Regulation 9VAC25-401-10, titled Sewage Treatment in the Dulles Area Watershed (also known as the Dulles Area Watershed Policy [DAWP]), governs discharge requirements to Broad Run, which is the immediate receiving stream for the Broad Run WRF, and the entire watershed. This watershed is located upstream of a major drinking water intake on the Potomac River. The DAWP was developed to protect both instream water quality and drinking water sources within the watershed and requires an extremely high level of treatment. Extensive pathogen control at the facility was also needed to protect downstream drinking water supplies from currently targeted and emerging pathogens such as pathogenic bacteria, viruses, and the parasitic protozoa *Giardia* and *Cryptosporidium*. Treatment processes at the Broad Run WRF also had to address other regional goals such as protecting the Chesapeake Bay through the control of nutrients.

Table 9-3 summarizes discharge limits for selected parameters for the Broad Run WRF. Monthly effluent concentration limits of 0.1 mg/L for total phosphorus (TP), 1.0 mg/L for total suspended solids, 1.0 mg/L for total Kjeldahl nitrogen, and 4 mg/L for total nitrogen (TN) are specified. A discharge limit of 10 mg/L is specified for chemical oxygen demand (COD), with no limit for 5-day biochemical oxygen demand (BOD_5). The COD limit is intended to ensure that a method for removing synthetic organic chemicals be provided, while the absence of a BOD_5 limit reflects the fact that compliance with the COD limit results in an effluent BOD_5 below detection limits. Annual effluent loading requirements are also specified for TP and TN.

Table 9-3 Permit requirements for the Broad Run Water Reclamation Facility

Parameter	Discharge Requirements
Total suspended solids, mg/L	1
Chemical oxygen demand, mg/L	10.0
Total Kjeldahl nitrogen, mg/L	1.0
Total nitrogen, mg/L (concentration limit for the calendar year)	4.0
Total phosphorus, mg/L (concentration limit)	0.1
Turbidity, ntu	0.5
Coliform, per 100 mL	<2
Total phosphorus, kg/y (annual loading of 0.1 mg/L)	1,519
Total nitrogen, kg/y (annual loading of 4.0 mg/L)	60,784

PLANT DESCRIPTION

The full-scale pretreatment configuration consists of influent screening (6 mm), grit removal, primary treatment, and fine screening (2 mm). Facilities for the equalization of screened primary effluent and storage of plant effluent are also provided. Biological treatment consists of a membrane bioreactor (MBR), configured as a five-stage Bardenpho reactor to maximize the system's nutrient removal capabilities. Post-treatment consists of upflow–downflow granular activated carbon (GAC) reactors for the removal of residual organics, followed by ultraviolet (UV) disinfection. Figure 9-8 is a generalized process schematic for the facility.

Loudoun Water made a special effort throughout the study and design period to involve local stakeholders, including regulators, other affected utilities, local political bodies, and the general public, in the planning and design process. A proactive outreach effort was also part of the final design process. Loudoun Water reserved a portion of the Broad Run WRF site for use in its efforts to increase the public's awareness of the importance of water resources and water reuse, as described below.

Plant effluent flows through a constructed, natural-looking stream and two effluent display ponds before being discharged to the receiving stream. The discharge stream area is tastefully landscaped and open to the public. Stormwater is captured and treated using natural "rain garden" techniques. A portion of the effluent flow is also diverted to a fountain at the entrance to the facility to further enhance the vision of the Broad Run WRF as a community asset that will provide long-lasting educational, aesthetic, and recreational value to Loudoun County, Virginia and the surrounding area. Loudoun Water also has constructed an interpretive visitor's center in its administration building to highlight the importance of water reclamation and explain the facilities of the Broad Run WRF to the general public.

MEMBRANE SYSTEM DESIGN SUMMARY

Twelve membrane trains are provided. The design is based on 10 in-service membrane trains and 2 out-of-service trains to accommodate the potential need to have one train out of service for in situ recovery cleaning of the membranes and one train out of service for mechanical repairs. Each train is fitted with four cassettes of GE Water & Process Technologies (formerly Zenon) Zeeweed 500d hollow-fiber membranes. Each cassette consists of 48 membrane modules of 340 ft2. A total of 783,360 ft2 of membrane area is provided.

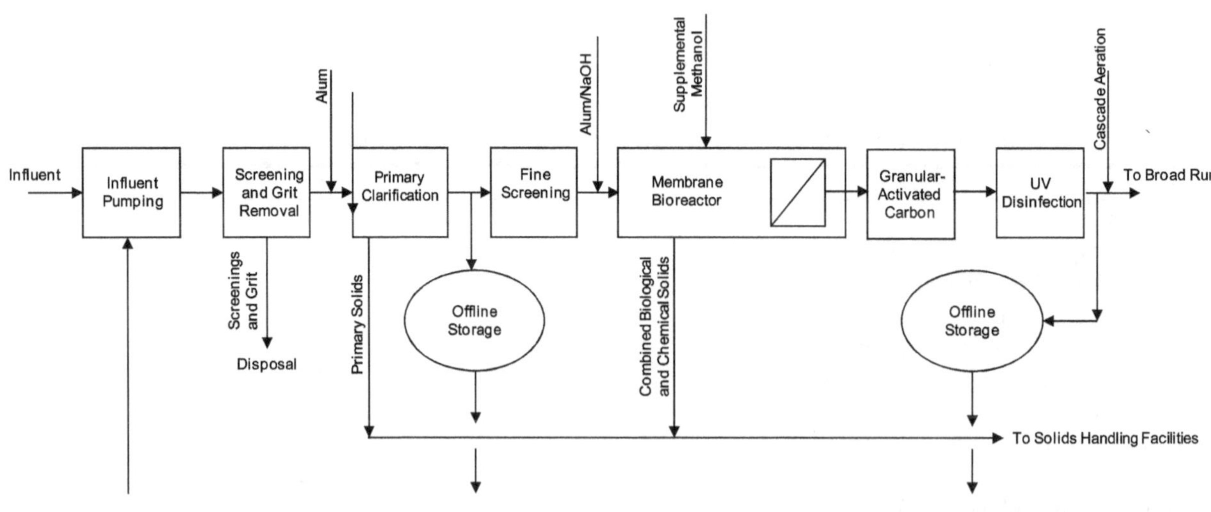

Figure 9-8 Process schematic for the Broad Run Water Reclamation Facility

GE Water & Process Technologies was selected as the membrane manufacturer/supplier for the Broad Run WRF. A preselection process was conducted concurrently with the preliminary design of the facility. After vendor selection, Loudoun Water entered into negotiations to finalize the scope of supply, services, and cost for the system. In addition to the scope of supply and services, issues such as risk, guarantees, warranties, replacement costs, financing, and other terms and conditions were also addressed during the negotiation process. This was done to protect the interests of all parties and to ensure that a long-term partnership between Loudoun Water and the vendor could be developed. Appendix B, Table B-1, provides a design summary for the MBR system.

The project was delivered using a conventional design–bid–build approach. Following an 18-month design period, construction was successfully completed within the specified 46-month time period to meet Loudoun Water's initial goal of being in operation by May 2008.

Start-up of this complex, green-field facility was successfully completed on schedule in accordance with Commonwealth of Virginia requirements. The plant has continued to operate well since the start-up period. However, influent loading characteristics are higher than originally anticipated. Appendix B, Table B-2, compares the design criteria for the influent and effluent with actual influent data and plant performance for the years 2011 and 2012.

One important "change" to the operational objectives for the facility is the greatly expanded emphasis on the reuse of plant effluent for commercial purposes. The initial planning effort (1999–2001) for the plant included a feasibility plan for the reuse of plant effluent. However, that planning effort focused on the potential for irrigation of golf courses, office parks, and similar areas. Since that time, an extremely strong market has developed for the reuse of effluent as cooling water for several large and recently constructed data centers located in the surrounding area. Loudoun Water now envisions a demand of up to 4.0 mgd for this purpose by the year 2015. Loudoun Water has subsequently invested approximately $12 million for the construction of reuse pipelines and a reclaimed water storage and pumping facility to serve only this market.

The new reuse will also lead to some market-specific changes to the treatment processes at the Broad Run WRF. For example, data centers do not need GAC-treated effluent for cooling water but do need chlorine residual in the supply pipelines to prevent biological growth in the distribution piping and subsequent biofouling of cooling equipment. Consequently, Loudoun Water will use a split-treatment option for reuse water to save operating costs by bypassing the GAC and UV disinfection process and by providing chlorine feed for the disinfection of this water only.

PLANT COSTS

Five construction projects were bid to maximize the use of local construction resources and complete the $190 million project on time. Two contractors were selected to deliver the five construction projects. In an effort to maintain the contract schedule, Loudoun Water provided an incentive fee equal to approximately 1 percent of the contract price to the construction contractors. Both the "errors and omissions" change order rate of 2 percent and the total change order rate of 4.5 percent were significantly below the industry averages.

The final negotiated price for one half of the membrane system, with all appurtenances, services, and other resources, was $4,421,948. During construction, Loudoun Water contracted with the vendor for the delivery of the remaining half of the system for an additional $2,800,000.

CONCLUSIONS

The Broad Run WRF is the result of an approximately 20-year planning, design, and construction effort by Loudoun Water. The project consisted of many phases, including detailed process evaluations, site studies, a pilot program, extensive permitting and public outreach efforts, design, construction, and start-up. The completed facility represents the first large-scale application that combines an MBR with GAC treatment and UV disinfection technologies to simultaneously meet extremely stringent nutrient limits while providing multiple pathogen barriers to protect public health. The completed Broad Run WRF has proven to be capable of meeting one of the most stringent discharge permits in the world. The facility was also the first treatment plant in Virginia to meet enhanced nutrient removal waste load allocation requirements of 4.0 mg/L of effluent TN for protection of the Chesapeake Bay.

In recognition of the innovative treatment processes and other features incorporated into its design, the Broad Run WRF has won several important awards, including the 2009 National Grand Prize (Design Category) for the Excellence in Engineering Awards of the American Academy of Environmental Engineers (AAEE) and the 2009 Grand Prizes for the Washington, D.C., area chapters of AAEE and the American Consulting Engineers Council.

Located in a campus-like setting that incorporates nature trails, rain gardens, and a display of the treated effluent to increase public understanding of water reclamation and reuse, the \$190 million, state-of-the-art, 11.0-mgd (41,600-m3/day) Broad Run WRF also provides a community amenity that will be of long-lasting educational, aesthetic, and recreational value to Loudoun County, Virginia.

PROJECT PHOTOS

Figure 9-9 Membrane bioreactor tanks and pipework

Figure 9-10 Granular activated carbon contactors and pipework

Figure 9-11 Ultraviolet light reactors

CASE STUDY—BIG SPRING RAW WATER PRODUCTION FACILITY: COLORADO RIVER MUNICIPAL WATER DISTRICT

By David Sloan, Freese and Nichols

The Permian Basin in the western part of Texas, like much of the western United States, has been subjected to an unprecedented period of drought since the mid-1990s. Reservoir levels have been very low for several years, and it appears that western Texas runoff response and reservoir yields are significantly different than records indicated when the principal reservoirs were originally planned on the Upper Colorado River. The Colorado River Municipal Water District has responded by pursuing new supplies and alternatives to continue providing a reliable and sustainable water supply to its member and customer cities. The treated wastewater currently discharged by cities in the district's service area has been identified as a sustainable source, and their first project to use this source began operation in April 2013.

The project is located in Big Spring, Texas, adjacent to the City of Big Spring's water reclamation plant (WRP), a conventional trickling filter–activated sludge facility that recently added cloth media filters. The district's facility diverts filtered secondary effluent prior to the final dechlorination step and pumps this water to the 1.0-mil gal source water storage tank at the district's raw water production facility. Following treatment (described below), the product water is pumped to an existing 42-in. raw water transmission pipeline, where it is blended with raw surface water and distributed to several municipal water suppliers. Each receiving system provides conventional surface water treatment of the blended water before distributing the finished water to individual users.

Reverse osmosis (RO) concentrate is discharged to an adjacent saline creek, which has been diverted downstream to reduce salt inflow to the Colorado River. Microfiltration (MF) backwash and other cleaning residuals are returned to the Big Spring WRP for retreatment.

This facility is believed to be the first in the United States to blend reclaimed water for municipal supply directly in a water transmission pipeline. Normal operating constraints limit the reclaimed fraction in the blended water to about 15 percent. The district has worked closely for several years with the Texas Commission on Environmental Quality to establish the case-specific requirements for pilot testing, permitting, design, operation, and monitoring. Public information has been made readily available, but no significant public opposition has been encountered.

PLANT DESCRIPTION

The treatment facility provides three major processes: MF, RO, and ultraviolet (UV) disinfection–advanced oxidation. Starting with up to 2.5 mgd of source water, the facility can produce up to 1.8 mgd of product water. A simplified process schematic is shown in Figure 9-12.

Microfiltration

Two hollow-fiber microfiltration systems operate in parallel to remove particles and protozoan pathogens from the source water. The systems are AP8 skid-mounted systems provided by Pall Corporation. Each system contains 78 Microza membrane modules designed to operate at a maximum flux of 32.9 gfd. The membranes operate in direct filtration mode with no coagulant or pretreatment other than the skid-mounted Amiad strainers.

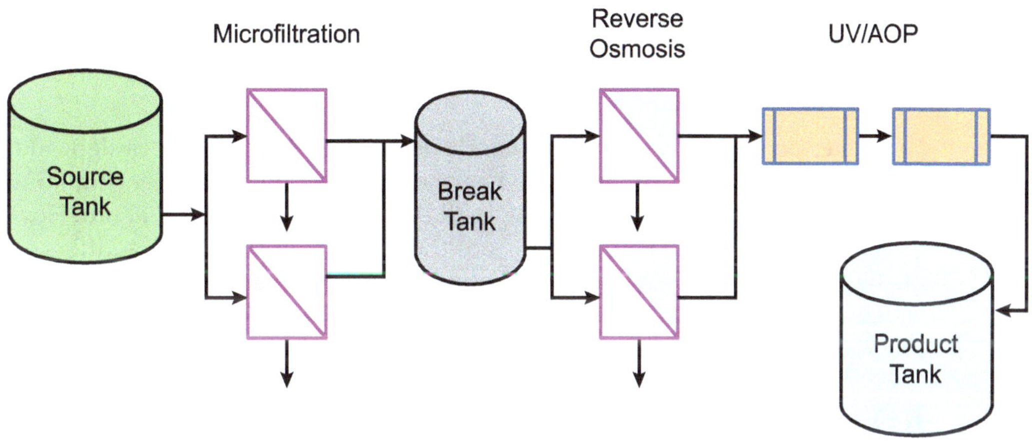

Figure 9-12 Process schematic for the Colorado River Municipal Water District raw water production facility in Big Spring, Texas

In order to better manage biological fouling within the MF units and, more importantly, the RO system, the district is in the process of implementing preformed chloramineaddition to the MF feedwater in order to provide a continuous monochloramine residual in both the MF and RO systems. Chloramination is scheduled to begin in June 2014 and will replace the use of antifoulant chemical and sodium bisulfite for biofouling control. Microfiltration design criteria are shown in Appendix C, Table C-1.

REVERSE OSMOSIS FEEDWATER PRETREATMENT

In addition to chloramination, RO feedwater pretreatment includes cartridge filtration and sulfuric acid and antiscalant dosing.

Reverse Osmosis

Two RO systems operate in parallel to remove dissolved salts, organics, viruses, and bacteria from the first stage filtrate. The systems, manufactured by Pall Corporation, are arranged in a 2-to-1 array, with 24 elements in the first stage of each system and 12 elements in the second stage. The elements are spiral-wound Toray thin-film composite membranes that operate at a flux of 10.5 gfd. The systems include an energy recovery turbine on the second-stage concentrate that recycles residual pressure in the concentrate to increase the pressure of the second-stage feed and reduce the energy consumption of the RO feed pump.

Ultraviolet Disinfection–Advanced Oxidation

Two Trojan UVPhox reactors operate in series to provide inactivation of viruses and other pathogens. The addition of hydrogen peroxide upstream of the reactors also causes the creation of hydroxyl radicals, establishing a highly potent oxidative environment and causing destruction of various organic molecules that could pass through the RO membranes. Each reactor contains 72 low-pressure lamps. The high UV dose and series configuration allow the process to achieve a documented 4-log virus inactivation.

WATER QUALITY

Feedwater

Feedwater quality has varied significantly during facility planning, design, and construction. Pilot testing was performed when the city WRP was experiencing failures of the old traveling bridge filter equipment, with turbidities approaching 10 ntu. However, feed quality has improved significantly since then, and with the recent addition of the cloth media filters, turbidity is typically around 1 ntu. Dissolved solids vary with the mix of water sources and with variations in precipitation. Plant feedwater quality for the period July 2008 through March 2013 is presented in Appendix C, Table C-2.

Treated Water Quality

Treated water quality goals and requirements are described below for each process.

Microfiltration

- Filtrate turbidity <0.15 ntu
- 4-log removal credit for *Cryptosporidium* and *Giardia*

Reverse Osmosis

- Nominal salt rejection of 99 percent
- Conductivity will be monitored for permeate consistency and will shut down the system when there is a 20 percent increase over the previous reading

Ultraviolet light–advanced oxidation

- 4-log virus inactivation; 2-log per reactor
- 1.4-log n-nitrosodimethylamine reduction
- 0.5-log 1,4-dioxane reduction

Product water (independent of individual processes)

- Nitrate <10, nitrite <1
- *Escherichia coli* 0 cfu/L

Although there is no significant operating history for this facility, initial treated water quality has been very good and consistent with or better than expected quality goals. An extensive program of water quality monitoring sponsored by the Texas Water Development Board is underway and will provide testing for an array of pathogens and regulated and unregulated contaminants.

PLANT COSTS

Construction cost for the facility was approximately $12 million, including source water diversion, pumping and piping, treatment building and equipment, storage tanks, product water pumping and pipeline, concentrate discharge, and site work. MF and RO membrane systems were preselected at a total cost of $2.75 million, included in the above total. The UV–advanced oxidation equipment was provided at a cost of $565,000, including the peroxide feed system.

PROJECT PHOTOS

Figure 9-13 Microfiltration system

Figure 9-14 Reverse osmosis system

CASE STUDY—EDWARD C. LITTLE WATER RECYCLING FACILITY

By Wyatt Won, West Basin Municipal Water District*

The West Basin Municipal Water District (West Basin) is a public agency in southern California that provides drinking and recycled water to customers in a 185-square mile service area. Faced with a declining water table from years of overpumping the groundwater basin, West Basin was formed in 1947 by voters to provide supplemental water to the area. West Basin purchases supplemental water from the Metropolitan Water District of Southern California (Metropolitan), which imports and treats water from the Colorado River and the Sacramento–San Joaquin Delta (State Water Project).

Prior to 1990, about 80 percent of the water used in West Basin was imported, with the remaining 20 percent pumped from the ground. The drought in the late 1980s and early 1990s pointed to a need to diversify water supplies and not depend so heavily on imported water. In the early 1990s, West Basin's Board of Directors approved development of a recycled water system and a water conservation program to help diversify the area's water supply. In the face of future water challenges, West Basin initiated its Water Reliability 2020 program to reduce dependence on imported water and provide more reliable, local supplies. The program will double West Basin's recycled water and conservation programs and eventually add 10 percent of future supplies from ocean water desalination, reducing its dependence on imported water to 33 percent by 2020.

West Basin's water recycling program delivers about 30,000 acre-ft per year (26.8 mgd) of recycled water to customers inside and outside of the service area. The program delivers five qualities of recycled water to meet its customers' water quality needs. The Edward C. Little Water Recycling Facility (ECLWRF) is the main facility in the recycled water program. Secondary effluent from the Los Angeles Hyperion Treatment Plant (Hyperion) is treated at ECLWRF through one of four treatment systems depending on the end use. The first system treats secondary effluent to tertiary standards using flocculation, sedimentation, monomedia filtration, and disinfection prior to distribution to more than 300 sites. Three satellite treatment plants located near industrial customers provide supplemental treatment for those industrial customers that require a higher-quality water than can be provided by the tertiary treatment processes.

The other three systems at ECLWRF consist of membrane processes and are the focus of this case study. The second system treats secondary effluent so that it can be injected into the ground to form a hydraulic barrier on the West Coast Basin Barrier Project (the Barrier) to prevent seawater from migrating inland to contaminate drinking water wells. The third and fourth systems at ECLWRF treat secondary effluent for industrial use. The third system produces water for use in low-pressure boilers and the fourth system for high-pressure boilers.

The membrane treatment processes at ECLWRF were built in phases to better match demand for recycled water. The phases are summarized below.

- *Initial construction:* completed in 1995 and capable of delivering 5 mgd of recycled water to the Barrier. Processes included lime clarification and trimedia filtration as pretreatment to the reverse osmosis (RO) units.
- *Phase II expansion:* completed in 1997 and increased the capacity of the recycled water Barrier system to 7.5 mgd. Microfiltration (MF) and additional RO units were added in this phase.
- *Phase III expansion:* completed in 2001. Additional MF and RO units were added so that recycled water could be used in an industrial customer's boilers. Also,

* Wyatt Won was an employee of the West Basin Municipal Water District at the time this manual was prepared.

1.7 mgd of single-pass low-pressure boiler feedwater and 2.6 mgd of double-pass high-pressure boiler feedwater were added.

- *Phase IV expansion:* completed in 2006 and increased the capacity of the recycled water Barrier system to 12.5 mgd. Additional MF and RO units were added, and the lime clarifier and trimedia filters were decommissioned.
- *Phase V expansion:* completed in 2013. This expansion increased the capacity of the Barrier recycled water system to 17.5 mgd and the capacity of the industrial single-pass RO system to 2.2 mgd. Ozone was also added to precondition the secondary effluent prior to microfiltration.

PLANT DESCRIPTION

A summary of the membrane facilities after the phase V expansion was completed is as follows:

- *Feedwater source:* secondary effluent from the Hyperion
- *Processes:*
  - Barrier treatment: ozone, MF, RO, advanced oxidation process (AOP), decarbonation, lime stabilization, and disinfection;
  - Low-pressure boiler feed treatment: ozone, MF, RO, and decarbonation; and
  - High-pressure boiler feed treatment: ozone, MF, RO, decarbonation, and second-pass RO.
- *Capacity:*
  - Barrier treatment: 17.5 mgd,
  - Low-pressure boiler feed treatment: 2.2 mgd, and
  - High-pressure boiler feed treatment: 2.6 mgd.

Feedwater quality is listed in Appendix D, Table D-1. A schematic of the processes is shown in Figure 9-15.

The different treatment processes are described below.

Ozone

Ozone was added as part of the phase V expansion. It is used to precondition the secondary effluent prior to all microfiltration at ECLWRF. Prior to the addition of ozone, the MF membranes experienced a severe decline in flux. The decline is thought to have been caused by organic fouling. Ozone was added to break down organic foulants and allow them to pass through the microfiltration membranes rather than fouling them. Design criteria for the ozone system are listed in Appendix D, Table D-2. The ozone injection system is shown in Figure 9-16.

Microfiltration

MF is used to pretreat the water prior to RO. The goal of MF is to remove as many particulates as possible prior to RO to prevent premature fouling. The secondary effluent is chloraminated prior to MF to help control biogrowth. Target combined chlorine residual is between 3 and 5 mg/L. The silt density index target for the MF effluent is less than 3.0. MF was added in four phases; phases II, IV, and V for Barrier treatment and phase III for industrial treatment. Design criteria for the MF systems are listed in Appendix D, Table D-3. Phase III CMF units are shown in Figure 9-17.

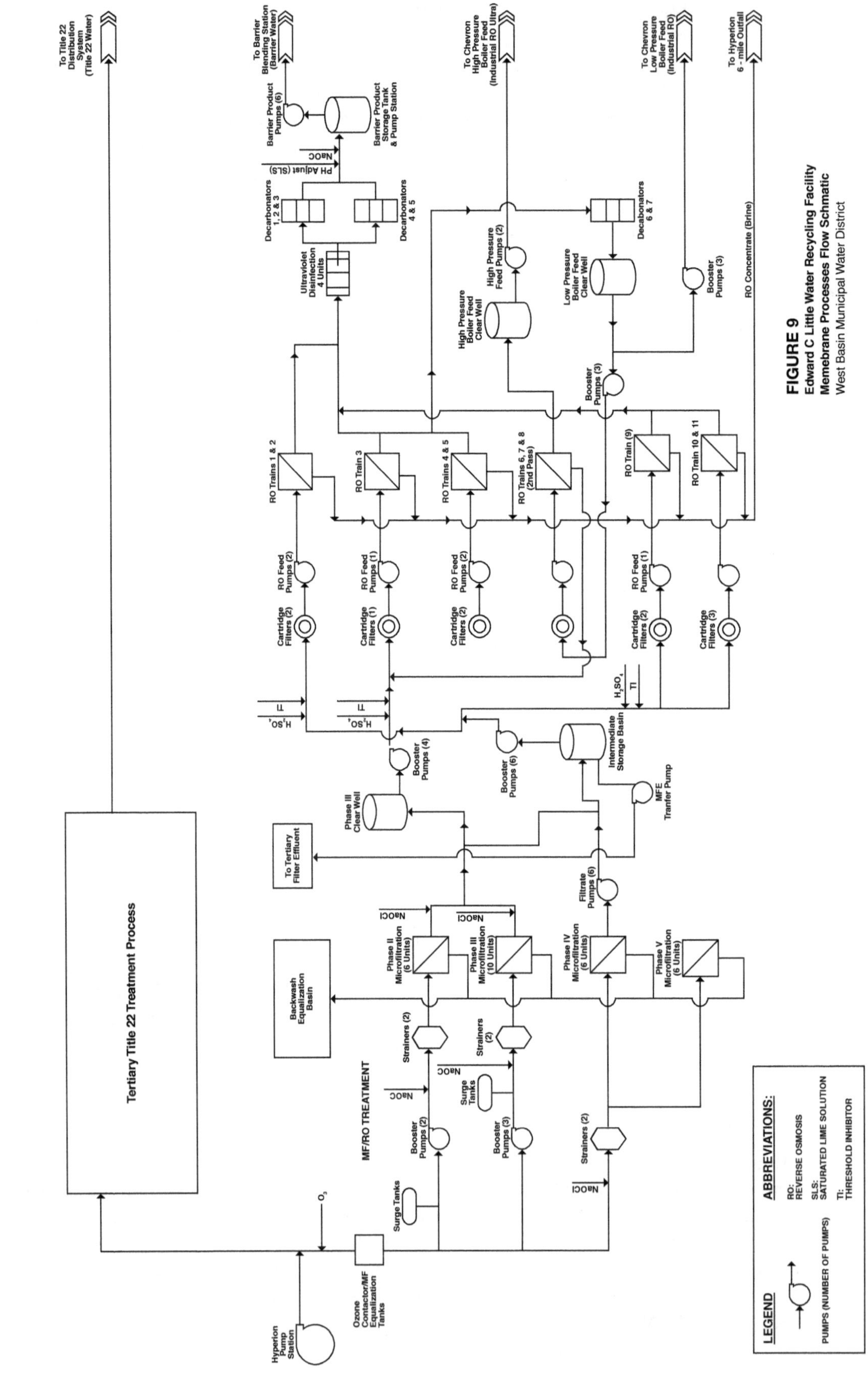

Figure 9-15 Process schematic of the Edward C. Little Water Recycling Facility

Figure 9-16 Ozone injection system

Figure 9-17 Phase III CMF units

Reverse Osmosis—Single Pass and Double Pass

RO is used to remove large molecular constituents and ions. RO was added in five phases; trains 1 and 2 were built in the first phase, train 3 in phase II, train 9 in the phase IV, and trains 10 and 11 in phase V for Barrier treatment. Trains 4, 5, 6, 7, and 8 were built in phase III for industrial treatment. Trains 4 and 5 are the first-pass trains for low-pressure boiler feed. Trains 6, 7, and 8 are the second-pass trains for high-pressure boiler feed. Design criteria for the different trains are listed in Appendix D, Table D-4.

Advanced Oxidation Process

The AOP treats RO permeate and consists of hydrogen peroxide addition followed by ultraviolet (UV) light. AOP provides disinfection of the barrier recycled water effluent, as well as destruction of n-nitrosodimethylamine and other unknown organic chemicals that may not be completely rejected by RO. Hydrogen peroxide is injected upstream of the UV reactors. Design criteria for the AOP are listed in Appendix D, Table D-5.

Decarbonation

All first-pass RO permeate for industrial use passes through decarbonators to release excess carbon dioxide. For the Barrier treatment process, most of the flow passes through the decarbonators. However, a portion of the permeate is bypassed to help control pH of the blended water after lime stabilization.

Lime Stabilization

Lime stabilization is not used on industrial-quality RO permeate. For the Barrier treatment process, a saturated lime solution is injected into the blend of decarbonated and bypassed RO permeate to reduce the water's corrosivity. Lime is added to increase the Langlier saturation index to values between −0.5 and +0.5.

Post-Disinfection

Post-disinfection is not used on industrial-quality RO permeate. Sodium hypochlorite is added to Barrier water prior to pumping to the injection wells. Sodium hypochlorite combines with ammonia in the water to form a chloramine residual, which is required by the Los Angeles County Department of Public Works because of concern about biofouling in the injection wells. The total chlorine residual target is between 3 and 4 mg/L.

TREATED WATER QUALITY PRODUCED

Quality of the recycled water delivered to the Barrier is described in Appendix D, Table D-6. In addition to the constituents listed in Table D-6, West Basin is required to test annually for disinfection by-products, volatile organic chemicals, nonvolatile synthetic organic chemicals, chemicals of concern, base/neutral extractables, pesticides, pharmaceuticals, endocrine disruptors, and tentatively identified compounds. Quarterly testing for radioactivity is also required. Levels of these contaminants are typically not detected; when they are detected, levels are below permit requirements.

Water quality of the single- and double-pass RO-based recycled water in 2012 is described in Appendix D, Tables D-7 and D-8.

CAPITAL COSTS

The membrane facilities at ECLWRF were built in phases, starting with initial construction in 1995. Phase V, the most recent membrane expansion, was completed in 2013. Total capital cost for improvements through phase V is approximately $225,000,000. These dollars are not escalated and are the sum of construction costs incurred. Costs include secondary effluent conveyance, MF, RO, membrane replacement, solids handling of microfiltration backwash, AOP, and brine discharge.

OPERATION AND MAINTENANCE COSTS

Expenditures to operate and maintain the membrane facilities at ECLWRF were $8,457,000 for fiscal year ending 2012; 10,421 acre-ft of Barrier recycled water, low-pressure boiler feedwater, and high-pressure boiler feedwater were produced. Costs include power, labor, chemicals, sludge disposal, laboratory work, supplies, equipment, and office support. Ozone was not online in fiscal year 2012, so costs associated with ozone production are not included.

CASE STUDY—SINGAPORE NEWATER FACILITIES

By Key Wee Ong, Singapore Public Utilities Board

As a small island city–state with only 710 square kilometers in land area and a population of 5.18 million, Singapore has to look for innovative ways to find new water resources to meet its growing need for clean water. As the national water agency of Singapore, the Public Utilities Board (PUB) manages the nation's water resources in an integrated manner, from the collection and management of raw water, the production and supply of potable water, and collection to the treatment and reclamation of wastewater.

SINGAPORE'S FIRST INITIATIVE ON RECYCLING: INDUSTRIAL WATER

All used water is collected through a comprehensive network of sewer reticulation systems and treated at centralized water reclamation plants to meet international standards before the treated effluent is discharged into the sea. The treated secondary effluent or treated wastewater is a viable potential as a water resource if it is harnessed through advanced water reclamation technology.

Reclamation of treated used water is not new. Since 1966, the Jurong Industrial Waterworks has been reclaiming secondary treated effluent from the upstream Ulu Pandan Water Reclamation Plant for industrial use. The plant uses a conventional treatment process that includes chemical coagulation, sand filtration, aeration, and post-chlorination. Most of the industrial water is used for cooling and washing. Some industries further treat the industrial water to produce high-grade water for nonpotable use.

In 1974, a study was done to determine the feasibility of producing drinking water from treated wastewater using membrane technology. Though the study concluded that production of drinking water was technically achievable, it was not implemented at that time because of the high cost of membrane technology. At that time, the membrane technology was also not reliable.

THE NEWATER STUDY

Results from research and pilot and demonstration testing as well as full-scale plant operation through the late 1990s confirmed that membrane technologies, including microfiltration (MF) and reverse osmosis (RO), had become more reliable and cost efficient to operate and maintain. In 1998, staff from PUB visited water reclamation projects in California and West Virginia. Their findings showed that water recycling was viable, and a study team was formed. This led to the construction of a 2.2-mgd (10,000-m^3/day) dual-membrane demonstration plant at Bedok in May 2000. This was followed by a two-year study to determine the suitability of the highly treated reclaimed water as a source of raw water to supplement Singapore's water supply.

The NEWater Demonstration Plant

The NEWater demonstration plant was built with dual-membrane (MF/ultrafiltration [UF] and RO) and ultraviolet (UV) disinfection technologies. Appendix E, Table E-1, provides selected technical details regarding the treatment processes.

NEWater demonstration plant feedwater was secondary-treated effluent from an activated sludge treatment process with the characteristics listed in Appendix E, Table E-2. The effluent was first microscreened (to 0.3 mm) and then microfiltered. The MF units used outside–in flow, hollow–fiber membranes, which present a physical barrier to particulates including most bacteria and some viruses. Chlorine was added before and after MF to control biofouling of both MF and RO membranes; chlorine reacts with ammonia in the effluent to form chloramines. The MF filtrate was then passed through RO membranes to remove dissolved substances. The RO membranes also served as a second barrier to any viruses, bacteria, or other pathogens that remained in the MF filtrate. The RO membranes were thin-film, polyamide flat sheets configured in spiral-wound elements. The membranes were operated in a cross-flow manner and in a three-stage array. The RO permeate was subsequently disinfected using UV irradiation as an added microbial inactivation (safety) barrier. The product water from the MF–RO–UV treatment train is referred to as NEWater.

The Expert Panel

As part of the NEWater demonstration study, an independent expert panel (comprised of local and foreign experts across multiple disciplines including engineering, water chemistry, toxicology, epidemiology, and microbiology) was convened to advise the team and evaluate the findings. Among other areas, the expert panel was tasked to review the sampling and analysis of NEWater and to conduct toxicological and carcinogenic risk assessments and other relevant health studies, Extensive water quality results generated from more than 20,000 tests carried out on close to 190 water quality parameters showed NEWater produced by the treatment train met all US Environment Protection Agency and World Health Organization drinking water standards. The expert panel also reviewed studies carried out on mice and fish and conducted to determine the toxicity and estrogenic effects of NEWater. These studies showed that short- and long-term exposure or consumption of NEWater concentrated 500-fold did not produce any tissue abnormalities or health effects in mice, and there was no evidence of carcinogenic or estrogenic effects on fish.

At the end of the study, the expert panel concluded that NEWater was safe for potable use (the review report is available at The panel also concluded that Singapore should consider the use of NEWater for indirect potable use (IPU) as it was a safe supplement to the existing raw (reservoir) water supply. First, the panel found that blending with reservoir water would result in trace minerals, which had been removed in the RO treatment process. Second, reservoir storage would provide an additional safety barrier beyond the membrane and UV technologies used to produce safe, high-quality NEWater, as the reservoir raw water (with the blended NEWater) will undergo further treatment at a water treatment plant to produce potable water. Last, blending NEWater with reservoir water will help to gain public acceptance.

PUBLIC EDUCATION

After proving the science and technology behind water reuse, the next challenge was to gain public acceptance for NEWater, overturn the psychological stigma associated with wastewater, and convince the public to embrace NEWater as a source of drinking water. This was a significant challenge because the public knew very little about NEWwater, unlike PUB potable water in which they had full confidence. The key was to shift attention away from the source of NEWater and instead focus it on the treatment processes used to purify and produce NEWater. It was critical to translate the technical information to the public in a simple manner. Although terms such as "wastewater" and "sewage" have acceptance with the technical and engineering profession, these terms have a negative connotation that remind the public of their source. A paradigm shift was needed to make the public understand that this water is technically not wastewater but used water, as it can be used and reused over and over, similar to how water recycles itself in nature.

In May 2002, the media were briefed, and journalists accompanied PUB staff on a trip to several advanced water reclamation facilities in the United States, including Water Factory 21 in Orange County, Calif., and the Water Campus in Scottsdale, Ariz. The purpose of the trip was to demonstrate that the indirect potable reuse is not a new phenomenon and that it has been produced in a safe and reliable manner for many years. A documentary that aired on local television, *Thirsting for New Sources*, looked at NEWater technology and experiences in the United States with advanced water reuse. With the help of members from expert panel, briefings were held with political leaders, members of parliament, grassroots leaders, and business groups to garner feedback. In addition, exhibitions were held at the community and school levels. In a greater attempt to reach out to the community, NEWater was bottled to allow the general public to sample the taste of NEWater.

Through these initiatives, the public was able to increase their understanding of water recycling and reuse and learn that NEWater was not only safe to drink but also a sustainable water supply. During the 2002 National Day Parade, close to 60,000 Singaporeans toasted to NEWater in a show of solidarity, demonstrating their support and confidence in NEWater. This set the stage for implementation of NEWater in Singapore. An independent survey by Forbes Research at the end of 2002 confirmed the success of the program. NEWater had garnered a 98 percent acceptance rate, with 82 percent of respondents indicating that they would drink NEWater directly and an additional 16 percent indicating they would drink it indirectly, after mixing with reservoir water.

The public education campaign continued in schools, community centers, and workplaces. Educational leaflets and posters were distributed to explain the concept of NEWater. In February 2003, the NEWater Visitor Center was opened as a primary avenue for public education on NEWater. At the center, which is integrated with Bedok NEWater Factory, visitors can learn about water technology and the experiences with water reuse around the world, and they can view the operation of the technologies used in the production of NEWater.

FULL-SCALE IMPLEMENTATION OF NEWATER

Following the success of the NEWater study and public education initiatives, NEWater became the fourth source of drinking water through water reuse and has been successfully implemented in Singapore nation-wide since February 2003.

NEWater is supplied for both nonpotable use and indirect potable reuse. It serves as an effective substitute for PUB tap water in manufacturing processes that require ultrapure water, such as wafer fabrication in the semiconductor industry. This frees up a large amount of PUB potable water. The supply of NEWater to wafer fabrication plants for the production of ultrapure water needed for the manufacturing process is the world's first. It demonstrates the quality and reliability of NEWater, as these industries need water that is of higher quality than that used for drinking purpose. Industries and commercial buildings have also reported savings in operation cost and less water consumption with NEWater because of its higher quality. As of January 2013, NEWater had been supplied to more than 440 customers. This includes wafer fabrication plants, petrochemical complexes and refineries, power stations, electronics companies, and commercial premises where NEWater is used for wafer fabrication processes, process cooling, air conditioning, boilers, and general washing purposes.

For IPU, a small percentage of NEWater (about 2 to 3 percent of Singapore's daily water consumption) is mixed with raw reservoir water before being treated at the treatment plant as potable water. This is strategically significant, as supply of NEWater can be used to boost water supply. NEWater is also not affected by fluctuations in rainfall and can enhance Singapore's resilience against prolonged periods of low rainfall or drought.

There are currently (in 2013) four NEWater factories with a total capacity of 532,000 m^3/day. By 2020, Singapore's NEWater capacity will be able to meet 40 percent of its total water demand.

Processes and Equipment

The general flow schematic for the NEWater factories is shown in Figure 9-18.

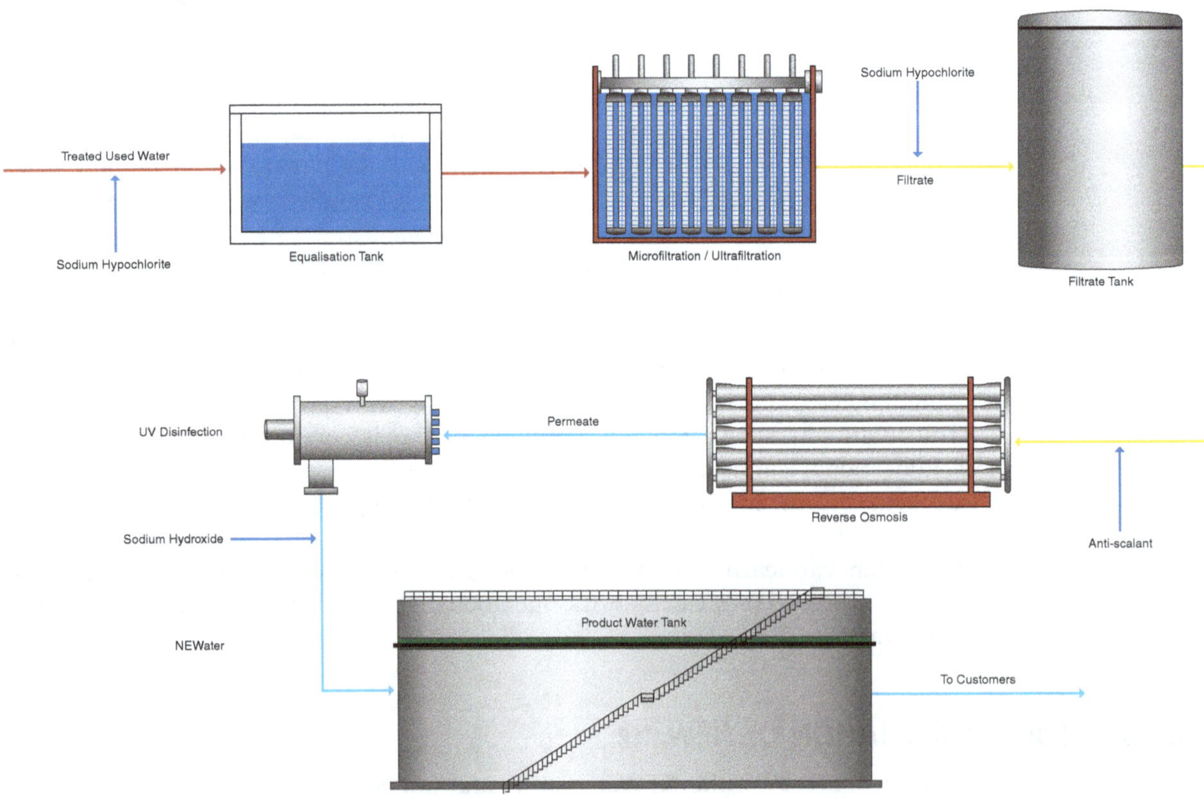

Figure 9-18 Flow schematic of NEWater factories

The first two NEWater factories, Bedok NEWater Factory and Kranji NEWater Factory, were commissioned in 2003. Each plant uses a dual-membrane treatment train (MF/UF and RO) followed by UV disinfection. Design criteria of the plant processes and equipment are summarized in Appendix E, Tables E-3 and E-4.

PROJECT PHOTOS

Figure 9-19 Zeeweed ultrafiltration at Bedok NEWater Factory, phase 1

Figure 9-20 Microza microfiltration at Bedok NEWater Factory, phase 2

Figure 9-21 Large-diameter (16-in.) reverse osmosis trains at Bedok NEWater Factory, phase 2

Figure 9-22 Memcor CMF-S system at Kranji NEWater Factory, phases 1 and 2

Figure 9-23 Reverse osmosis units at Kranji NEWater Factory, phases 1 and 2

To meet the growing demand for NEWater, additional NEWater factories were built. Beginning in 2004, PUB engaged private sector participation in NEWater supply by implementing a public–private partnership approach in which the private sector would design, build, own, and operate (DBOO) the NEWater plant, with PUB serving as the exclusive purchaser of NEWater via a long-term contract (usually 20- or 25-year contract). This arrangement allowed PUB to focus on acquiring the private sector's services at the stipulated level of quality and the most cost-effective price while encouraging greater innovation to enhance production and operational efficiency, and, in the longer term, develop the Singaporean water industry into a more vibrant, innovative, and export-oriented sector.

Two NEWater factories were constructed and are operated under the DBOO model: the Ulu Pandan NEWater Factory (2007) and the Changi NEWater Factory (2010). Details on these facilities are provided in Appendix E, Table E-5.

The capital cost of all NEWater factories and the first-year NEWater tariff for the 2 DBOO NEWater factories are provided in Table 9-4.

Design Criteria and Water Quality

The process units are designed to produce treated water quality that meets the criteria listed in Appendix E, Table E-6. Typical NEWater quality is listed in Appendix E, Table E-7.

Table 9-4 Cost information for NEWater factories

Plant	Bedok NEWater	Kranji NEWater	Ulu Pandan NEWater (DBOO)	Changi NEWater (DBOO)
Capital cost (Singaporean dollars)	Phase 1: $34 million, includes civil structure for phases 1 and 2 Phase 2: $25 million, all 16-in. reverse osmosis membranes were offered at $1 in the tender bids	Phase 1: $36.4 million, includes civil structure for phases 1 and 2 Phase 2: $33.6 million	Capital costs are not available to PUB as under the DBOO arrangement, the respective concession company will DBOO the NEWater factory and sell the NEWater exclusively to PUB under a long-term water agreement	
First-year NEWater tariff (Singaporean dollars)	Not applicable	Not applicable	$1.14/kgal	$1.14/kgal

Abbreviations: DBOO, design, build, own, and operate; PUB, Public Utilities Board.

CASE STUDY—BUNDAMBA 1A ADVANCED WATER RECLAMATION FACILITY: SOUTHEAST QUEENSLAND WATER

By Vasu Veerapaneni

South East Queensland (SEQ) in Australia is the fastest growing metropolitan region that faced a severe drought. The rapidly growing population coupled with an unusually long drought period exerted tremendous stress on water resources in this region. The Queensland government responded by initiating several long-term projects to ensure a secure long-term water supply. One such initiative, a recycled water project, was commissioned to reclaim wastewater for industrial and agricultural use and is intended to reduce the current demand on the region's water supply. As part of this, three advanced water treatment plants (AWTPs) were constructed to treat municipal effluent from several wastewater plants. The design and construction of the first advanced wastewater treatment facility is described here. The initial stage 1A consists of a 30-ML/d wastewater recycle facility that was commissioned in 2007. Within a few months after commissioning stage 1A, a second stage expansion was implemented (stage 1B) for an additional 36-ML/d capacity. This facility treats water from four of the seven wastewater plants in the region.

The treatment scheme consists of chloramination, coagulation, flocculation, clarification using plate settlers, microfiltration (MF), reverse osmosis (RO), ultraviolet (UV) irradiation and hydrogen peroxide addition, and post-treatment. Chloramination is practiced to limit biofouling. Because of the potential for calcium phosphate scaling, particularly when treating water from three wastewater treatment plants (WWTPs) that have high phosphate concentrations, provisions for removal of phosphate through coagulation and clarification are also provided. MF pretreatment is used as it has been proven to be very effective in controlling fouling of the RO membranes in municipal effluent reuse applications. Advanced oxidation with UV and hydrogen peroxide is used to decompose various emerging contaminants such as *N*-nitrosodimethylamine (NDMA), endocrine-disrupting chemicals, and pharmaceutically active compounds that may have gone through the previous barriers at trace levels, as described later in this case study. Post-treatment involves disinfection using chlorine and stabilization of the water using lime and carbon dioxide. The waste flow from the MF process is recycled back to the head of the AWTP. The RO concentrate is further treated through nitrification using moving bed biofilm reactors, chemical precipitation, and denitrification filters to remove phosphorus and nitrogen compounds. At the time of its construction, Bundamba 1A was the largest facility in the world using 18-in.-diameter RO elements.

The facility was designed and constructed in record time. The design–build contract for stage 1A was awarded in June 2006 and the facility was commissioned in September 2007. Stage 1B was designed and constructed in a similar time period. The duration of the operation contract is 15 years. The AWTP constitutes 28 percent of the total water reclaimed by this facility and two other facilities.

While in service, the facility operated well and met all treated water quality goals. In December 2010 heavy rains occurred that filled freshwater catchments. As a result, the Bundamba and other Western Corridor AWTPs were placed in standby mode. The plants will remain in standby mode until combined dam levels drop to 60 percent, at which time the Bundamba AWTP can be brought back online within a two-week period.

WATER QUALITY

Phases 1A and 1B are designed to treat effluent from four wastewater treatment facilities. The quality and quantity of water to be treated from these facilities varies significantly. The anticipated maximum percentage of feed flow from each facility as well as the water quality are listed in Appendix F, Table F-1.

The facility is designed to treat the effluent to meet the Australian Drinking Water Guidelines for Water Recycling (phase II) standards: Augmentation of Drinking Water Supply and Water Supply (Safety and Reliability) Act 2008. Pertinent parameters are listed in Appendix F, Table F-2.

In addition to the treated water quality goals, the RO concentrate is also required to meet discharge requirements related to nutrient and suspended solids, as shown in Appendix F, Table F-3. Although all wastewater facilities currently discharge their effluent to a river, as part of this project, the SEQ government intended to lower the nutrient loading on to the Brisbane River.

Treatment

AWTP treatment components consist of the following:

- Flow balancing,
- Chloramination,
- Pretreatment (flocculation and clarification using plate settlers),
- Screening (in-line strainers),
- MF,
- RO,
- Advanced oxidation using UV and hydrogen peroxide,
- Post-treatment (lime and carbon dioxide addition for stabilizing treated water and chlorination),
- RO concentrate treatment,
- Treated water storage, and
- High-service pump station.

A brief description of various treatment components is given below. A simplified process schematic is shown in Figure 9-24.

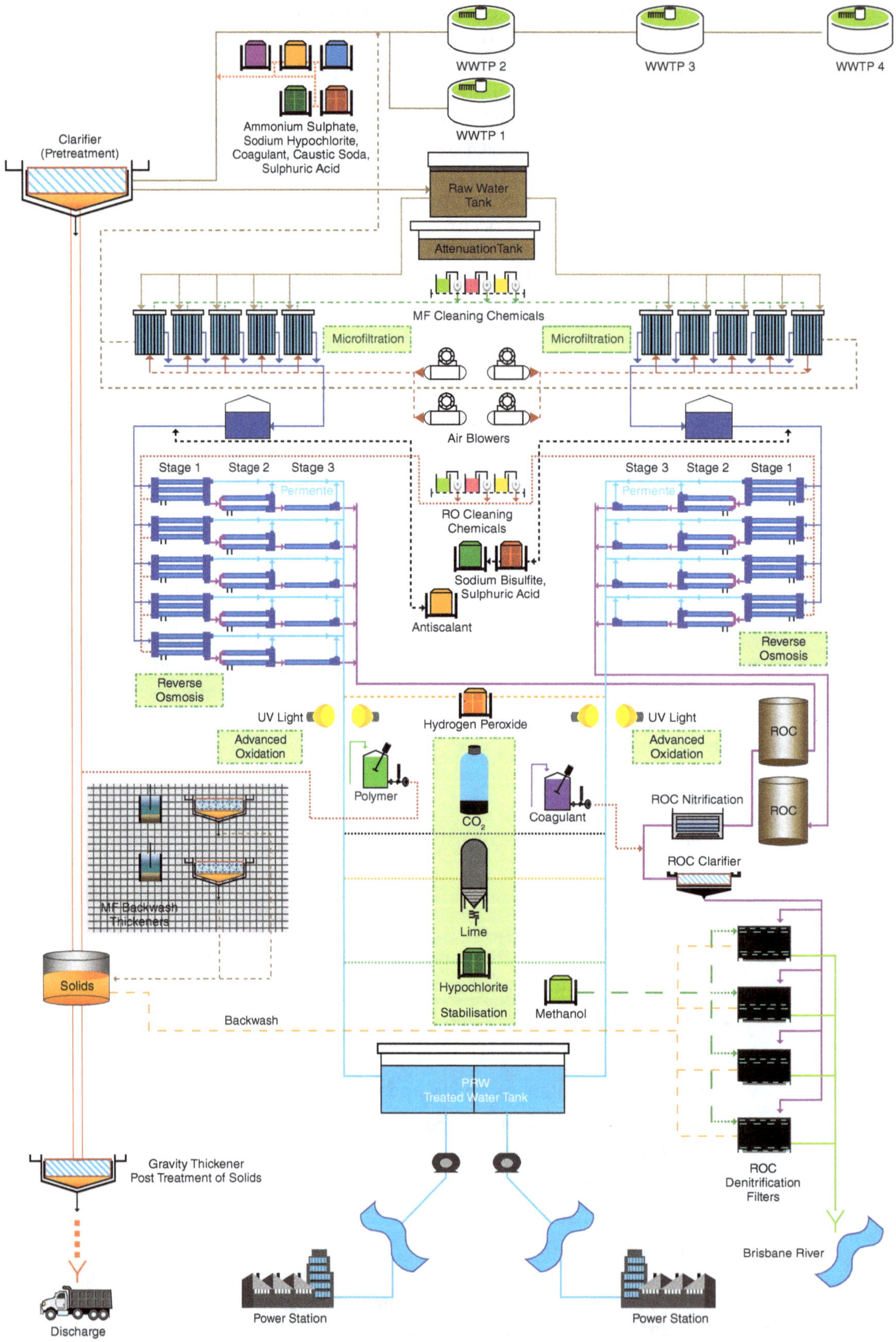

Figure 9-24 Process flow diagram for Bundamba advanced wastewater treatment plant

Flow Balancing and Clarification

As part of stage 1A, a raw water storage tank was constructed with a capacity of 15 ML. No clarification was provided during that stage, given the low concentrations of phosphate in the Bundamba WWTP effluent (stage 1A was designed to treat primarily Bundamba WWTP effluent). As part of stage 1B, pretreatment sedimentation tanks were constructed when the facility began treating water from wastewater treatment facilities with higher phosphate concentrations in their effluent. The clarification process has two trains, each consisting of two flocculation tanks in series and a sedimentation tank equipped with inclined plate settlers and a sludge collection system. The design hydraulic loading rate for plate settlers is 0.6 m/h.

Microfiltration

An encased (pressurized) microfiltration system is provided to ensure that the RO feedwater has consistently low turbidity and low fouling potential. Pretreated water is pumped to encased microfiltration trains using a bank of feed pumps. The membranes were provided by Siemens (now Evoqua; Memcor). For stage 1A, four duty and one standby trains are provided, each with a capacity of 7.6 ML/d. For stage 1B, five duty trains, each identical to the stage 1A trains, are provided.

Reverse Osmosis

Effluent from MF membranes is collected in an RO feed tank. Water from the feed tank is drawn by the high-pressure RO feed pumps. Four RO trains were provided for the stage 1A facility, each with a capacity of 7.75 ML/d. Five additional duty trains were provided for stage 1B. Each train has a dedicated RO feed pump. Large-diameter (18-in.) RO membrane elements are used to minimize the footprint and to reduce costs. Each train has three stages, with 13 pressure vessels arranged in a 7:4:2 configuration. Each pressure vessel houses five elements connected in series. The nominal design permeate flux is 20.5 L/m^2/h. The first stage permeate flux is limited to less than 22 L/m/h. The system is designed for a typical recovery of 85 percent. However, the pretreatment is designed to provide sufficient RO feedwater when the system operates at 82 percent recovery while producing finished water of 30 ML/d.

Advanced Oxidation

The RO permeate is treated with UV irradiation and hydrogen peroxide to provide 1-log NDMA and 0.5 log 1,4-dioxane removal. Two UV trains are provided for each stage of construction, each with a power of 75 kW.

Stabilization

Lime and carbon dioxide are added to reduce the corrosiveness of RO permeate and to stabilize the treated water.

Waste Treatment

Residuals from the pretreatment plate settlers are pumped by the pretreatment sludge pumps to a gravity thickener. The thickened solids are pumped to sludge holding tanks. The backwash water from the MF unit is recycled to the plant feedwater pipeline. The backwash from the strainers is also directed to this recovery tank.

The RO concentrate is pumped to a moving bed biofilm reactor for nitrification and then to a solids contact clarifier. Effluent from this clarifier flows by gravity to the denitrification filters for nitrate and total suspended solids removal to meet the RO concentrate discharge limits.

OPERATIONAL PERFORMANCE

Availability

The facility was required to have an availability of 90 percent, defined as the fraction of days when production exceeded rated capacity. During the first three months of operation, the availability was 96 percent, accounting for disruptions due to tie-ins required for implementation of stage 1B expansion.

Microfiltration Performance

The MF design criteria and selected performance requirements are listed in Appendix F. The facility operational performance data are provided in Figure 9-25. All the required treatment goals were met. Specifically, the filtered water silt density index (SDI) value was significantly better than the required value for RO feedwater of 3. SDI is a parameter that indicates the fouling potential for the downstream RO process. Higher values indicate higher fouling potential. The low SDI value of the microfiltered water indicates low fouling potential, which is reflected by the stable RO performance data (permeability) presented in Figure 9-26. The MF system operated at an average flux of 27 L/m/h and consistently achieved more than 5-log removal of pathogens larger than 3 μm. Further, only two clean-in-place processes were performed for the operating period shown below, indicating a cleaning interval of longer than 30 days.

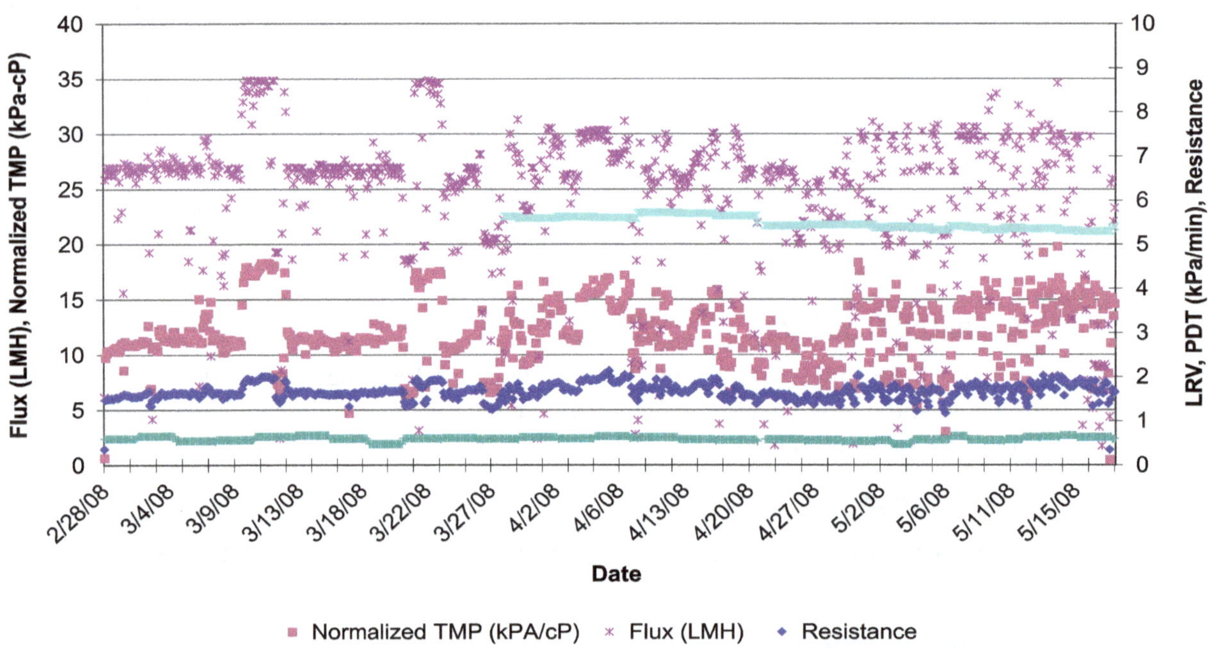

Figure 9-25 Microfiltration unit 2 operating performance at the Bundamba advanced wastewater treatment plant

Abbreviations: TMP, transmembrane pressure; LRV, log removal value; PDT, pressure decay testing; LMH, liters per m^2 of membrane per hour.

Reverse Osmosis Performance

Given the high quality of MF filtered water, the RO performance also exceeded design goals. One treatment goal for RO was rejection of nitrogen compounds. SEQ required that the AWTP meet the nitrogen concentrations in the receiving body. The selection of the RO membranes was largely based on their ability to reject nitrogen compounds, specifically nitrate. The membranes' permeability for the first six months of operation is shown for three trains. The initial decline in permeability is anticipated for wastewater applications, primarily due to plugging of any imperfections in the membrane surface and due to sorption of some foulants onto the membrane surface. After the initial decline in flux, the permeability remained steady, indicating that no further membrane fouling was occurring.

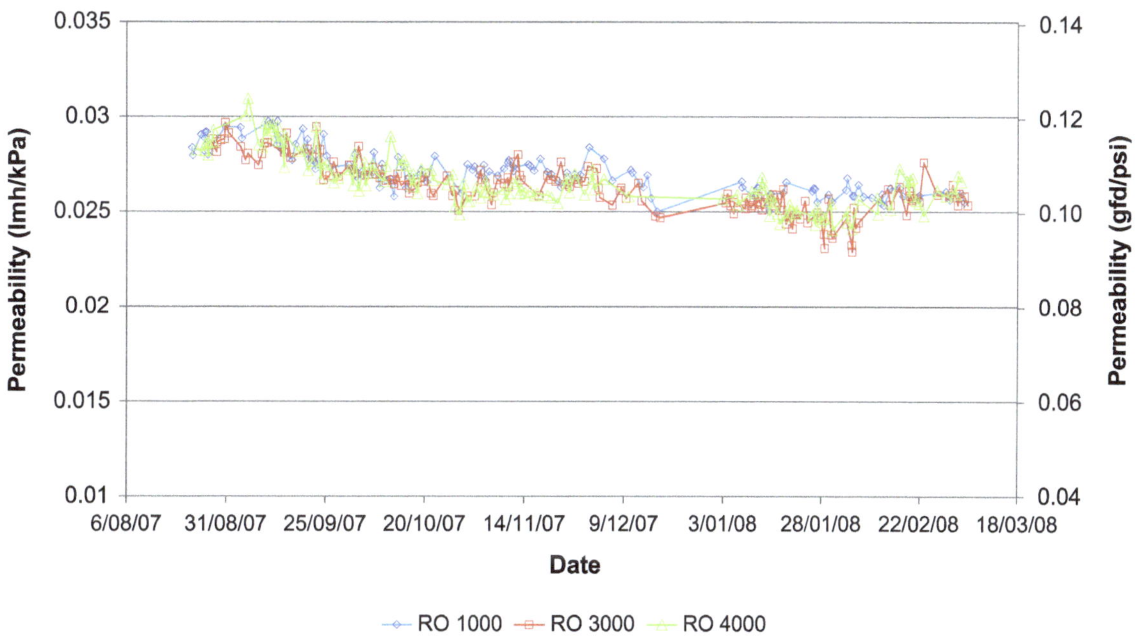

Figure 9-26 Permeability trends for reverse osmosis units at the Bandamba advanced wastewater treatment plant

Abbreviation: RO, reverse osmosis.

Table 9-5 Reverse osmosis performance at Bumdamba stage 1A

Parameter	Design	Actual
Recovery	82 to 85 percent	85 percent
Total nitrogen concentration in permeate (after breakpoint chlorination)	<0.8 mg/L average	0.68 mg/L average 0.87 mg/L maximum
CIP interval	No specific goal	The reverse osmosis units have operated for the first six months without requiring a CIP

Abbreviation: CIP, clean-in-place.

Treated Water Quality Goals

During its operation between commissioning in September 2007 and placing it in standby in 2011, the facility met all treated water quality goals. Because the drought has ended and supply reservoirs are at near capacity, demand has declined, and the facility has been operating at reduced capacity in recent years.

Specific Energy Consumption

The facility was designed to operate at a specific energy consumption of 1.0 kWh/m^3 of water produced. Since the facility was designed in two stages, the energy consumption during the operation of only stage 1A was close to the design estimate (see Figure 9-27 data until July 2008). Upon commissioning of stage 1B, the facility's energy consumption ranged from 0.8 to 0.9 kWh/m^3 (Figure 9-27). The total power consumption, including raw water conveyance, pretreatment, UF, RO, advanced oxidation, concentrate treatment, and distribution to a power station that is 6 km away, is 1.77 kWh/m^3.

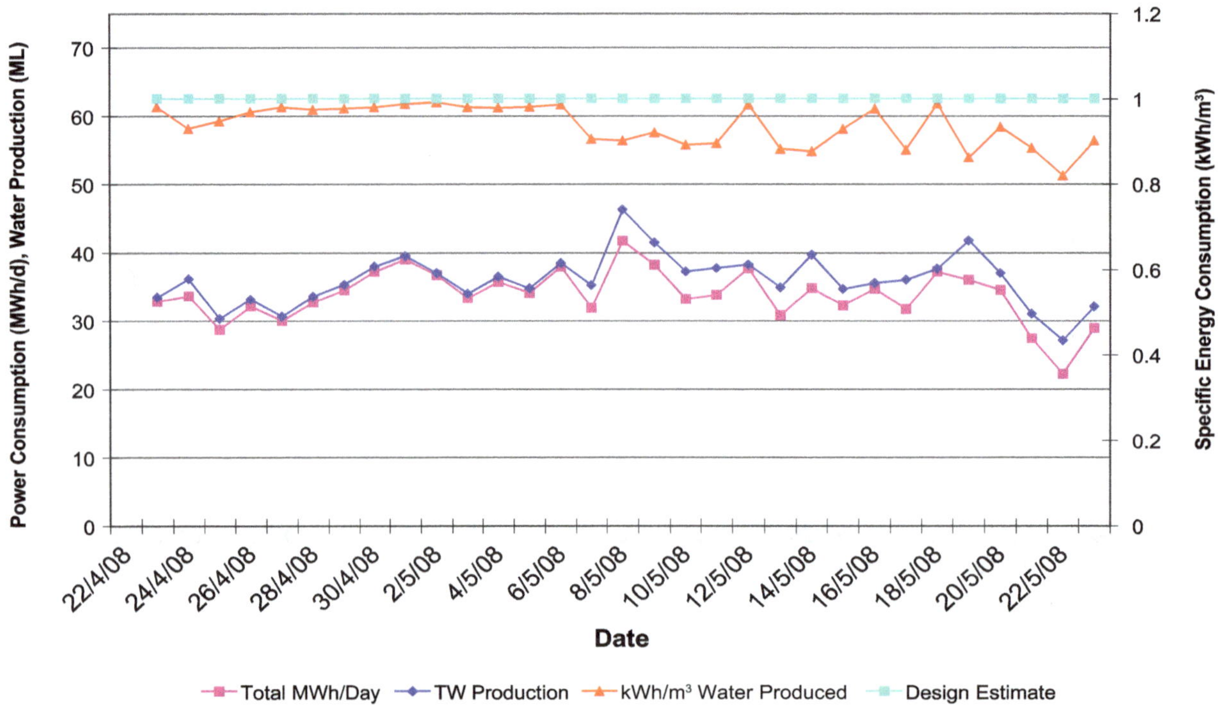

Figure 9-27 Energy consumption at the Bundamba advanced wastewater treatment plant

Abbreviation: TW, total water.

CASE STUDY—GIBSON ISLAND ADVANCED WATER TREATMENT PLANT: SOUTH EAST QUEENSLAND WATER

By Andrew Findlay, P.E., MWH

The Queensland government developed the Western Corridor Recycled Water Scheme to provide an alternative source of water in response to a period of prolonged drought in South East Queensland, Australia. The Western Corridor project consisted of three advanced water treatment plants (AWTPs), Gibson Island, Luggage Point, and Bundamba; 12 pump stations; and more than 200 km of pipeline. The project was originally designed to deliver purified reclaimed water to Southeast Queensland's main water supply reservoir, Lake Wivenhoe, for indirect potable use and to various industrial users, predominately coal-fired power stations. The Gibson Island AWTP achieved substantial completion in 2008.

PLANT DESCRIPTION

The Gibson Island AWTP was designed to treat secondary effluent from the Gibson Island and Luggage Point wastewater treatment plants (WWTPs), with the capacity to produce 100 ML/d (26 mgd) of treated water. Figure 9-28 shows a schematic of this process. The feedwater design basis is listed in Appendix F, Table F-5.

The Gibson Island AWTP was designed to meet the Australian Drinking Water Guidelines (ADWG 2004), as well as additional parameters required by the Queensland government, which is the owner. Appendix F, Table F-6, shows the key treated water parameter limits.

Subsequent to the design of the Gibson Island AWTP, the Australian Guidelines for Water Recycling (2000) for augmentation of drinking water supplies were issued, and the Queensland Public Health Regulation (2005) was updated to include requirements for recycled water for the augmentation of drinking water supplies. These documents set standards for water quality parameters and for the removal of pathogens for recycled water for indirect potable reuse.

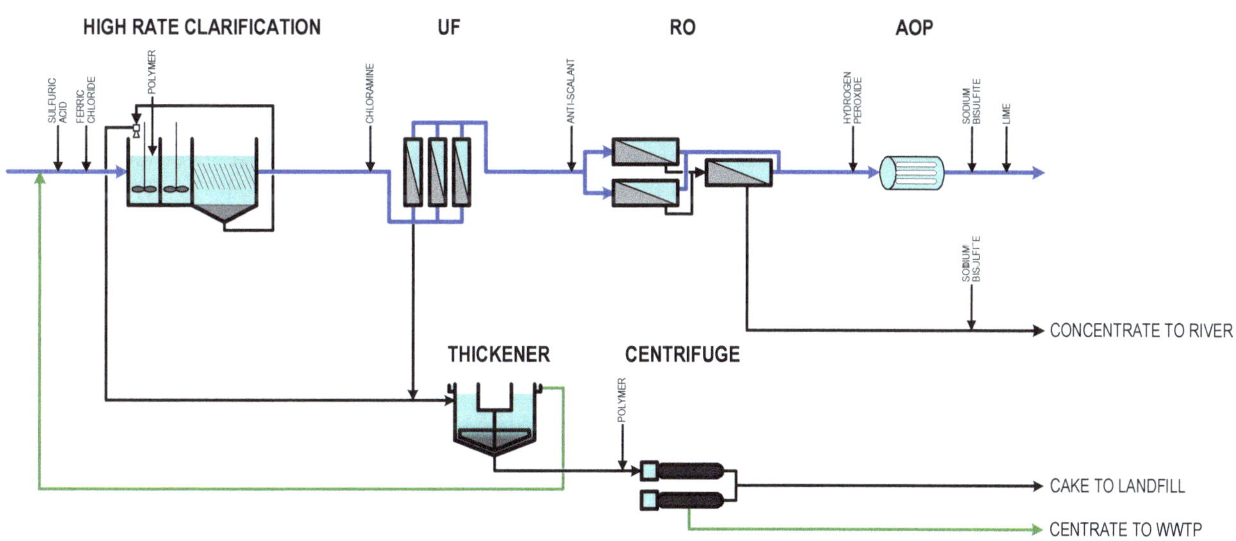

Figure 9-28 Gibson Island advanced wastewater treatment process schematic

In recognition of the high level of pathogen removal required by public health regulators to allow discharge of recycled water into drinking water supplies, each AWTP in the Western Corridor Project was required to use a multiple-barrier approach for pathogen removal. Additionally, the requirement to reduce total dissolved solids, total nitrogen, and endocrine-disrupting compounds (that is, estrone, 17-beta estradiol) drove the design to be based on reverse osmosis (RO), as RO is used for the removal of dissolved compounds and as a physical barrier to microbial material. For the Gibson Island AWTP, the barriers to pathogens included (in sequence) ultrafiltration (UF), RO, advanced oxidation (AO), and chlorination. A process schematic (see Figure 9-30) summarizes the major processes. Additionally, the design incorporated a high-rate clarification process (Actiflo) as pretreatment to the UF system, primarily for phosphorus removal, and gravity thickening of solids from the clarification process and ultrafiltration backwash prior to dewatering by centrifugation. The plant was required to achieve 82 percent recovery, that is, treated water output must be 82 percent or greater of the wastewater flow to the plant, and an availability, that is, percentage of time the plant is capable of producing at capacity, of 98 percent.

Because phosphorus is rejected efficiently by RO membranes, the phosphorous concentration increases on the feed side of the membranes. This RO concentrate, which is discharged to the Brisbane River, was required to have a total phosphorus content of 4 mg/L or less. However, it was calculated that the resultant RO concentrate would have a total phosphorus content in excess of 25 mg/L. Therefore, to achieve the discharge limit for the RO concentrate, phosphorus removal was required. Raw feedwater was pretreated for phosphorus removal, instead of removing phosphorus from the RO concentrate stream, due to the added benefit of a reduction in the likelihood of calcium phosphate scaling during RO.

Phosphorus removal is the first step in recycled water processing at the Gibson Island AWTP. Initially, ferric chloride is added to the raw feedwater to precipitate phosphate. After pH adjustment to 6.5, water flows into the clarification process, which was designed to achieve an effluent dissolved phosphorus concentration of 0.5 mg/L, where microsand and polymer are added and mixed to create heavy floc particles that rapidly settle in the clarification zone. Clarified water then flows by gravity to the UF feed tank. Sludge drawn from the clarification zone is recycled to a hydrocyclone where the relatively light ferric phosphate precipitates are separated from the microsand by centrifugal force. The precipitates flow to solids handling, and the microsand is returned to the clarification process.

As clarified water flows into the UF feed tank, chloramine is added. Chloramine is dosed as preformed monochloramine to ensure formation of monochloramine (versus dichloramine) and to reduce the likelihood of *N*-nitrosodimethylamine (NDMA) formation. A chloramine residual of approximately 3.5 mg/L as Cl_2 is targeted in the UF feed to control biological growth in the downstream membrane systems. Water is pumped from the UF feed tank through 500-µm strainers to the UF system.

UF is provided for removal of suspended and colloidal solids as pretreatment for the downstream RO system and to provide a barrier for microbial material. The UF system used at the Gibson Island plant is a CP360 system by Siemens (now Evoqua; Memcor) designed to produce filtrate with turbidity less than 0.1 ntu and a silt density index (SDI) of 3 or less. The system was also designed to achieve a minimum log removal value of 4 for *Giardia* and *Cryptosporidium*. Design criteria are shown in Appendix F, Table F-7.

During early operation of the plant, the UF system experienced significant fouling caused by polymer carryover from the upstream clarification process. Early attempts to resolve polymer fouling resulted in fouling of the UF system by solids, which also caused integrity issues with the membranes. These two problems were resolved with the addition of a flushing step in the backwashing of the UF (to reduce solids fouling) and the addition of coagulant dosing (aluminium chlorohydrate) to the clarification process effluent (to mitigate polymer fouling). Resolution of the UF solids fouling problem also resolved the membrane integrity issues (Findlay & Layson 2012).

Filtrate from the UF system flows to the RO feed tank where it is pumped by low-pressure forwarding pumps. Antiscalant and sulfuric acid are added (target pH in the RO feed is 6.3) prior to filtration through 5-μm cartridge filters. Filter effluent then flows to the high-pressure feed pump of each RO unit. The RO system used at the Gibson Island AWTP was supplied by GE. Design criteria for the RO system are listed in Appendix F, Table F-8.

Concentrate generated by the RO system is treated with sodium bisulfite to provide dechlorination prior to discharge to the Brisbane River. The RO permeate flows to the AO system. This system, which consists of hydrogen peroxide dosing and a UV system, is used to remove low-molecular-weight compounds and other compounds that pass through the RO membranes (such as NDMA), as well as for the inactivation of pathogens. The AO system is designed to achieve 1-log removal of NDMA and 0.5-log removal of 1,4-dioxane. The UV component consists of four units in a three-duty, one-standby arrangement. Each UV unit is equipped with six reactors, each housing 72 low-pressure, high-output lamps.

Downstream of the AO system, sodium bisulfite is added to quench residual hydrogen peroxide prior to addition of sodium hypochlorite for disinfection (Findlay, Poinapen, & Walker 2012). Lime and carbon dioxide are also added to the treated water to achieve the pH, alkalinity, hardness, and calcium carbonate precipitation potential requirements.

Residuals from the process are treated as follows. Solids from the clarification process and UF backwash are combined in a balance tank prior to being pumped to a combined inclined-plate clarifier–gravity thickener. Supernatant from the clarifier is returned to the head of the AWTP for reprocessing, and thickened solids are pumped to centrifuges for dewatering. The resultant sludge cake is trucked off site for disposal in a landfill. Centrate and neutralized clean-in-place waste from the UF and RO units are pumped to the Gibson Island WWTP as trade waste.

As discussed previously, high levels of pathogen removal are required for augmentation of drinking water supplies with recycled water (Australian Guidelines for Water Recycling 2008). The-log removal credits assigned to each barrier in the process are shown in Table 9-6. It should be noted that the WWTP upstream of the AWTP is credited with some of the log removal required.

CAPITAL COSTS

Table 9-7 provides the overall plant and equipment costs for the membrane systems.

Table 9-6 Gibson Island advanced wastewater treatment plant log-removal values

	Log Removal Value			
Process	Viruses	Bacteria	Protozoa	Helminths
Wastewater treatment plant	2	2	1	1
Ultrafiltration	0	3.5	3.5	3.5
Reverse osmosis	2	2	2	2
Ultraviolet disinfection–advanced oxidation	4	4	4	2
Chlorine disinfection	2	2	0	0
Total LRV provided	10	13.5	10.5	8.5
Required LRV (augmentation of drinking water supplies)	9.5	8	8	8
Required LRV (other uses)	6.5	5	5	5

Abbreviation: LRV, log-removal value.

Table 9-7 Gibson Island advanced wastewater treatment plant costs

Gibson Island advanced water treatment plant	$320M AUD
Ultrafiltration system	$15M AUD
Reverse osmosis system	$30M AUD

CONCLUSION

The Gibson Island AWTP treats secondary wastewater effluent to meet ADWG, Australian Guidelines for Water Recycling, and the Queensland Public Health Regulation. Appendix F, Table F-9, provides a comparison of the key treated water parameter limits defined by the ADWG and the owner's requirements with plant performance. Appendix F, Table F-10, provides a comparison of some of the requirements of the Australian Guidelines for Water Recycling and the Queensland Public Health Regulation with plant performance (Gibson Island Proving Period Report). The Gibson Island AWTP meets or exceeds all stipulated requirements.

During implementation of the Gibson Island AWTP, the Queensland government indicated that treated water from the AWTPs would not be directed to Lake Wivenhoe unless water levels in the lake were below 40 percent. This was a change from the original plan, which had no restriction on discharge of treated water to Lake Wivenhoe based on water level.

In 2011, significant rainfall filled Lake Wivenhoe, which resulted in the Queensland government deciding to "mothball" the Gibson Island AWTP. Currently, the plant is decommissioned and is being managed so that it can be brought back on line within six months.

PROJECT PHOTOS

Figure 9-29 Gibson Island advanced wastewater treatment plant ultrafiltration unit

Figure 9-30 Gibson Island advanced wastewater treatment plant reverse osmosis unit

Figure 9-31 Gibson Island advanced wastewater treatment plant. In the foreground are ultraviolet units, on the left are ultrafiltration units, and in the back are reverse osmosis units.

REFERENCES

Australian Drinking Water Guidelines. 2004. National Health and Medical Research Council, National Resource Management Ministerial Council.

Australian Guidelines for Water Recycling. 2008. Augmentation of Drinking Water Supplies, May. Environment Protection and Heritage Council, National Health and Medical Research Council, Natural Resource Management Ministerial Council.

Findlay, A.S. & Layson, A.J. 2012. Overcoming Membrane Performance Issues Resulting from Upstream Polymer and Ferric Chloride Addition. WEFTEC 2012 Conference Proceedings.

Findlay, A.S.; Poinapen, J.; & Walker, T. 2012. Changing Hydrogen Peroxide Quenching in an AOP to Reduce Chlorate Levels in Purified Recycled Water. WEFTEC 2012 Conference Proceedings. New Orleans, LA.

Gibson Island Recycled Water Management Plan

Gibson Island Proving Period Report

Queensland Public Health Regulation. 2005. Reprint 2A effective 30 October 2009, Office of Queensland Parliamentary Counsel.

CASE STUDY—LUGGAGE POINT ADVANCED WATER TREATMENT PLANT: SOUTHEAST QUEENSLAND WATER

By Larry Schimmoller and Jim Lozier, CH2M HILL

To provide an alternative source of water in response to a period of prolonged drought in Southeast Queensland, Australia, the Queensland government developed the Western Corridor Recycled Water Scheme. The Western Corridor project consisted of three advanced water treatment plants (AWTPs): Bundamba, having an initial treated water capacity of 20 ML/d, subsequently expanded to 66 ML/d; Gibson Island, having a treated water capacity of 100 ML/d; and Luggage Point (LP), having a treated water capacity of 70 ML/d. This case study describes the Luggage Point AWTP, on which work began in November 2006* and handover was completed in June 2011 following a two-year plant demonstration period.

INFLUENT CHARACTERISTICS AND TREATED WATER QUALITY REQUIREMENTS

The LP AWTP was designed to treat secondary effluent from the LP wastewater treatment plant (WWTP). The AWTP feedwater design basis is provided in Appendix F, Table F-11, based on historical quality of effluent from the WWTP.

In addition to complying with the Australian Drinking Water Guidelines (2004), treated water from LP AWTP required compliance with water quality parameters shown in Appendix F, Table F-12, which included parameters for limiting the concentration of inorganic nutrients (nitrogen and phosphorus) for control of eutrophication in the receiving catchment (Wivenhoe Dam), minimizing corrosivity (stability) of the treated water during conveyance, and addressing the emerging health-related issues associated with microconstituents (for example, *N*-nitrosodimethylamine [NDMA] and endocrine-disrupting compounds).

TREATMENT PROCESS

The LP AWTP includes the following major treatment processes:

- Rapid mix and flocculation,
- Inclined plate clarification,
- Microfiltration (MF),
- Cartridge filtration,
- Reverse osmosis (RO),
- Ultraviolet light/advanced oxidation (UV/AO) with hydrogen peroxide, and
- Stabilization.

A process flow diagram for the plant, as originally designed, is shown in Figure 9-32. During the plant demonstration period, an MF backwash treatment system that consisted of a packaged flocculation and inclined-plate settler unit was installed to reduce the solids returned to the raw water tank from this residuals stream.

The plant was required to achieve 82 percent recovery and an availability, that is, percentage of time the plant is capable of producing at capacity, of 90 percent.

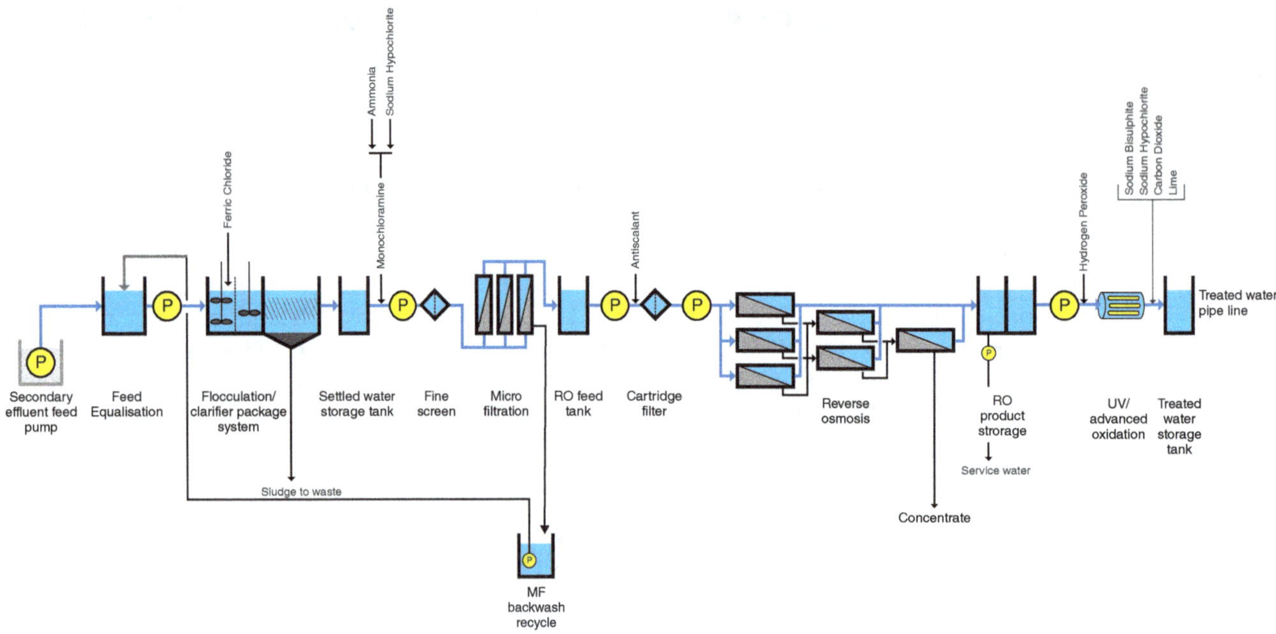

Figure 9-32 Process flow diagram for the Luggage Point advanced water treatment plant

Abbreviations: MF, microfiltration; RO, reverse osmosis.

Chemical Clarification and Chloramination

Rapid mix, flocculation, and clarification were provided to reduce the level of phosphorus in the WWTP effluent to a target of 0.5 mg/L. This was necessary for two reasons: to reduce the phosphorus concentration in the RO concentrate and to maximize recovery of the RO system. The RO concentrate is discharged to the Brisbane River, which required phosphorus levels to be less than 4 mg/L. Assuming nearly 100 percent phosphorus rejection by RO, this required a RO feed phosphorus level of less than 0.6 mg/L at the RO design recovery of 85 percent. In addition, if unreduced, RO recovery would have been limited by calcium phosphate scaling to 75 percent. To achieve the target phosphorus levels in the clarified water, ferric chloride dosing of between 15 and 120 mg/L is required. Solids discharged from the clarifier are conveyed to a gravity thickener. Supernatant is recycled back to the raw water storage tank, and thickened solids are pumped to the centrifuges. Centrate is returned to the WWTP.

The clarified water is dosed with preformed monochloramine to control biological (bacterial) fouling in the RO system; however, it also benefits such fouling control in the MF and cartridge filtration systems. Monochloramine is preformed by dosing both aqueous ammonia and sodium hypochlorite into carrier water (RO permeate) at pH 9.0. Preformation at this pH with a controlled chlorine-to-ammonia molar ratio maximizes the formation of monochloramine and minimizes the formation of dichloramine. This, in turn, minimizes the formation of disinfection by-products, most notably, NDMA, which have been shown to be formed by the reaction between dichloramine and nitrogenous compounds in wastewater effluent (Luggage Point AWTP Performance Test Report). Table 9-8 lists the avoidance of NDMA formation using preformed monochloramine, as measured during pilot testing conducted during the predesign phase of the LP AWTP project.

RO biofouling control has been demonstrated with as little as 0.5 mg/L of monochloramine in the RO feedwater. However, a dose of 2.0 mg/L monochloramine was used for LP AWTP design.

Table 9-8 *N*-nitrosodimethylamine levels prior to and following preformed chloramination

	N-nitrosodimethylamine Concentration, ng/L				
Location	Feb. 21, 2008	Feb. 26, 2008	Feb. 28, 2008	Mar. 4, 2008	Mar. 6, 2008
Feed	10	14	10	5	7
Settled water	10	13	10	< 5	5
Microfiltration feed*	9	13	9	< 5	7
Reverse osmosis feed	10	12	10	5	6

* Monochloramine addition point.

Microfiltration

The MF system comprises 11 Pall Microza valve/module racks and associated systems for fouling management, including reverse filtration pumps and air scour blowers and acid (using both citric and sulfuric acid) and caustic–chlorine clean-in-place (CIP) systems. CIP systems are also used for enhanced flux maintenance. The MF system is designed to produce up to 85 ML/d with only the duty racks in service. Figure 9-33 A shows two of the module racks.

MF is provided for removal of suspended and colloidal solids in order to minimize particulate fouling on the downstream RO system and to provide a barrier to the passage of all pathogens except viruses. The Pall Microza MF system is designed to produce filtrate with turbidity less than 0.1 ntu and a silt density index of < 3 (90th percentile). The system was also designed to achieve a log-removal value of 5 or greater for *Giardia* and *Cryptosporidium*. Basic design criteria for the MF system are shown in Appendix F, Table F-13.

During the predesign pilot testing phase, the MF system experienced more rapid fouling than anticipated. This was determined to be caused by iron (and manganese) accumulation that was not being effectively controlled by the manufacturer's recommended procedure of daily chlorine and weekly acid-enhanced flux maintenance (A-EFM). Modifications, including increasing coagulation pH to *greater than or equal to* 6.5 to minimize soluble iron carryover in the clarified water, modifying the A-EFM regime to acid–acid–chlorine on three-day intervals, and permeating a portion of the A-EFM solution through the membrane to dissolve intrapore deposits, were effective in minimizing fouling and allowing long periods (greater than 60 days) of operation between CIPs (Mueller, Foster, & Morgan 2008). The impact of these modifications on MF permeability is shown in Figure 9-34.

MF filtrate flows to the RO feed tank (flow balancing) and is then pumped through 5-μm cartridge filters and into the suction of the high-pressure pumps. The filtrate is dosed with antiscalant prior to entering the cartridge filters. Although the cartridge filters were designed strictly as "fail-safe" filters to protect the RO element in case of significant particulate ingress through the RO feed tank or catastrophic failure of the MF system, the filters were shown to retain an appreciable amount of iron solids, most likely from iron precipitation post-MF.

Reverse Osmosis System

The RO system is designed to remove a variety of contaminants in the microfiltered effluent, including nutrients (nitrogen and phosphorus), bulk dissolved organics, total dissolved solids, pathogens, and microcontaminants. The RO system (manufactured by Doosan, Korea), includes four identical RO skids along with CIP and flushing systems. Figure 9-35 shows one of the RO skids. Design criteria for the RO system are listed in Appendix F, Table F-14.

Figure 9-33 Pall Microza microfiltration racks

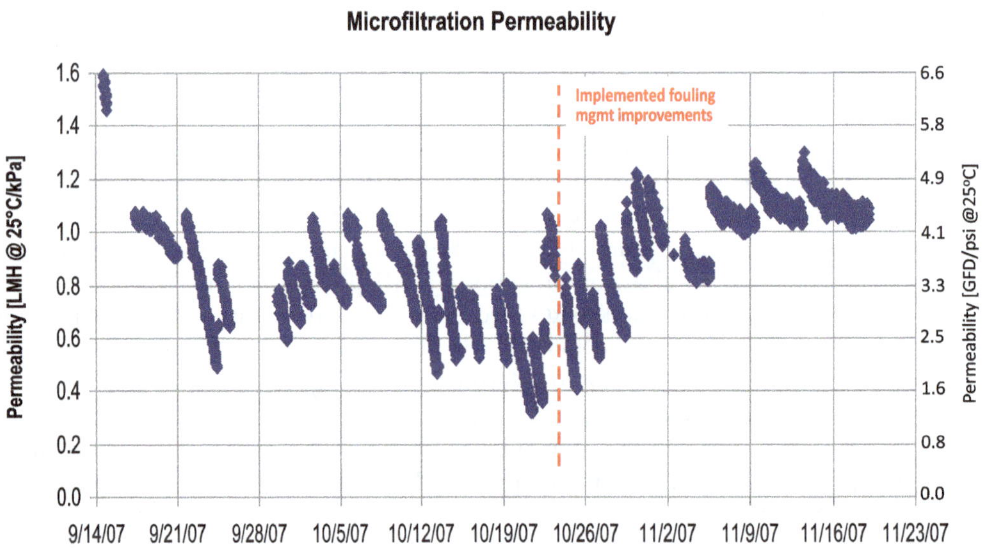

Figure 9-34 Beneficial impact of fouling management modifications on microfiltration permeability

Figure 9-35 Three-stage reverse osmosis skid. The hydraulic turbocharger is shown on bottom left.

Concentrate generated by the RO system is treated with sodium bisulfite to extinguish residual chloramines prior to discharge to the Brisbane River. The RO permeate flows to the UV–AO system.

Performance of RO train 1 from the period of start-up through Sept. 3, 2010 (705 days), is shown in Figures 9-36, 9-37, and 9-38. Normalized product flow during this period was very stable, indicating very little fouling. Although variable, based on the impact of variations in both feedwater temperature and salinity that are not effectively corrected by normalization, normalized salt passage exhibited no net increase in the nearly two-year operating period. Some increases in normalized differential pressure were observed. These increases were effectively reduced by chemical cleaning using a combination of sodium hydroxide and sodium ethylenediaminetetraacetic acid. Together with metals analysis of the spent cleaning solution, it was shown that the particulate fouling was attributed primarily to co-deposition of dissolved organics and iron. Differential pressure increases were greatest in stage 3 where organics and iron concentrations were highest. The minimal increases in stage 1 suggest biofouling was well controlled by chloramination.

Ultraviolet–Advanced Oxidation System

The UV–AO system includes a hydrogen peroxide storage and dosing system followed by a closed-vessel UV system. It is used to oxidize low-molecular-weight and weakly charged microcontaminants that are poorly rejected by the RO process (for example, NDMA) and are present in the RO permeate at levels greater than the treated water requirements. The UV portion of the system is also used to inactivate pathogens that remain following MF and RO. Microcontaminants are removed by both photolysis and oxidation via hydroxyl radicals that form when hydrogen peroxide reacts with UV light. The system is designed to achieve 1-log removal of NDMA and 0.5-log removal of 1,4-dioxane. The UV system consists of three Trojan UV PHOX units in a 3 duty and 1 standby arrangement. Figure 9-39 shows one of the UV units. Each UV unit is equipped with six reactors, each housing 72 low-pressure, high-output lamps.

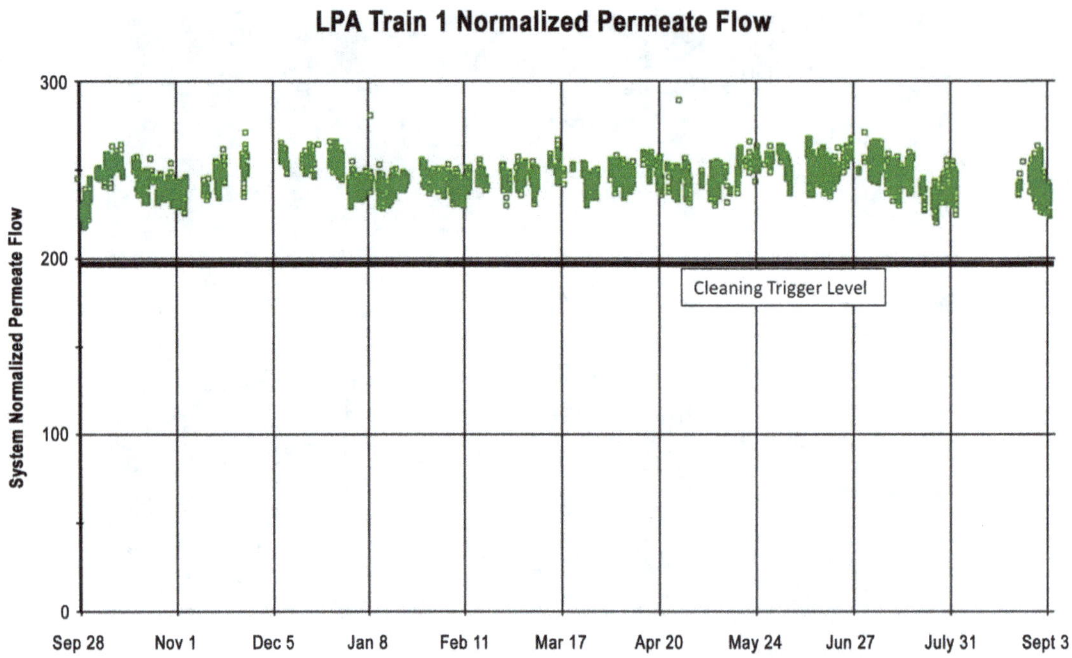

Figure 9-36 Normalized permeate flow for the Luggage Point advanced water treatment plant reverse osmosis train 1 from start-up to Sept. 3, 2010

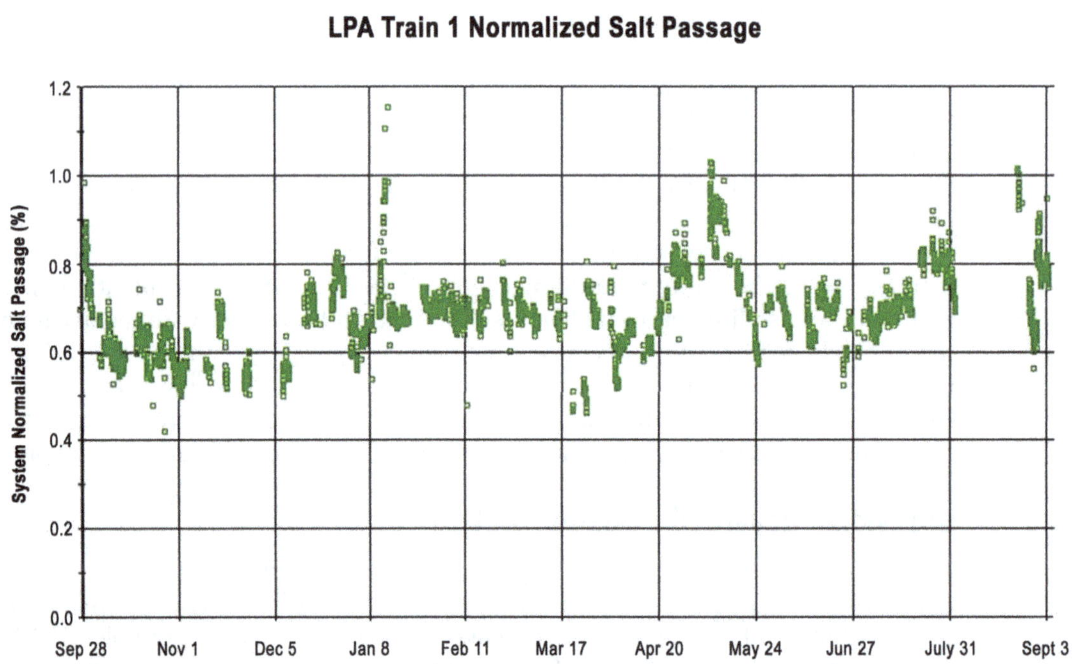

Figure 9-37 Normalized salt passage for the Luggage Point advanced water treatment plant reverse osmosis train 1 from start-up to Sept. 3, 2010

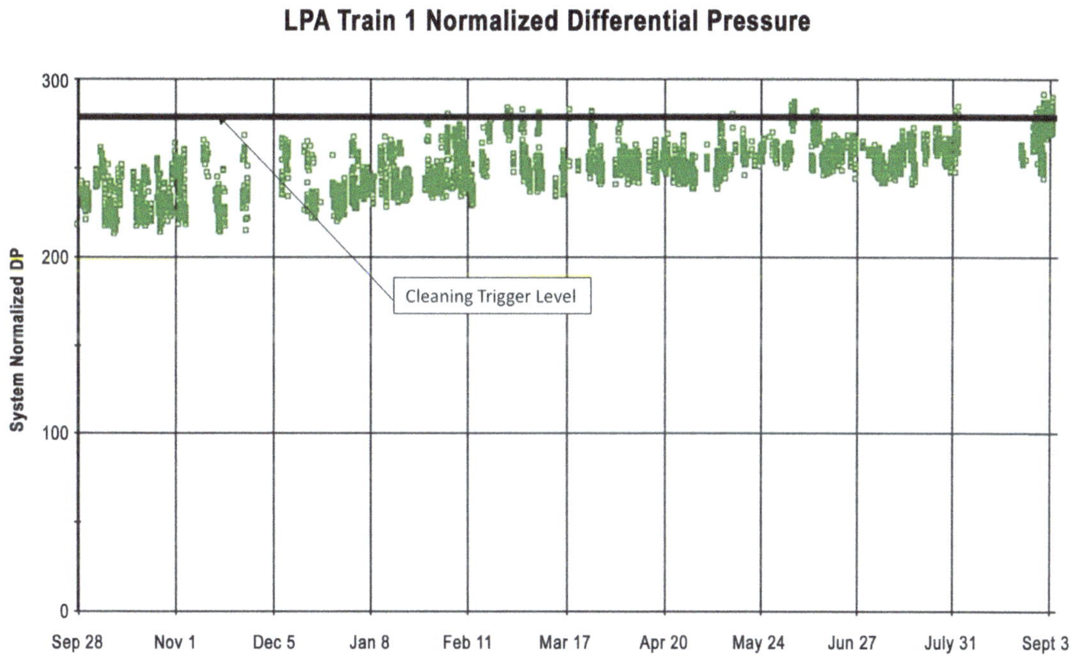

Figure 9-38 Normalized differential pressure for the Luggage Point advanced water treatment plant reverse osmosis train 1 from start-up to Sept. 3, 2010

Figure 9-39 One of the four Trojan ultraviolet PHOX units

Sodium bisulfite is added to the UV effluent to quench residual hydrogen peroxide prior to addition of sodium hypochlorite for disinfection. Lime and carbon dioxide are also dosed to the UV effluent for stabilization, that is, to achieve the required treated water pH, alkalinity and hardness, and calcium carbonate precipitation potential requirements of the treated water as listed in Appendix F, Table F-12.

PLANT PERFORMANCE RELATIVE TO TREATED WATER REQUIREMENTS

Following commissioning of the LP AWTP, a 30-day performance test was completed to demonstrate the treatment system's ability to comply with treated water quality requirements. The results from the performance test (Sedlak & Kavanaugh 2006), summarized in Appendix F, Table F-15, indicate that all treated water requirements were met with the exception of one exceedance for calcium carbonate precipitation potential (a measure of corrosivity) out of 28 samples.

COSTS

The capital cost for the LP AWTP, in Australian dollars, was $310 million.

SUMMARY

The 70-ML/d LP AWTP was designed, constructed, and commissioned. A 30-day performance test was completed in January 2010 in which the plant successfully met all treated water and production requirements. Successful plant handover occurred in June 2011. The plant was operated on a near-continuous basis through February 2011 when, following heavy rains in December 2010, area dams were filled to capacity and the need for recycled water production by the Western Corridor scheme was significantly reduced. A decision was made by the scheme operator (SEQwater) to place the Luggage Point, Bundamba, and Gibson Island AWTPs in standby mode, where they will remain until levels in the area dams drop to 60 percent of capacity.

During the two-plus years of operation, the LP AWTP reliably produced up to design capacity with both MF and RO systems experiencing stable performance and minimal fouling.

REFERENCES

Australian Drinking Water Guidelines. 2004. National Health and Medical Research Council, National Resource Management Ministerial Council.

Luggage Point AWTP Performance Test Report.

Mueller, P.A.; Foster, L.; & Morgan, R. 2008. *Pretreatment to Optimize MF/RO Design and Operation in a Municipal Secondary Effluent Application.* WEFTEC 2008 Conference Proceedings. WEFTEC Conference, Chicago, IL.

Sedlak, D. & Kavanaugh, M. 2006. *Removal and Destruction of NDMA and NDMA Precursors during Wastewater Treatment.* Alexandria, VA: Water Reuse Foundation.

CASE STUDY—CAPCO INDUSTRIAL WASTEWATER RECLAMATION FACILITY

By Joseph M. Wong, Brown and Caldwell

The China American Petrochemical Co., Ltd. (CAPCO) purified terephthalic acid (PTA) plant, located in Lin Yuan Petrochemical Industrial Complex, Kaohsiung County, Taiwan, is one of the largest PTA producers in the world, producing more than 1.4 million metric tons per year. CAPCO is a joint venture of BP and CPC Corporation, Taiwan. The Lin Yuan PTA plant consumes 12.7 mgd of freshwater for its operations—approximately 52 percent for cooling tower makeup, 40 percent for feed to the ion exchange–based deionization (DI) system, which produces pure water for boiler feed and PTA process production, and the rest for miscellaneous plant uses. The plant also generates approximately 5.3 mgd of wastewater from the manufacturing processes, cooling tower blowdown, DI system regeneration, and miscellaneous plant operations. The wastewater is treated and discharged to a local adjoining (or co-located) wastewater treatment plant (WWTP) for ocean discharge with the approved quota of 5.8 mgd.

In the 1990s, PTA was in high demand and was the most profitable petrochemical worldwide. Dry periods occurred almost every year and often caused water rationing in the Kaohsiung area that led to PTA production curtailment. During dry periods, the total dissolved solids (TDS) content of the water supplied to the plant increased to the point that the DI system could not produce enough DI water to meet plant demands. The DI beds required regeneration every 4 to 5 hours instead of the normal 8 hours due to high influent TDS loadings. Another problem was that the WWTP effluent discharge sewer was overloaded during heavy storm events. Government authorities sometimes ordered the plant to cease discharging effluent and to store it in onsite ponds. This was problematic when the ponds were already full due to an extended period of nondischarge. Due to local conditions, water and wastewater capacities were the main limitations if CAPCO wanted to expand its PTA production. Finally, there was constant pressure on the government to drastically increase water and wastewater fees as a way to curtail water consumption.

CAPCO's consultant conducted a water resource management study to evaluate alternatives to alleviate these water-related problems. The study concentrated on improvements to the two major water uses. Many water conservation and minimization opportunities were identified, evaluated, and recommended (Wong 1995). Some improvements had been implemented; however, the study's key finding was the identification of a major wastewater recovery project at the plant. The plant's main WWTP was well designed and well operated, with relatively good and stable effluent quality. The plant also consumed a large quantity of high-purity DI water. This presented a unique opportunity for using an advanced membrane-based treatment system to supply predemineralized water to the DI system as makeup. The project appeared to be economically attractive, with a payback of approximately five years, and could solve many of the water-related problems. After the study, CAPCO conducted a comprehensive pilot testing program to verify process design criteria and then designed and constructed a 1.75-mgd membrane-based (ultrafiltration–reverse osmosis [UF–RO]) in-plant water reclamation system with a recovery of 73 percent. The reclamation system was commissioned in 2000 and has been operating successfully, except for a six-month period in 2008 due to economic reasons. The UF–RO reclamation system is the first of its kind for the worldwide petrochemical industry and the first significant membrane-based water reuse system in Taiwan.

WASTEWATER RECLAMATION SYSTEM DESCRIPTION

Figure 9-40 shows how the wastewater reclamation system relates to the plant's water and wastewater systems (Wong 1998). The three main effluent streams from the petrochemical plant include cooling tower blowdown (CTB), organic wastewater treatment system (OWWTS) effluent, and inorganic wastewater treatment system (IWWTS) effluent. The wastewater reclamation system treats a combination of OWWTS effluent and CTB and recycles the reclaimed water to supply the DI system. Rejects from the membrane-based treatment system are discharged with the IWWTS effluent to the regional (joint) WWTP for ocean disposal. The IWWTS effluent is not reclaimed because of its high TDS and silica concentrations. Reusing the reclaimed water to supply the DI system makes the project economically feasible because of the high quality of the RO permeate in comparison to the normal DI system feedwater from the municipal water company.

Typical characteristics of the two reclaimed effluent streams compared to the effluent discharge standards are presented in Appendix G, Table G-1 (Wong 1999). The individual effluent streams meet the discharge requirements but require comprehensive treatment before they can be reused as DI system feedwater. However, the two streams appear to complement each other in producing a suitable blended feed for membrane treatment. For example, the CTB is high in hardness and silica concentrations, but the OWWTS effluent is low in these two constituent concentrations; thus, a high RO recovery could be achieved from the blended feed. The OWWTS effluent contains cobalt and manganese because these are used as catalysts in the manufacturing process.

Figure 9-41 shows a block flow diagram of the wastewater reclamation and treatment system (Wong 1999). The OWWTS effluent and CTB are blended prior to chemical oxidation to precipitate heavy metals such as cobalt, manganese, and iron. The dual-media filter (DMF) removes the precipitates and suspended solids. The granular activated carabon (GAC) filter removes residual organics to protect downstream membrane systems and to control the chemical oxygen demand (COD) concentration in the RO reject stream so that it remains within the discharge limit of 100 mg/L. The cartridge filter removes particles larger than 10 μm. UF removes suspended and colloidal particles larger than 0.01 μm, including colloidal silica and large-molecule organics, to protect the RO membrane and enhance the RO recovery rate. The UF selected is the hollow-fiber type, which can be backwashed periodically to renew flux. The UV reactor disinfects the RO feedwater when necessary. The spiral-wound RO is the heart of the treatment system and removes dissolved inorganics and organics, silica, and residual colloidal materials. After degasification to remove carbon dioxide formed in the acidification of the RO feedwater, the reclaimed water, which is of superior quality compared to the water supplied by the water company, is supplied to feed the DI system. The DI system regeneration frequency is reduced by 70 to 80 percent by switching to this water source because of the low TDS and silica concentrations. This reclamation project reduces both water purchase from the water purveyor and wastewater discharge loading.

ECONOMIC ANALYSIS FOR THE REUSE PROJECT

CAPCO wanted to determine the reclamation project's economic feasibility prior to its implementation. The net investment cost and annual operation and maintenance cost of a 15,000 4.0 mgd (m3/d) (influent) reclamation project were estimated to be approximately $20 million and $2.5 million, respectively. Tables 9-9 and 9-10 present the cost breakdowns (Wong 1998). Using reclaimed water to supply the DI system would provide several major direct cost savings for the plant, including cost reductions in water purchase, DI system regeneration chemical (acid and caustic) and water usage, wastewater discharge fees, and IWWTS chemicals. The amount of water recycled to the DI system replaces the same amount of water that was purchased from the water purveyor. In addition, the DI system consumes approximately 25 to 30 percent of its own product water for the regeneration process. The reduction in regeneration frequency further reduces the amount of water to be purchased and the amount of inorganic (regeneration) wastewater to be treated. Table 9-9 shows a project cost analysis for the reclamation project (Wong 1998). The payback would be 4.6 years at the present water purchase cost and would be shorter at increased water cost. This payback does not include indirect cost savings such as avoiding production cutbacks due to water rationing, concomitant expansion of the ion-exchange-based DI system, and alleviation of wastewater discharge overloading during heavy storms. Hence, this project is economically very attractive.

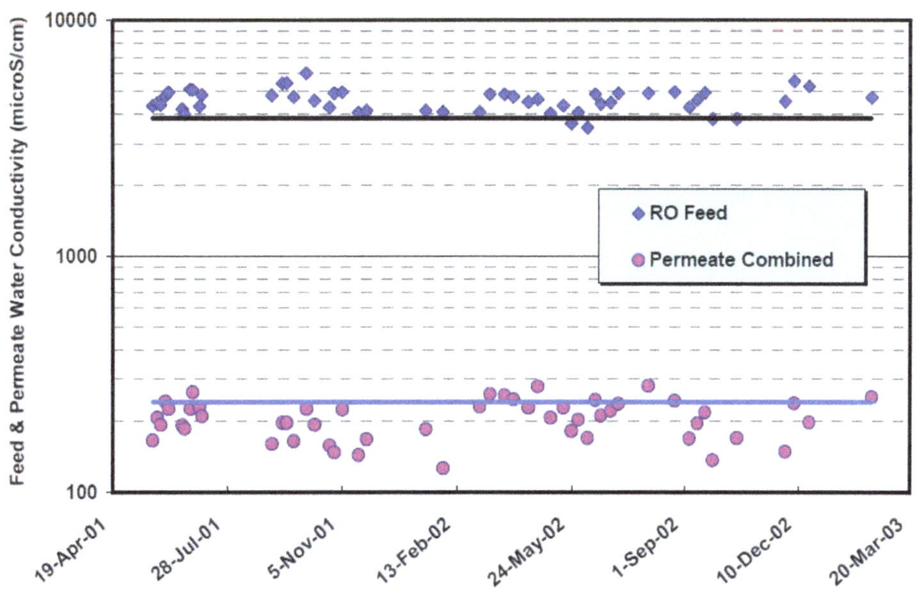

Figure 9-40 Relation of wastewater reclamation system with plant water systems

Abbreviations: DI, deionization; RO, reverse osmosis; UF, ultrafiltration; WWTS, wastewater treatment system; C.T., cooling tower.

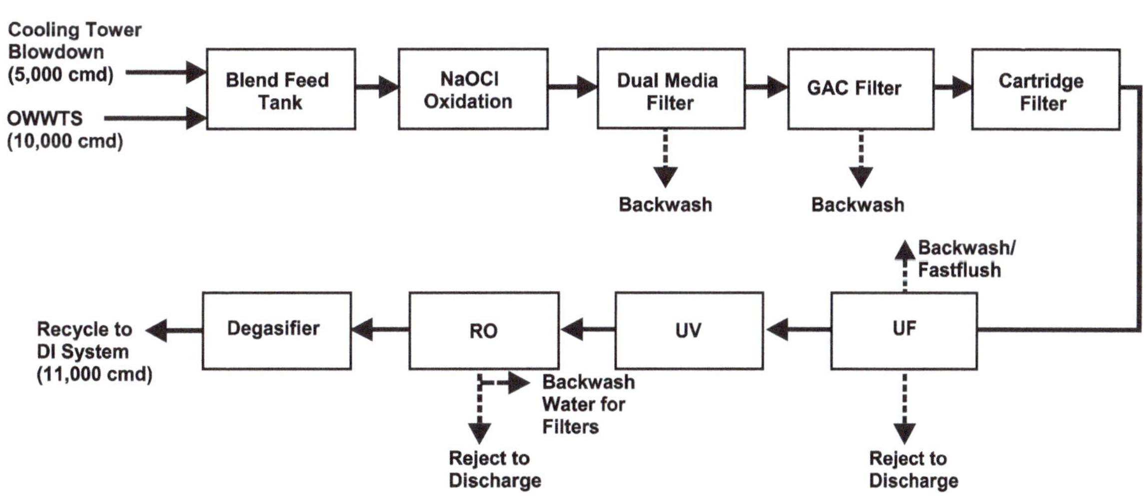

Figure 9-41 Wastewater recovery treatment system block flow diagram

Abbreviations: GAC, granular activated carbon; RO, reverse osmosis; UF, ultrafiltration; UV, ultraviolet.

Table 9-9 Estimated capital cost of the wastewater reclamation system: 3.0 mgd of product water, 1998

Water Reclamation System	Cost, US dollars	Subtotal	Running Total
Equipment			
Horizontal dual-media filters	400,000		
Granular activated carbon filters	500,000		
Cartridge filters	100,000		
Hollow-fiber ultrafiltration system	2,500,000		
Ultraviolet system	150,000		
Reverse osmosis system	2,000,000		
Degasifier	100,000		
Pumps	500,000		
Tanks	500,000		
Total major equipment cost		6,750,000	6,750,000
Equipment installation at 100 percent	6,750,000		
Piping	1,000,000		
Instrumentation and control	500,000		
Building	1,000,000		
Electrical	1,000,000		
Total construction cost		10,250,000	17,000,000
Contingency at 10 percent	1,700,000		
Legal, administrative, and engineering at 20 percent	3,400,000		
Total capital cost			22,100,000
Tax credit at 10 percent			<2,210,000>
Net investment cost			20,000,000
Unit capital cost ($/gal)			6.67

Table 9-10 Estimated annual operations and maintenance cost of the wastewater reclamation system: 3.0 mgd of product water, 1998

Item	Cost, US dollars
Electrical power	500,000
Chemicals	
Sulfuric acid	480,000
Sodium hypochlorite	100,000
Polymer	25,000
Antiscalant	60,000
Ultrafiltration–reverse osmosis cleaning chemicals	40,000
Reverse osmosis membrane replacement at four-year life	170,000
Ultrafiltration membrane replacement at five-year life	270,000
Ultraviolet lamp replacement at one-year life	15,000
Cartridge filter replacement	60,000
Granular activated carbon regeneration/makeup	600,000
Other maintenance and analytical materials	80,000
O&M labor	100,000
Total annual O&M cost	2,500,000
Unit O&M cost ($/1,000 gal)	2.28

Abbreviation: O&M, operations and maintenance.

Table 9-11 Project cost analysis of the wastewater reclamation system: 3.0 mgd of product water, 1998

Net investment cost	$20,000,000
Total annual savings at present water/wastewater costs	$6,885,000
Annual operations and maintenance cost	$2,500,000
Net annual savings (present cost)	$4,385,000
Simple payback (present cost)	4.56 years

FULL-SCALE IMPLEMENTATION

Based on favorable economic feasibility and technical verification by long-term pilot testing, CAPCO decided to implement a full-scale project using advanced UF–RO technologies. The initial system treats an influent flow of 9,000 2.4 mgd (m3/d) and produces 6,600 1.75 mgd (m3/d) of pure reclaimed water for feed to the DI system (Wong 1999). The system was designed in accordance with the process design criteria developed in the pilot tests. The treatment system was designed to be expandable in the future by 50 percent with minimal facility additions. Major equipment in the full-scale plant includes a large oxidation retention tank; three horizontal pressure DMF units; three vertical pressure GAC filters; five hollow-fiber UF skids with 109 cartridges each (Koch PM 100, 6 in. × 48 in., design flux of 55 gfd); a 254-nm UV reactor unit (WEDCO); three 5-μm cartridge filters; eight RO skids with TFC polyamide membranes (FILMTEC BW-30-400) in a two-stage array (design flux of 13 gfd); a fiberglass reinforced plastic degasifier; a UF clean-in-place (CIP) system and an RO CIP system; a large RO feed and filter backwash tank; a large backwash settling tank; and associated pumping, piping, control, chemical feed and storage, and electrical facilities (Wong 1999). Due to site constraints, the major treatment equipment is housed in a four-story concrete building.

The design–build project required approximately two years to complete. The total project cost was approximately $15 million. A start-up and commissioning period of two months was allocated during project scheduling. However, the start-up effort turned out to be much more complicated than anticipated due to some deficiencies in design, construction, and operational oversight (Wong 2000). After correcting the deficiencies, the water reclamation treatment system was commissioned in 2000 (Dow Chemical Company 2004).

A BRIEF REVIEW OF THE 13 YEARS OF OPERATION

During the past 13 years of operation, the reclamation plant produced several billion gallons of recycled water that met the guaranteed water quality requirements for manganese, cobalt, COD, and conductivity. The final product water quality is better than that obtained during pilot testing. Figure 9-42 shows the conductivity of RO feed and combined permeate from all trains during the first three years of operation (Dow Chemical Company 2004). These values are typical for the last 13 years of operation except when the RO elements were near the end of life. The low conductivity in the feedwater to the DI system drastically reduced the regeneration frequency and chemical cost. During the first month of operation, chlorine inadvertently reached the RO elements; the elements in five out of the eight trains were replaced. This problem was corrected, and the RO system operated very well with infrequent CIP (once a year). One CIP was initiated because of a high RO influent silt density index, which was caused by damage to the UF cartridges upstream. The average life of the RO elements was four years, which is the same as that predicted during the economic evaluation. The RO elements were replaced due to significant conductivity rise in permeate.

The UF system uses the Koch PM 100 membrane, which is made of polysulfone and has an anisotropic pore structure with nominal molecular weight cutoff of 100,000 Daltons. During pilot testing, the 5-in.-diameter cartridges were used. However, because 6-in.-diameter cartridges were available during project design, the 6-in.-diameter cartridges were installed. The average life of the UF cartridges was six years compared to the warrantied life of five years by the vendor. Before UF cartridges needed to be replaced, larger-diameter cartridges (8- and 10-in.) became available, and Koch discontinued the 6-in.-diameter cartridges. CAPCO and Koch worked together to design and convert the UF system to use 10-in.-diameter cartridges while preserving the original frames and headers (Wong 2012). This conversion leaves significant room for expansion when needed, as the 10-in.-diameter cartridges have much higher packing density for filtration surface than the 6-in.-diameter cartridges.

A review of the reclamation plant's past 13 years of operation indicates that, in spite of some start-up and operation problems, this was a very successful project. It helped alleviate the water shortage and wastewater discharge limitation problems for CAPCO and was economically profitable. This project was beneficial to the local community as more freshwater was available for domestic consumption, especially during water shortages. Furthermore, this was the first major recycled water of its kind in Taiwan and has served as a model for water reclamation for industrial plants on the island.

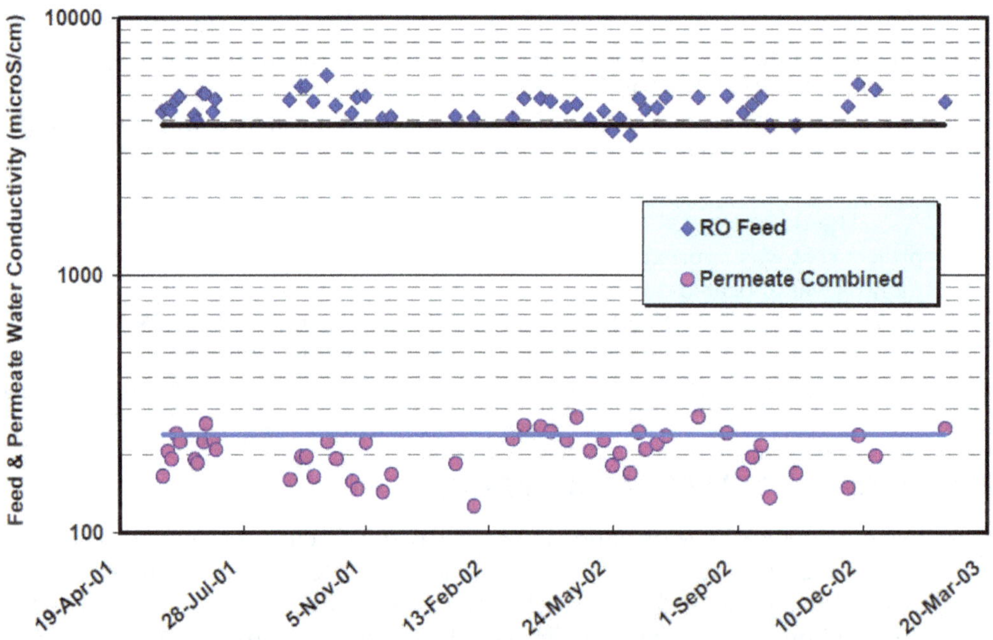

Figure 9-42 Conductivity of reverse osmosis feed and combined permeate from all trains from April 2001 to March 2003

REFERENCES

Dow Chemical Company. 2004. *FILMTEC Elements Key to Success of Integrated Membrane-Based Reclamation System in Major Petrochemical Plant in Taiwan*. FILMTEC Membranes Case History, Form No. 609-00451-0904.

Wong, J.M. 1995. *Wastewater Reclamation and Reuse in a Petrochemical Plant*. Proceedings 50th Purdue Industrial Waste Conference, Ann Arbor Press, Chelsea, MI; pp. 683–693.

———. 1998. *Can a Major Industrial Wastewater Reuse Project be Economically Feasible?* Proceedings WateReuse Symposium XIII, Oakland, CA.

———. 1999. *Testing and Implementation of an Advanced Wastewater Reclamation and Recycling System in a Major Petrochemical Plan*t. Proceedings International Association of Water Quality Conference on Wastewater Minimization and End of Pipe Treatment in Chemical and Petrochemical Industries, Merida, Yucatan, Mexico; pp. 119-124.

———. 2000. *Lessons Learned from the Design and Startup of an Advanced Industrial Wastewater Reclamation System*. Proceedings WEFTEC 2000, Water Environment Federation, Anaheim, California.

———. 2012. *The History and Present Status of the World's First Major Membrane-Based Water Reuse System in the Petrochemical Industry*. Proceedings WEFTEC 2012, Water Environment Federation, New Orleans, Louisiana.

Wong, J.M. & Weddle, S.C. 1997. *Pilot Testing of UF/RO for a Major Water Reuse Project*. Proceedings Membrane Technology Conference, American Water Works Association, New Orleans, LA; pp. 441–460.

This page intentionally blank.

Chapter 10

Future Technology Trends and Contaminants of Emerging Concern

The first part of this chapter presents some of the technological advances in low-pressure membranes (microfiltration–ultrafiltration [MF–UF]), new reverse osmosis (RO) cleaning techniques, novel membrane materials, and emerging membrane processes that can remove dissolved constituents in the applications for water reuse. The technological advances described in this chapter are by no means exhaustive. The latter part of this chapter presents a discussion of contaminants of emerging concern (CECs) and the ability of membrane processes to remove them. CECs are a major public health concern in potable water reuse (direct and indirect), which is becoming an important way to augment existing supplies in many parts of the world with limited freshwater resources and long-term drought conditions.

ADVANCES IN LOW-PRESSURE MEMBRANES

The use of MF/UF in drinking water treatment increased significantly in the mid-1990s. Since the late 1990s, MF–UF has become standard pretreatment (in place of granular media filtration) for RO in advanced water reuse applications. Over the last 20 years, most of the advances in MF–UF have been in modifying polymer chemistry for membranes to increase permeability and improve fouling resistance and fiber strength and in producing larger modules to reduce footprint and capital cost. For the most part, polymeric membranes in a hollow-fiber configuration have been used in the water reuse tertiary filtration applications. While polyvinylidene difluoride (PVDF) is the most popular membrane material used for this application, polypropylene is still used in the original 82-mgd groundwater replenishment system in Orange County, Calif. (see Chapter 9) and is also being used in its expansion. Some vendors have claimed that alternative membrane materials such as polytetrafluoroethylene and modified polyethersulfone have advantages over PVDF. However, membranes made of these materials have not gained a foothold in reuse applications. Polymeric membranes are expected to be used for the foreseeable future, and new MF–UF membrane materials will continue to become more robust, with advancements to improve permeability, improve fouling resistance and cleanability, and reduce energy consumption.

Ceramic membranes use alternative materials that are more rigid and stronger than polymeric membranes and are usually much more expensive. However, recent advancements in ceramic membranes are making them competitive with polymeric membranes for certain situations. Another recent advancement in MF–UF technology is the universal membrane skid. These advancements are discussed as follows.

CERAMIC MEMBRANES

Ceramic membranes are monolithic ceramic filter modules that serve as a filtration barrier. Made of rigid material (typically alumina, zirconia, and titania), ceramic filter modules are housed in vessels and plumbed together to create a membrane filtration system. Most advantages of ceramic membranes are related to operations (Freeman & Shorney-Darby 2011). Ceramic membranes experience fewer failures in terms of breaches to the filtration barrier than polymeric membranes. They are expected to have a service life of at least 20 years compared to 7 to 10 years for polymeric membranes, making membrane replacement cost lower. Another advantage is the tolerance of ceramic membranes to a wider variety of chemicals and chemical concentrations, which is helpful in dealing with severe fouling situations. Ceramic membranes can also endure higher pressures than polymeric membranes. This means that higher operating pressure can be used during operation, allowing for much higher flux operation (for example, greater than 100 gfd); however, there is the potential for a higher fouling rate with this higher pressure. Higher pressure tolerance also results in more efficient and much less frequent backwashing. This leads to a higher recovery, typically higher than 98 percent, compared to 93 to 96 percent for polymeric membranes. The main disadvantage is higher capital cost; the membrane elements are more costly to manufacture and the systems are equipment-intensive (Freeman & Shorney-Darby 2011). Ceramic membrane system suppliers are working to reduce capital costs so as to compete with polymeric membrane suppliers on a total life-cycle cost basis.

Tubular ceramic membranes, which have been available for several decades, are used in specific industrial applications because of their higher costs and lower packing density (AWWA 2005). However, a new generation of ceramic membranes was introduced to the drinking water market by NGK of Japan in the late 1990s (Freeman & Shorney-Darby 2011). Many ceramic membrane systems have been built in Japan, including a 10-mgd drinking water plant in Oshio and a 45-mgd plant in Yokohoma. In 2007 NGK and Fuji Electric merged to form Metawater, which markets ceramic membrane modules as well as systems. PWN Technologies of the Netherlands has researched a new ceramic membrane system configuration (CeraMac) that houses multiple Metawater ceramic membrane elements in a single vessel. The company designed and constructed a 32-mgd facility for its Andijk treatment plant, which began operation in 2014. PWN has pilot tested projects with possible applications in the United Kingdom, Singapore, Vietnam, and Australia. A pilot project is being conducted in Melbourne, Australia, to evaluate the efficacy of combining ozone and ceramic membranes for water reuse. Several successful pilot projects have been conducted in the United States, and a 10-mgd ceramic membrane drinking water plant is being constructed in Parker, Colo. (Bookout et al. 2014). Another 7-mgd Metawater ceramic MF system will be constructed for surface water treatment at the Butte-Silver Bow (Mont.) Public Works Department's Basin Creek Water Treatment Plant. Metawater offered a full 20-year replacement warranty for the ceramic modules and won the 20-year present worth competition with another qualified polymeric membrane vendor.

As ceramic membranes become more cost competitive in drinking water treatment, their use in water reuse is expected to increase, especially in industrial applications where wastewater is more difficult to treat. A specific application is the treatment of fracking flowback and produced water for reuse (Drewes et al. 2008; Zhang, Zhong, & Xing 2013).

UNIVERSAL MF–UF MEMBRANE SKID

A unique and, at times, challenging characteristic of MF–UF systems is the fact that each MF–UF membrane module is unique, with varying module dimensions and different operations and cleaning schemes. Each system requires different types and sizes of ancillary equipment. Therefore, it has been common practice to preselect a MF–UF supplier at the initial stage of a project and to complete the detailed design based on the preselected membrane system. To date, this approach has been successful for many projects. However, the end user has to use the same module over the life of the plant, even if the module's performance is not satisfactory over time. If the end user wants to try a different membrane module, the entire train has to be replaced. This may result in increasing the life-cycle cost of the MF–UF membrane system (Hyung et al. 2014). In recent years, several manufacturers have introduced the universal membrane skid, which is designed to accommodate membrane modules from various MF–UF manufacturers with minor system modifications.

Examples of universal membrane skids include the FiberFlex system (H_2O Innovation) and the Smartrack system (Infilco Degremont). The FiberFlex system can accommodate UF modules supplied by CSM, Dow, Econity, Hydranautics, and Toray. The Smartrack system can accommodate UF modules supplied by Asahi, Dow, GE, and Toray (Westerling 2014). All of the UF modules are of the pressure type. Some small-scale installations include universal MF–UF membrane skids that can accept at least two membranes. The membrane supplier provides only these membranes. When vendors that currently offer both membranes and the system start offering just membranes, implementation of the universal systems is anticipated to become commonplace. The first full-scale universal MF–UF skid system was installed at the Santa Barbara (Calif.) El Estero wastewater treatment plant.

ADVANCES IN REVERSE OSMOSIS MEMBRANE CLEANING

RO membrane fouling occurs in all RO plant operations. Effective cleaning of RO membranes is essential to ensure efficient plant operation, reduce energy usage, and ensure longer membrane life; all of which will lead to reduced overall costs of an RO system. There are two notable advances in this area—the direct osmosis–high salinity (DO–HS) cleaning method and the microbubble cleaning method.

DO–HS Cleaning

The DO–HS cleaning technique uses direct osmosis (a natural process) to draw water from the permeate side to the feed side by injecting a high-salinity solution into the feed stream for a few seconds during the RO train operation. Because of the high salinity gradient, the permeate flows back into the feed side, reversing the normal direction of flow and resulting in actual backwash. The reversible flow helps to dislodge any foulants and scaling on the membrane surface. As a result, online cleaning is achieved without having to shut down the RO train.

Three full-scale RO plants for desalination of brackish groundwater have been implemented using the DO–HS technique (Liberman & Liberman 2005). Pilot testing of the DO–HS technique was conducted in Singapore to simulate full-scale NEWater production (Qin et al. 2010). The results indicated that DO–HS significantly reduced clean-in-place (CIP) frequency at stage 2 of the RO train without impacting product water quality in terms of total organic carbon and conductivity. The DO–HS technique has been used at the Kranji NEWater plant (2.1 mgd) since May 2012 and at a petrochemical industrial water reuse RO system (2.9 mgd) with a membrane bioreactor (MBR) as pretreatment for RO since February 2011. DO–HS technology is becoming more popular and has been endorsed by several RO membrane vendors.

Microbubble Cleaning

The microbubble cleaning technique uses proprietary acidic or alkaline cleaning compounds combined with air induction to generate microbubbles that agitate deposits on the membrane surface of RO membranes. This agitation, which uses a high concentration and wide distribution of bubble sizes, is used to clean a variety of deposits in different industries. The cleaning effect occurs "when bubbles expand and collapse close to boundaries, a shear flow is generated which is able to remove particles from the surface, thus locally cleaning it" (Al-Bastaki & Abbas 2001). The key to effective cleaning of RO–NF membranes using bubbles is to produce very small, refined bubbles that do not damage membranes and are evenly distributed over the membrane surface in a pulsed fashion. A UK company (Genesys International) developed two specially formulated cleaners—low pH (cleaner A) and high pH (cleaner B)—that work with air induction to give off microbubbles for RO membrane cleaning. Cleaners A and B contain detergents, chelants, effervescent, surfactants, and ionic-strength builders (Chester et al. 2014). Cleaner A is used to remove some mineral scales and metal deposits, and cleaner B is used to remove clay, biofilm, and organics. During experiments conducted by Genesys International, it was found that when used at a 1 to 2 percent solution in conjunction with the microbubble generator, the cleaners had a profound effect on bubble size distribution and also imparted pulsing on the cleaning solutions after exit from the physical generator device (Chester et al. 2014).

The Genesys microbubble cleaning method was tested in a 0.12-mgd RO system at the water reuse plant of a salad producer the eastern United Kingdom (Chester et al. 2014). The wastewater treatment and reuse plant use an advanced MBR system ahead of RO to treat and purify water for reuse. The RO system, which was commissioned in 2004, had been prone to fouling during periods of peak production. Even with various improvements, the RO membranes required CIP every 7 to 14 days, and the membranes were replaced every 18 to 24 months. The results of testing in 2013 indicated that using cleaner B with microbubbles increased reduction in differential pressure and a trending of continuous improvement. It was also found that this method greatly improved the ability to clean the rapidly fouling membranes in a significantly shorter time. No loss of salt rejection had been detected, and permeate flow improved after each clean. The frequency of cleaning has declined from every 10.5 days in 2013 to more than 30 days at the beginning of 2014 and later to every 50 to 70 days (Chester & Armstrong 2014).

NEW MEMBRANE MATERIALS FOR DESALINATION

While polyamide thin film composite (TFC) RO membranes are the state of the art in desalination, several other novel materials have been or are being developed to improve flux, reduce energy consumption and fouling, and increase durability. The notable ones include carbon nanotubes, nanocomposite membranes, graphene-based membranes, and aquaporin membranes. These new membrane materials are still under the research and development stages, and full-scale implementation for water reuse applications has not occurred. These materials are discussed below.

Nanotube Membranes

Carbon nanotubes that range in diameter from 6 to 11 Å have been evaluated for desalination because of their fast water transport properties, large surface area, and ease of functionalization (Majumder et al. 2005; Hummer, Rasaiah, & Noworyta 2001; Humplik et al. 2011; Corry 2008). Energy consumption is expected to be low because of the atomic smoothness of the pores and because water molecules pass through them in one-dimensional single-file procession. Ion rejection by carbon nanotube membranes is governed by the steric effects between nanopores and the hydrated diameter of ions, along with the Donnan equilibrium of the membrane surface. Compared to conventional membranes, membranes made of carbon nanotubes are expected to have longer life due to excellent mechanical properties.

Aligned nanotube membranes have significant advantages over conventional membranes because of their reduced hydraulic driving pressure and lower energy costs, which have been reported using theoretical modeling studies. However, productivity is limited by osmotic pressure via thermodynamic restriction (Song et al. 2003). It is also uncertain if the nanotubes can be aligned with a high packing density to obtain the projected permeability. Carbon nanotubes are producible in large quantities. However, fabrication of large surface areas after incorporation of nanotubes is a key step to their commercialization (Pendergast & Hoek 2011). In addition to the uncertainty of aligning nanotubes with a high packing density, manufacturing on a large-scale and rejection properties in larger applications are challenges for further development of nanotube membranes.

NANOTUBE MEMBRANES

Advantage

- High permeability and salt rejection when compared to commercial RO membranes

Disadvantages

- Packing density of nanotubes on substrate not known for practical applications
- Limited experimental data with real feedwater sources
- Rejection of specific contaminants not known

Technology Provider

- Not commercially available for desalination applications

Nanocomposite Membranes

The thin film nanotechnology (TFN) membrane incorporates Linde-type A zeolite nanoparticles or other molecular additives such as alkaline earth metals into the membrane's thin film layer to enhance water permeability while maintaining salt rejection (Jeong et al. 2007). These membranes are commercially available for seawater desalination applications. However, membranes are currently being developed for water reuse and brackish desalination applications. TFN technology uses an interfacial polymerization process with benign nanoparticles dispersed in one or more of the monomer solutions in order to create the nanocomposite membranes. The increased flux through TFN membranes compared to TFC membranes can be as much as 40 percent (Jeong et al. 2007).

The TFN membranes have a measurably smoother, more hydrophilic, and more negatively charged surface compared to a typically pure polyamide thin film membrane due to the presence of the nanoparticle pores (Jeong et al. 2007). These surface properties improve the membrane's permeability, essentially creating molecular, hydrophilic tunnels across the membrane matrix that water preferentially passes through. The greater negative charge on the nanoparticle pore walls enhances ion exclusion and thus maintains salt rejection (Pendergast, Ghosh, & Hoek 2013). The highest TFN element flux reported is a permeate flow rate of 13,700 gpd and a minimum sodium chloride rejection of 99.7 percent (NanoH_2O, Inc. 2011).

Hofs et al. (2013) compared TFN and TFC membrane elements at pilot scale and reported that TFN elements exhibited 1.4 times higher permeability than TFC elements. In this study, a TFN membrane element from NanoH_2O, Inc. (SW365ES) was compared with a TFC membrane element from DowFilmtec (SW30XHR-440i). Although the TFN membrane exhibited higher water permeability while maintaining similar salt rejection, boron rejection and low-molecular-weight nitrosamines rejection was lower when compared to the TFC membrane. Thus, it is important to compare the rejection of TFN with TFC membranes for water reuse applications.

TFN technology has been developed for commercial use by LG-NanoH_2O, Inc., with four membrane elements currently in production and available for purchase (NanoH_2O, Inc. 2012). TFN membranes have been installed at several locations around the world, and multiple pilot installations have been using these elements for desalination, processing more than 130,610 ft^3 of seawater and operating more than 28,000 combined hours (Kurth et al. 2011). Membranes for brackish water applications are currently being developed by LG-NanoH2O.

THIN FILM NANOTECHNOLOGY

Advantages

- Higher water permeability compared to TFC membranes
- Reduced feed pressure requirement compared to TFC membranes
- Lower specific energy consumption compared to TFC membranes

Disadvantages

- More expensive membrane elements compared to commercially available TFC membranes
- Lower boron and nitrosamines rejection compared to TFC membranes
- Retrofit of existing plants may require a variable-speed drive on pumps in order to reduce operating pressure

Technology Provider: NanoH_2O, Inc. www.nanoh2o.com

Graphene-based Membranes

Graphene-based membranes are being developed for desalination and water reuse applications due to their fast water transport properties and good mechanical properties. Similar to the water permeation mechanism in carbon nanotubes, two-dimensional graphene nanocapillaries allow a low-friction flow of monolayer water with size exclusion as the dominant sieving mechanism (Nair et al. 2012; Xue et al. 2013; Choi et al. 2013; Mi 2014). Graphene-based membranes are prepared by dispersing graphene through sonication and forming laminates using spray-coating or spin-coating and forming graphene oxide (GO) sheets. GO sheets are vacuum tight in the dry state but swell and act as molecular sieves when immersed in water. These sheets reject all solutes with a hydrated radii larger than 4.5 Å (Nair et al. 2012). Nanocapillaries are opened up in the hydrated state and result in a low frictional flow of water. A one-atom thick carbon allotrope of graphene called graphyne has been shown to reject all common ions in seawater while maintaining water permeability twice that of commercially available polyamide membranes (Xue, Qiu, & Gui 2013). Selectivity of GO membranes can be tuned through the generation of subnanometer pores. There have been several attempts to fabricate GO membranes on polyamide TFC RO membranes, which resulted in increased hydrophilicity and reduced surface roughness of the RO membrane (Choi et al. 2013). This also resulted in enhanced resistance to protein fouling and increased chlorine resistance.

Graphene can be mass produced, and it is possible to make laminates with high mechanical strength and flexibility. However, water productivity, as with carbon nanotubes, will be limited by osmotic pressure via thermodynamic restriction. Commercialization of graphene-based membranes for desalination applications will depend on the ability to synthesize large quantities of graphene material and the mechanical strength of the nanolayers when an applied hydrostatic pressure is present.

GRAPHENE-BASED MEMBRANES

Advantages

- High permeability and salt rejection compared to commercial RO membranes
- Potential resistance to fouling and higher chlorine tolerance

Disadvantages

- Limited experimental data with real feedwater sources
- Impact of hydrostatic pressure on nanolayer stability is not known
- Rejection of specific contaminants is not known

Technology Provider

- Not commercially available for desalination applications

Aquaporin Membranes

Aquaporins are protein channels that control water flux across biological membranes. They are found widely in human tissues; they provide rapid, passive transport of water molecules across cell membranes (Agre et al. 1993). Water movement in an aquaporin is mediated by selective, rapid diffusion and caused by osmotic gradients (Agre 2003). Aquaporin-1, with selective extracellular and intracellular vestibules at each end, allows water molecules to pass rapidly in a single-file line while excluding proteins and ions by an electrostatic tuning mechanism (Agre 2003; Sui et al. 2001). This leads to only water molecules being transported through the aquaporin channels and charged ions being rejected (Sui et al. 2001).

Aquaporin membranes are considered to be 100 times more permeable than commercial RO membranes (Kaufman, Berman, & Freger 2010). Membranes that are based on aquaporins show promise for desalination where the driving mechanism is an osmotic pressure gradient (salt concentration) rather than a mechanically applied pressure gradient as in RO (Pendergast & Hoek 2011). With 75 percent coverage of aquaporins on a membrane, an order of magnitude increase in hydraulic permeability has been predicted (Kaufman, Berman, & Freger 2010). Due to the absence of applied pressure, energy consumption is expected to be substantially lower compared to RO membranes. Studies have been performed where aquaporins were deposited onto commercially available membranes. A good review of desalination using aquaporin membranes and their status of development has been provided by Tang et al. (2013). One difficulty to be overcome with aquaporin membrane synthesis is that they need to withstand high operating pressures and harsh chemical cleaning. Because it is difficult to attain large quantities of proteins and produce large areas of membrane material, aquaporin-based membranes are not widely available for commercialization (Pendergast & Hoek 2011). Only small membrane areas have been fabricated using highly specialized nanofabrication techniques, and scale-up to commercial size membrane elements is still an issue.

AQUAPORIN MEMBRANES

Advantages

- High permeability (an order of magnitude higher than commercial RO membranes)
- 100% rejection of solute molecules

Disadvantages

- Synthetic approach to producing and purifying aquaporin in large quantities is essential
- Limited experimental data with real feedwater sources
- Chemical resistance of aquaporin is unknown

Technology Providers: Applied Biomimetic, http://www.appliedbiomimetic.com/

NOVEL MEMBRANE PROCESSES FOR DESALINATION

Because RO requires high energy usage for pumping, other novel membrane processes have been developed or redeveloped to reduce energy usage. Two notable processes are forward osmosis and membrane distillation, which are discussed below.

Forward Osmosis

Forward osmosis (FO) has been recognized as a practical commercial process since the 1930s (Semiat et al. 2008). It has been evaluated for a wide range of applications and shows promise in decreasing the energy required and, therefore, the cost to desalinate water. In the FO process, instead of using hydraulic pressure as in conventional RO desalination, a concentrated draw solution is used to generate high osmotic pressure. This pressure pulls the water across a semipermeable membrane from the feed solution (McCutcheon, McGinnis, & Elimelech 2005). The draw solutes are then separated from the diluted draw solution to recycle the solutes and to produce a final product water. A mixture of ammonia and carbon dioxide gas is the draw solution in most instances (McCutcheon, McGinnis, & Elimelech 2006). When ammonia and carbon dioxide are mixed in the right proportion, a solution with a high osmotic pressure is formed. This solution is used for drawing water saline feeds. The advantage of using this mixture is that it can be regenerated when heated and reused for the FO process. FO can be considered a combination of membrane and thermal processes.

While a variety of osmotic agents can be used, the draw solutions must be nontoxic, stable, have a near neutral pH, be highly soluble to avoid precipitation, and be removed from water at a low cost using existing technology (Cath et al. 2006; Liu et al. 2010; Achilli, Cath, & Childress 2010; Yen et al. 2010; Klaysom et al. 2013). In addition to choosing the optimal osmotic agent, a variety of membrane types and configurations will affect FO performance. Because internal and external concentration polarization can significantly affect performance, FO membranes should be designed with a high-density active layer that will achieve an elevated solute rejection, a thin support layer with minimal porosity that will limit internal concentration polarization, and hydrophilic properties to maximize flux and minimize fouling (Coday et al. 2013). The total energy used by the FO process is approximately 25 to 45 percent of the thermal energy needed for multi-effect distillation (Cath et al. 2009). FO uses heat at a much lower or higher temperature than multi-effect distillation processes. The FO process can use heat as low as 40°C and as high as 200 to 250°C. Cath et al. (2009) reported specific energy consumption of less than 0.9 kWh/1,000 gal for the membrane portion of FO. However, after considering the energy consumption associated with draw solution recovery, the specific energy consumption was similar to that for RO (approximately 13.3 kWh/1,000 gal) (McGovern & Lienhard 2014). However, the FO process has the advantage of less fouling propensity when compared to RO due to the absence of a hydraulic driving pressure (Mi & Elimelech 2010).

In an innovative approach to reducing energy consumption, Cath et al. (2009) used FO in combination with RO to form a hybrid process. Recycled water (tertiary-treated effluent) was passed through an FO system, with seawater used as the draw solution. The seawater was diluted by the recycled water within the FO process. The diluted seawater was then passed through an RO system where the feed pressure requirement was lowered by dilution of the seawater, and energy consumption was lowered for the seawater desalination process. The concentrate (brine) from the RO process was further treated through a second-stage FO process, and the final seawater brine was discharged to the ocean. By using a combination of FO and RO, seawater was desalinated with lower energy consumption, and the recycled water was simultaneously treated through two physical barriers—FO and RO.

Using the same approach, Yangali-Quintanilla et al. (2011) desalinated Red Sea water using FO and RO. Secondary wastewater effluent was used as the feedwater, and Red Sea water was used as the draw solution. The system was reported to consume only 50 percent of the energy (approximately 5.7 kWh/1,000 gal) compared to a high-pressure seawater RO system desalinating the same seawater. Although the FO–RO process had lower energy consumption, cost analysis suggested that a minimum average flux of 6.2 GFD was necessary to compete with a conventional RO process that uses a flux of 10 to 16 gfd for water reuse applications.

FORWARD OSMOSIS

Advantages

- Multiple applications when combined with other separation processes such as RO
- Can accommodate higher salinity compared to RO processes
- Potential for lower energy consumption and fouling potential due to absence of applied pressure

Disadvantages

- Difficult to choose optimal osmotic agent
- Recovery of draw solution consumes additional energy
- External and internal concentration polarization causes empirical flux rates to be much lower than theoretical flux rates

Technology Providers

Modern Water, http://www.modernwater.com/membrane-processes
HTI, http://www.htiwater.com/technology/forward_osmosis/
Porifera, http://porifera.com/

Membrane Distillation

Membrane distillation (MD) is a thermally driven, membrane-based process (El-Bourawi et al. 2006; Adham et al. 2013) that combines membrane technology and evaporation processing in one unit. The temperature difference across a membrane results in a vapor pressure differential that leads to the transfer of the water vapor is transferred through the hydrophobic membrane to the condensation surface. Membrane materials used for MD include polytetrafluoroethylene, polyvinylidene fluoride, polyethylene, and polypropylene (Alkhudhiri, Darwish, & Hilal 2012). Porosity values are 0.06 to 0.85; pore sizes are 0.2 to 1.0 mm; and thickness is 0.06 to 0.25 mm. MD, considered an alternative to conventional desalination technologies such as multistage flash (MSF) and RO, has not been competitive due to low flux rates and lack of commercial designs. Both MSF and RO involve high energy and high operating pressure, respectively, which result in excessive operating costs. MD can operate at atmospheric pressure and low temperatures (30 to 90°C), with the theoretical ability to achieve 100% salt rejection.

Typical configurations used for MD are direct contact membrane distillation, air gap membrane distillation, vacuum membrane distillation, and sweeping gas membrane distillation (Evans & Miller 2002). Different configurations of the MD process are used to improve efficiency during desalination. The selection of membrane material is critical. The selected membrane should be porous and hydrophobic (should not be wetted by liquids) and must not have capillary condensation inside the pores. Only vapor should be transported through the pores, and the membrane must not alter the vapor equilibrium of process liquids (Tomaszewska 2000). The specific energy consumption for MD systems has been reported to be about 162 kWh/1,000 gal of water produced (Walton et al. 2004). However, this energy consumption will be substantially lower when waste heat is available to heat the source water that requires treatment. A major drawback of MD is the low flux obtained, less than 5 gfd.

There are several pilot- and demonstration-scale studies being proposed to evaluate MD for brackish water treatment and treatment of wastewater from oil and gas fields. In a study conducted by the National Center for Excellence in Desalination (Australia), an MD system was proposed for treatment of brackish water at Tjuntjuntjara in a remote corner state of Western Australia where the primary contaminant is nitrate (*New York Times* 2011). At the Singapore Membrane Technology Center, MD is being evaluated for treatment of water contaminated with oil (*New York Times* 2011). In both studies, vacuum gap multi-effect membrane distillation is being used. Warm water is passed through a series of MD effects working at progressively lower pressures. This increases the gain output ratio (GOR) to the level that it could deliver large units with a GOR of 10, comparable to commercially available multi-effect distillation systems. Because of the vacuum, water boils at a much lower temperature (50 to 80°C). IBM developed a method for internal heat recovery and recycling (*New York Times* 2011).

Research in MD needs to focus on creation of membrane material that has a structured morphology to facilitate mass transfer and enhance flux. Gethard and colleagues (2011) studied desalination using carbon nanotube–enhanced MD. Carbon nanotubes were immobilized in the pores of a hydrophobic membrane that favorably altered the water–membrane interactions to promote vapor permeability while preventing liquid penetration into the membrane pores. With the incorporation of carbon nanotubes, the flux and salt rejection was enhanced from 1.85 to 15 times with seawater (Gethard, Sae-Khow, & Mitra 2011).

MEMBRANE DISTILLATION

Advantages

- Absence of applied pressure
- High rejection capacity with the removal of volatile organic compounds
- Lower operating temperatures when used in a vacuum
- Possibility of using plastic material to avoid corrosion
- Minimal effect of performance due to feed salt concentrations
- Possibility of using different energy sources such as waste heat, solar energy, and geothermal

Disadvantages

- Low permeate flux
- Only bench-scale or pilot-scale studies have been performed
- Lack of membranes and modules designed specifically for MD
- Requires pretreatment to prevent organic fouling of hydrophobic membranes

Technology Providers
MemSys, http://www.memsys.eu/home.html
Xzero, http://www.xzero.se/technology.html

CONTAMINANTS OF EMERGING CONCERN

In this section, contaminants of emerging concern (CECs) and the application of membrane technologies for their removal from water are discussed. Contaminants of emerging concern includes a broad spectrum of anthropogenic compounds found in groundwater, surface water, drinking water, and wastewater. Many of these compounds are found in almost every surface water or wastewater sample across the world. While there is well-documented evidence of the effects of some of these compounds on aquatic and amphibious organisms, the human health risks are not well understood. Many agencies approach the risks to human health with an abundance of caution and have chosen to implement treatment for the reduction of the CECs to the maximum extent possible.

Membrane technologies provide high levels of treatment for traditional contaminants and are gaining wide acceptance for removal of CECs. However, not all membrane technologies are well suited for this application. Prediction models for the removal of CECs using membrane technologies are only now beginning to take useable form, and pilot testing is still necessary to properly select and design a treatment system.

Defining CECs

CECs include tens of thousands of chemicals. Many are frequently detected in natural waters and wastewaters. Understanding the uses, occurrence, and properties of these chemicals helps researchers focus on those chemicals of most importance.

Categories. CECs can be categorized in general terms by their use. CECs have been categorized into the following categories (US Bureau of Reclamation 2009): hormones; pharmaceutically active compounds (PhACs); personal care products; insecticides, herbicides, and pesticides; industrial and household products; flame retardants, and disinfection by-products. Illegal substances can be added to this list.

Personal care products are often included with PhACs into a grouping referred to as pharmaceuticals and personal care products. Endocrine-disrupting chemicals are a subset of many of these categories including hormones, industrial chemicals, disinfection by-products (DBPs), and pesticides.

Hormones. Hormones include both naturally occurring and synthetic hormones and are included in the category of PhACs. Some of the more common hormones found in water include:

- Estrone,
- 17α estradiol,
- 17β estradiol,
- Estriol,
- Testosterone,
- Progesterone,
- 17α ethynylestradiol,
- Equilenin, and
- Cholesterol.

Pharmaceutically active compounds. There are more than 10,000 PhACs in use worldwide. Because their chemical properties are diverse, there is no single treatment technology that is suitable for all PhACs (US Bureau of Reclamation 2009). Table 10-1 lists many of the PhACs discussed in literature and their rank in terms of number of doses distributed per year.

Table 10-1 Pharmaceutically active compounds and rank of use

Name	Use	Rank of Use
Amoxicillin	Antibiotic	3
Atenolol	Beta blocker	7
Ibuprofen	Analgesic	14
Flouxetine	Antidepressant	19
Carbemazepine	Anticonvulsant	168
Diclenofac	Antiinflammatory	80
Gemfibriziol	Anticholesterol	82
Sulfamethoxazole	Antibiotic	62
Meprobamate	Antianxiety	Not ranked
Estradiol	Hormone replacement	81
Propanolol	Beta blocker	78

Source: AwwaRF (2010).

Personal care products. Table 10-2 lists some of the commonly used personal care products found in water and wastewater.

Table 10-2 Commonly used personal care products

Name	Use
Caffeine	Stimulant
Butylparaben	Antibacterial
Ethylparaben	Antibacterial
Methylparaben	Antibacterial
Nonylphenol	Detergent metabolite
Octylphenol	Detergent metabolite
Oxybenzone	Sun screen
Benzophenone	Sun screen
Triclosan	Antimicrobial
3-Phenolpropionate	Fragrance
Acetophenone	Fragrance
Musk ketone	Fragrance
1,4-Dioxane	Shampoo

Pesticides, herbicides, and insecticides. Table 10-3 lists some of the commonly used pesticides, herbicides, and insecticides found in water and wastewater.

Table 10-3 Pesticides, herbicides, and insecticides found in water and wastewater

Name	Use
Dimethyltoluamide (DEET)	Insecticide
Hydroxyatrazine	Pesticide
Bentazone	Pesticide
Dicamba	Pesticide
Aminotriazole	Pesticide
Methoxychlor	Pesticide
Atrazine	Herbicide
Propazine	Herbicide
Cyanazine	Herbicide

Industrial and household products. The list of industrial and household products, including cleaning products, is extensive. A few compounds that are discussed in the research literature for wastewater treatment include:

- Perfluorooctanoic acid (PFOA),
- Bisphenol A,
- Bis(2-ethylhexyl) phthatate (DEHP),
- Nonylphenols (associated with detergents),
- Benzene, and
- Butylphenol.

Flame retardants. Flame retardants can be particularly resistant to biological and chemical oxidation. A common flame retardant found in water and wastewater is tri (2-chloroethyl) phosphate.

Illegal substances. Illegal substances that have been studied include the following:

- Cannabis (tetrahydrocannabinol),
- Lysergic acid diethylamide (LSD),
- Diamorphine (heroine),
- 3,4-methylenedioxy-methamphetamine (MDMA or ecstasy), and
- Amphetamine.

Disinfection by-products. Many DBPs are already regulated for drinking water; by definition, they are not emerging contaminants. Though not exclusively a DBP, *N*-nitrosodimethylamine (NDMA) is one of the most studied compounds and one of the few that appears close to being regulated nationwide. California has already set a public health goal of 3 ng/L and an action level of 10 ng/L for drinking water.

Endocrine disruptors. Endocrine disrupting compounds (EDCs) include chemicals from a number of groups, including the steroidal and hormone supplement products. Some more well-known EDCs include the following:

- Dichlorodiphenyltrichoroethane (DDT),
- Polychlorinated biphenyl (PCB),
- Bisphenol A,
- Organochlorine phthalates,
- Phthalates,
- Akylphenolic ethoxylates,
- Polycyclic aromatic hydrocarbons,
- Halogenated aromatic hydrocarbons, and
- Natural and synthetic estrogenic hormones (US Bureau of Reclamation 2009).

Metabolites. Metabolites include the intermediate by-products of CECs, particularly PhACs. The environmental and health risks associated with metabolites of CECs are even more poorly understood than their original forms.

CEC Characteristics

An understanding of the characteristics of CECs will aid in the development of treatment strategies. By comparing the physical and chemical characteristics of a variety of compounds, groups of contaminants can be formed and surrogate parameters can be selected to represent those groups. This can help reduce the cost of analyses for studies and ongoing monitoring of treatment performance. The chemical characteristics most commonly referenced are grouped into the three categories shown in Figure 10-1.

In the size category, the molecular depth, width, length, and volume are rarely referenced. While the molecular volume is probably a more appropriate parameter for comparing removal efficiencies by membranes, mole weight is by far the more convenient and universally accepted parameter. Many UF and NF membranes have ratings in terms of molecular weight cutoff (MWCO). The relationship of a contaminant's molecular weight and membrane molecular weight cutoff is central to much of the recent research in the removal of CECs, as discussed later in this chapter.

The hydrophobic properties of contaminants include the octanol–water partition coefficient $\log K_{ow}$, which is the ratio of ionized species of a compound in equilibrium concentration in a solution of octanol versus its concentration in a solution of water. In general, compounds with a $\log K_{ow}$ greater than 3 are considered hydrophobic, and those with a $\log K_{ow}$ less than 2 are considered hydrophilic (Yangali-Quintanilla 2009). The logD is similar to the $\log K_{ow}$ but differs in that logD accounts for both ionized and unionized forms of the compound. In most research, $\log K_{ow}$ is reported instead of logD.

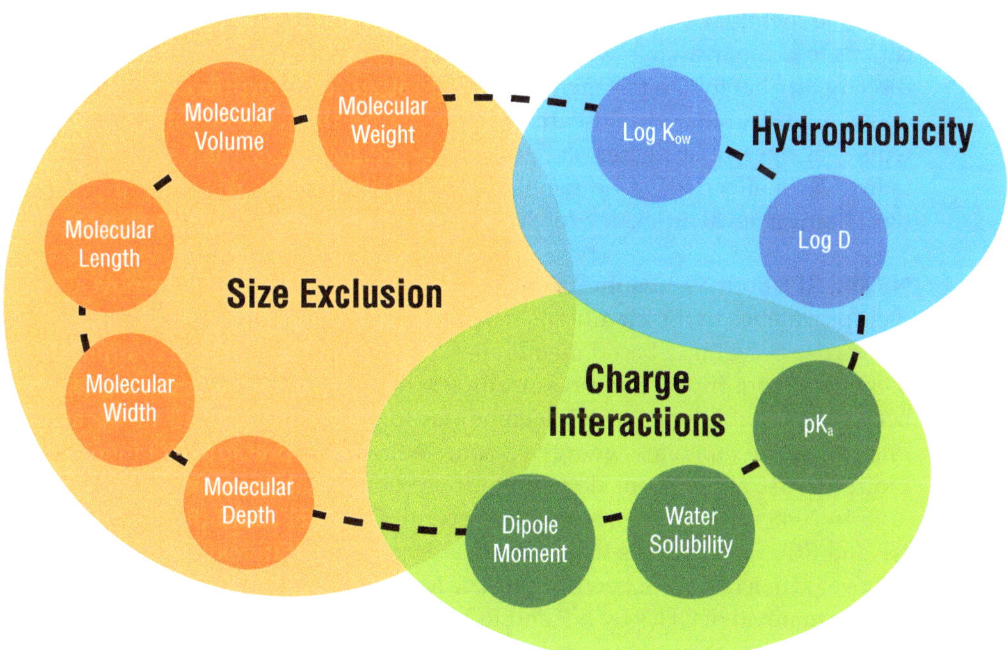

Source: Brown and Caldwell.

Figure 10-1 Categories of constituents of emerging concern by chemical characteristics

Ionization parameters, including water solubility, the negative log of the acid dissociation constant (pKa), and dipole moment, have been reported in the literature for the purpose of evaluating removal efficiencies of NF and RO membranes. The acid dissociation constant is the most widely referenced.

In addition to these parameters, some studies include the $\log K_{oc}$, or soil sorption coefficient. This is generally used for transport and fate analyses of contaminants in groundwater and natural water sediments, but may also apply to sorption in activated sludge systems such as MBRs. Values for these parameters are available in a number of chemical databases. The Secondary/Emerging Constituents Report (US Bureau of Reclamation 2009) includes a summary of the key chemical properties for 36 common CECs.

Membrane Treatment Technologies for CEC Removal

While membrane separation technology has been in use for more than 50 years, it has only been in the past 25 years that these technologies have become widely used for water and wastewater treatment. The more common membrane technologies include MF, UF, NF, and RO. Membrane bioreactors use MF or UF in conjunction with activated sludge. The capabilities of membrane technology have led to their use for removing CECs from water. Other membrane technologies that have been described in the literature for the removal of CECs are electrodialysis membranes and FO.

MF–UF. At typical pore sizes of 1 to 0.01 μm, MF–UF as a standalone unit process is not effective for the removal of CECs. MF and UF have been used in conjunction with other processes such as enhanced coagulation, powdered activated carbon, ozone, and other advanced oxidation processes (AOPs) and as pretreatment to NF and RO.

MBR. The MBR, which is a hybridization of the activated sludge process with membrane filtration, has been shown to be effective for the removal of a number of CECs. MBRs use the same membrane material that is used for drinking water and tertiary filtration applications and are available with both MF and UF pore sizes. The removal of CECs is not directly attributable to the membrane, but rather to adsorption to activated sludge particles and biodegradation. Researchers have found that conventional activated sludge systems that operate at higher sludge ages (typically longer than seven days) also remove many CECs effectively. The membranes contribute to consistent removal rates through their high removal efficiency for suspended solids. Table 10-4 summarizes.

NF and RO. Both NF membranes and RO membranes are used to remove CECs. Reflecting their prevalence in the membrane marketplace, most of the research and full-scale installations use polyamide-based membranes. For indirect potable reuse applications, most RO membranes are in the low-pressure brackish groundwater product category. There has been significant interest in NF membranes for the removal of CECs because these membranes require much less energy to operate than low-pressure RO membranes and most indirect reuse applications do not require a substantial decrease in total dissolved solids.

In 2008, Drewes et al. stated that no single model could predict removal of CECs by NF and RO. Some models that provide at least a qualitative or approximate prediction of removal efficiency have been proposed. However, a reliable, quantitative model for predicting removal of CECs by NF and RO membranes has not been developed.

Table 10-4 Observed average efficiencies of membrane bioreactors in the removal of various constituents of emerging concern

Compound	Reported Removal Efficiency, %	Source
Acetaminophen	>95	Erdal et al. 2010
	99.8	Sipma et al. 2010
	100	Tambosi et al. 2010
Atenolol	71	Sipma et al. 2010
Bisphenol-A	74	Kim et al. 2009
	90	Ngheim, Tadkaew, & Sivakumar 2009
Caffeine	75	Erdal et al. 2010
Carbemazepine	<20	Erdal et al. 2010
	0	Sipma et al. 2010
DEET	>90	Erdal et al. 2010
2,4-Dichlorophenol	78	Kim et al. 2009
17α-estradiol	90	Erdal et al. 2010
	>73	Le-Minh et al. 2010
17α-ethinylestradiol	>92	Le-Minh et al. 2010
Flouxetine	25	Erdal et al. 2010
	98	Sipma et al. 2010
Gemfibrozil	45	Erdal et al. 2010
Ibuprofen	>90	Erdal et al. 2010
Iopromide	59	Sipma et al. 2010
Meprobamate	>25	Erdal et al. 2010
Naproxen	>80	Erdal et al. 2010
	86–89	Tambosi et al. 2010
Oxybenzone	>65	Erdal et al. 2010
Progesterone	85	Erdal et al. 2010
Propanolol	72	Sipma et al. 2010
Sulfamethoxazole	<25	Erdal et al. 2010
	50	Ngheim, Tadkaew, & Sivakumar. 2009
	75	Le-Minh et al. 2010
	55–64	Tambosi et al. 2010
Tri (2-chloroethyl) phosphate	<5	Erdal et al. 2010
Testosterone	>74	Le-Minh et al. 2010
Triclosan	<50	Erdal et al. 2010

Size exclusion, charge interactions, and hydrophobicity play significant roles in the rejection of CECs by NF and RO membranes. Sieving of compounds larger than the membrane's MWCO is a major removal mechanism for NF membranes but only provides a rough estimate of removal efficiency (WRRF 2006). With the exception of a few compounds such as NDMA, the molecular weight of most CECs is greater than 150, which is a conservative estimate of the MWCO for RO membrane. Some more popular NF membranes have MWCO values of 200 to 300. Rejections of ionic compounds have been found to be higher than for nonionic or neutral compounds due to interactions between the membrane surface and charged solute (Yangali 2010). Low rejections have been observed for compounds with molecular weights that are lower than the membrane MWCO and of a hydrophobic nature (generally defined as pK_{ow} greater than 2) (Zraik et al. 2013).

Zraik et al. (2013) performed several experiments to evaluate a simplified model previously proposed by Verliefde et al. (2007). This model categorizes contaminants into eight groups according to the properties of molecular weight, hydrophobicity (log K_{ow}), and the charge as reflected by the acid dissociation constant (pKa). One distinction between this model and the quantitative structure-activity relationship (QSAR) model is that it pertains to NF and RO. The model provides only qualitative results; however, these results provide a framework for evaluating the usefulness of membranes in any treatment scheme and help in tailoring a pilot study program. This model is similar to the one developed by Bellona et al. (2005). The results from Zraik et al. showed that the model served as an accurate predictor of CEC removal by NF and RO, though not a precise one. Table 10-5 reproduces the approach proposed by Verliefde et al. (2007).

Figure 10-2 is a flowchart for qualitatively predicting rejection based on solute and membrane properties (WRRF 2006). The researchers compared the predicted rejection against observed values and demonstrated good qualitative predictive results.

These models can help narrow the scope of research or aid in the selection of appropriate membranes. However, accurate values for the rejection of CECs must be determined by pilot testing with selected membranes, with the same wastewater, and under the same environmental conditions to be expected by the full-scale plant. Table 10-6 summarizes the efficiency of removing several CECs reported in studies from numerous sources. What is not listed in Table 10-6 is the operating recovery of the NF and RO processes. Drewes et al. (2006) reported that other studies have shown that operating conditions, including pressure and recovery, affect the rejection of some solutes.

Table 10-5 Predictions of removal of constituents of emerging concern by nanofiltration and reverse osmosis

Category	Hydrophobicity, log K_{ow}	Molecular Weight	Charge, pKa relative to pH of solution	Removal Mechanism	Qualitative Removal Prediction
1	<2 (HL)	MW < MWCO	pH<pKa	Steric hindrance	Moderate
2	<2 (HL)	MW > MWCO	(uncharged)	Steric hindrance	Moderate to high
3	>2 (HB)	MW < MWCO		Hydrophobic interactions	Low
4	>2 (HB)	MW > MWCO		Hydrophobic interactions	Moderate
5 and 7	HL or HB	MW < MWCO	pH >pKa	Charge repulsion	High
6 and 8	HL or HB	MW > MWCO	(charged)	Charge repulsion	Very high

Abbreviations: HB, hydrophobic; HL, hydrophilic; MW, molecular weight; MWCO, molecular weight cutoff (pore size of membrane).

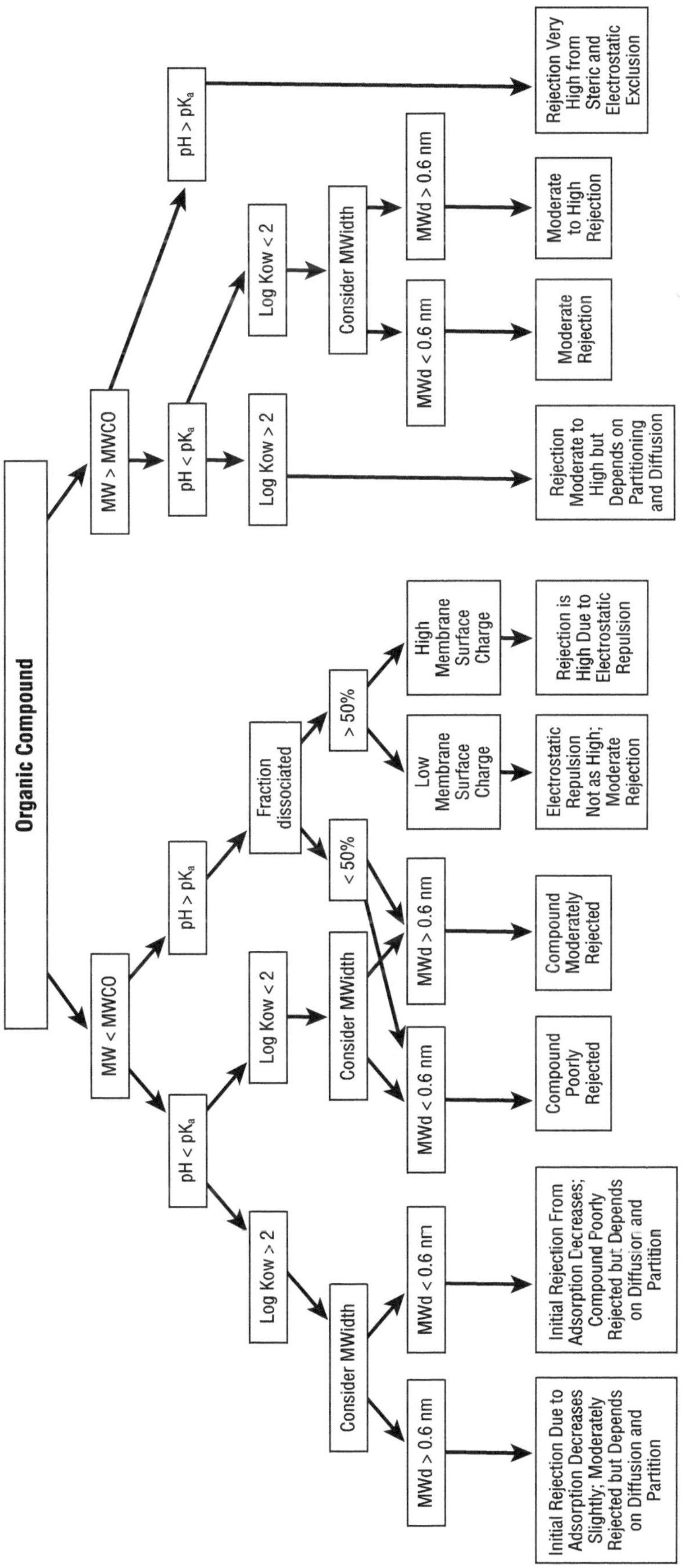

Adopted from WateReuse Research Foundation (2006).

Figure 10-2 Rejection diagram for organic micropollutants during membrane treatment

Abbreviations: MW, molecular weight; MWCO, molecular weight cutoff

Table 10-6 Efficiencies of nanofiltration and reverse osmosis in the removal of various constituents of emerging concern

Compound	Membrane Type	Reported Removal Efficiency, %	Source
Acetaminophen	RO	>90	US Bureau of Reclamation 2009
	RO	60	Alexander et al. 2009
	NF	22–78	Yangali-Quintanilla et al.2010
Bisphenol-A	NF	40	Yangali-Quintanilla et al.2009
	NF	55–90	Yangali-Quintanilla et al.2010
Caffeine	NF	40	Yangali-Quintanilla et al.2009
	NF	40–72	Yangali-Quintanilla et al.2010
	RO	62	Alexander et al. 2009
	RO	>99	US Bureau of Reclamation 2009
Carbemazepine	NF	80–92	Yangali-Quintanilla et al.2010
	RO	96.6	Alexander et al. 2009
	RO	>99	US Bureau of Reclamation 2009
Clozapine	NF	99	Yangali-Quintanilla et al.2009
DEET	RO	>95	US Bureau of Reclamation 2009
Diclofenac	NF	80	Yangali-Quintanilla et al.2009
	RO	>97	US Bureau of Reclamation 2009
17α-Estradiol			
17α-Ethinylestradiol	NF	89–94	Yangali-Quintanilla et al.2010
Flouxetine	RO	>96	US Bureau of Reclamation 2009
Gemfibrozil	NF	90–96	Yangali-Quintanilla et al.2010
	RO	90.9	Alexander et al. 2009
	RO	>99	US Bureau of Reclamation 2009
Ibuprofen	NF	70	Yangali-Quintanilla et al.2009
	NF	89–97	Yangali-Quintanilla et al.2010
	RO	79.7–82	Alexander et al. 2009
	RO	>99	US Bureau of Reclamation 2009
Iopromide	RO	>99	US Bureau of Reclamation 2009
Meprobamate	RO	98.6–99.7	Alexander et al. 2009
	RO	>99	US Bureau of Reclamation 2009
Naproxen	NF	70	Yangali-Quintanilla et al.2009
	NF	91–95	Yangali-Quintanilla et al.2010
	RO	>99	US Bureau of Reclamation 2009
N-nitrosodimethylamine	RO	5–44	Alexander et al. 2009
	RO	1–26	Bates et al. 2009
Oxybenzone	RO	>93	US Bureau of Reclamation 2009
Risperidone	NF	99	Yangali-Quintanilla et al.2009
Sulfamethoxazole	NF	50	Yangali-Quintanilla et al.2009
	NF	84–96	Yangali-Quintanilla et al.2010
	RO	92.2	Alexander et al. 2009
	RO	>99	US Bureau of Reclamation 2009
Tri (2-chloroethyl) phosphate	RO	>91	US Bureau of Reclamation 2009
Trimethoprim	NF	90	Yangali-Quintanilla et al.2009
	RO	>99	US Bureau of Reclamation 2009
Triclosan	RO	>97	US Bureau of Reclamation 2009

Abbreviations: NF, nanofiltration; RO, reverse osmosis.

Other Membrane Processes

Recently, electrodialysis and FO have been evaluated for their ability to remove CECs. Electrodialysis and electrodialysis reversal use alternating stacks of anion-selective and cation-selective membranes. The driving force for separation of ionized components is electric current rather than pressure as with RO and NF. One study of the removal of hormones using electrodialysis found that hormones adsorb into the membranes and can desorb depending on the electrostatic interactions with the membrane and solution pH. Given the wide variation in the chemical properties of CECs, it seems unlikely that electrodialysis or electrodialysis reversal would be applicable for most waters that contain a heterogeneous mixture of CECs.

FO works by using a solution of high osmosis pressure to draw clean water across a semipermeable membrane. The draw solution is then desalinated or the agent otherwise separated by other means in a much less energy intensive process (see more FO descriptions earlier in this chapter). The membranes are similar to RO membranes. In one study, the rejection of CECs was modest compared to rejection for RO. However, when coupled with an RO system for recovery of the draw solution, the rejection of the system was comparable to rejection for traditional RO systems.

Advanced Oxidation

No membrane process can completely remove all CECs commonly found in reclaimed water. AOPs act to destroy CECs or at least break them up into more biodegradable forms. When coupled with membrane processes, AOP becomes an effective tool in removal for many of the more persistent CECs. The combination of membrane filtration, reverse osmosis, and UV/AOP is a widely adopted model for advanced water reclamation where final product CEC goals are very stringent.

REFERENCES

Achilli, A.; Cath, T.Y.; & Childress, A.E. 2010. Selection of Inorganic-based Draw Solutions for Forward Osmosis Applications. *J. Membr. Sci.*, 364:233–241.

Adham, S.; Hussain, A.; Matar, J.M.; Dores, R.; & Janson, A. 2013. Application of Membrane Distillation for Desalting Brines From Thermal Desalination Plants. *Desalination*, 314:101–108.

Agre, P. 2003. Aquaporin Water Channels, Nobel Lecture. Stockholm, SE.

Agre, P.; Preston, G.M.; Smith, B.L.; Jung, J.S.; Raina, S.; Moon, C.; Guggino, W.B.; & Nielsen, S. 1993. Aquaporin CHIP: The Archetypal Molecular Water Channel. *Am. J. Physiol. Ren. Physiol.*, 265:F463–F476.

Ahadian, S. & Kawazoe, Y. 2009. An Artificial Intelligence Approach for Modeling and Prediction of Water Diffusion Inside a Carbon Nanotube. *Nanoscale Res. Lett.* 4:1054–1058.

Ahn, C.H.; Baek, Y.; Lee, C.; Kim, S.O.; Kim, S.; Lee, S.; Kim, S-H.; Bae, S.S.; Park, J.; & Yoon, J. 2012. Carbon Nanotube-based Membranes: Fabrication and Application to Desalination. *J. Ind. Eng. Chemistry*, 18:1551–1559.

Al-Bastaki, N. & Abbas, A. 2001. Use of Fluid Instabilities to Enhance Membrane Performance, A Review. *Desalination*, 136:255–262.

Alexander, K. L.; Boysen, R. E.; Nunez, A.; Lee, B. W.; & Matta, J. 2009. Full Scale Evaluation of Antiscalant and NDMA Removal at Scottsdale Water Campus. Proceedings of 24th WateReuse Symposium. Seattle, Washington, Sept. 13–16; Alexandria, VA: WateReuse Association.

Alkhudhiri, A.; Darwish, N.; & Hilal, N. 2012. Membrane Distillation: A Comprehensive Review. *Desalination*, 287:2–18.

AWWA. 2005. M53, *Microfiltration and Ultrafiltration Membrane for Drinking Water.* Denver, CO: AWWA.

Bates, J.; Hemken, B.; Hunter, G.; Wallis-Lage, C.; & Zavlanos, V. 2009. The Rise and Fall of Micro-constituents at the Bundamba AWTP. Proceedings of 82nd Annual Water Environment Federation Technical Exposition & Conference; Orlando, Florida, Oct. 11–14. Alexandria, VA: Water Environment Federation.

Bolong, N.; Ismail, A. F.; Salim, M. R.; & Matsuura, T. 2009. A Review of the Effects of Emerging Contaminants in Wastewater and Options for Their Removal. *Desalination*, 239:229–246.

Bookout, K.; Perry, B.; Neeman, J.; Freeman, S.; & Hansen. A. 2014. Evaluating and Testing Ceramic Membranes to Expand Treatment Capacity. Proceedings 2014 AWWA/AMTA Membrane Technology Conference, Las Vegas, NV, March 10–14.

Cath, T.Y.; Adams, V.D.; & Childress, A.E. 2004. Experimental Study of Desalination Using Direct Contact Membrane Distillation: A New Approach to Flux Enhancement. *J. Membrane Sci.*, 228:5–16.

Cath, T.; Childress, A.; & Elimelech, M. 2006. Forward Osmosis: Principles, Applications, and Recent Developments. *J. Membrane Sci.*, 281:70–87.

Chester, S. & Armstrong, M. 2014. Bubble Power Enhancing a Weapon for RO Membrane Cleaning. *WaterWorld*, 29:3.

Chester, S.; Wilson, R.; Armstrong, M.; & Fazel, M. 2014. What's New in RO Membrane Cleaning? Micro-Bubbles and High Ionic Strength Effervescent Cleaners Enhance RO Membrane Foulant Removal. Proceedings 2014 AWWA/AMTA Membrane Technology Conference, Las Vegas, NV, March 10–14.

Choi, W.; Choi, J.; Bang, J.; & Lee, J.H. 2013. Layer-by-Layer Assembly of Araphene Oxide Nanosheets on Polyamide Membrane for Durable Reverse-Osmosis Applications. *ACS Applied Materials and Interfaces*, 5:12510–12519.

Coday, B.D.; Heil, D.M.; Xu, P.; & Cath, T.Y. 2013. Effects of Transmembrane Hydraulic Pressure on Performance of Forward Osmosis Membranes. *Environ. Sci. Technol.*, 47: 2386–2393.

Comerton, A.M.; Andrews, R.C.; & Bagley, D.M. 2009. The Influence of Natural Organic Matter and Cations on the Rejection of Endocrine Disrupting and Pharmaceutically Active Compounds by Nanofiltration. *Water Research*, 43:613–622.

Corry, B. 2008. Designing Carbon Nanotube Membranes for Efficient Water Desalination. *J. Phys.Chem.B*, 112:1427–1434.

Corry, B. 2011. Water and Ion Transport Through Functionalized Carbon Nanotubes: Implications for Desalination Technology. *Energy Environ. Sci.*, 4:751–759.

Drewes, J.; Cath, T.; Xu, P.; Graydon, J.; Veil, J.; & Snyder, S. 2008. An Integrated Framework for Treatment and Management of Produced Water. Golden, CO: RPSEA Forum.

El-Bourawi, M.S.; Ding, Z.; Ma, R.; & Khayet, M. 2006. A Framework for Better Understanding of Membrane Distillation Separation Process. *J. Membrane Sci.*, 285:4–29.

Erdal, U.; Shyamasundar, V.; Wilson, K.; Schimmoller, L.; & Daigger, G. 2010. Are Membrane Bioreactors Better Than Conventional Activated Sludge Systems to Remove Compounds of Emerging Concern? Proceedings Water Environment Federation Membrane Applications Conference, Anaheim, California. Alexandria, VA:

Evans, L.R. & Miller, J.E. 2002. Sweeping Gas Membrane Desalination Using Commercial Hydrophobic Hollow Fiber Membranes, SAND 2002-0138. Albuquerque, N.M.: Sandia National Labs Report.

Fatta D.; A. Achileos, A. Nicholaou, S. Meric. 2007. Analytical Methods for Tracing Pharmaceutical Residues in Water and Wastewater. *Trends in Analytical Chemistry*, 26(6):515–533.

Foreign Press Center/Japan. 2014. "City of Yokohoma: Japan's Largest Membrane Filtration Plant Completed," March 13.

Freeman, S. & Shorney-Darby, H. 2011. Manager to Manager–What is the Buzz about Ceramic Membranes. *Journal of the American Water Works Association*, 103: 12.

Gai, J.G. & Gong, X.L. 2014. Zero Internal Concentration Polarization FO Membrane: Functionalized Graphene. *Journal of Materials Chemistry A*, 2:425–429.

Gai, J.G.; Gong, X.L.; Wang, W.W.; Zhang, X.; & Kang, W.L. 2014. An Ultrafast Water Transport Forward Osmosis Membrane: Porous Graphene. *Journal of Materials Chemistry A*, 2:4023–4028.

Gethard, K.; Sae-Khow, O.; & Mitra, S. 2011. Water Desalination Using Carbon-Nanotube-Enhanced Membrane Distillation. *Appl. Mat. Interf.*, 3:110–114.

Hilder, T.A.; Gordon, D.; & Chung, S. 2009. Salt rejection and water transport through boron nitride nanotubes. *Small*, 5:2183–2190.

Hofs, B.; Schurer, R.; Harmsen, D.J.H.; Ceccarelli, C.; Beerendonk, E.F.; & Cornelissen, E.R. 2013. Characterization and Performance of a Commercial Thin Film Nanocomposite Seawater Reverse Osmosis Membrane and Comparison With a Thin Film Composite. *J. Membrane Sci.*, 446:68–78.

Holt, J.K.; Park, H.G.; Wang, Y.M.; Stadermann, M.; Artyukhin, A.B.; Grigoropoulos, C.P.; Noy, A.; & Bakajin, O. 2006. Fast Mass Transport Through Sub-2-Nanometer Carbon Nanotubes. *Science*, 312:1034–1037.

Hoover, L.A.; Phillip, W.A.; Tiraferri, A.; Yip, N.Y.; & Elimelech, M. 2011. Forward With Osmosis: Emerging Applications for Greater Sustainability. *Environ. Sci. Technol.*, 45:9824–830.

Hummer, G.; Rasaiah, J.C.; & Noworyta, J.P. 2001. Water Conduction Through the Hydrophobic Channel of a Carbon Nanotube. *Nature*, 414:188–190.

Humplik, T.; Lee, J.; O'Hern, S.C.; Fellman, B.A.; Baig, M.A.; Hassan, S.F.; Atieh, M.A.; Rahman, F.; Laoui, T.; Karnik, R.; & Wang, E.N. 2011. Nanostructured Materials for Water Desalination. *Nanotechnology*, 22:292001.

Hyung, H.; Wetterau, G.; You, E.; & Cutler, D. 2014. Application of the Universal MF/UF Membrane Skid in a Full Scale Wastewater Reclamation Plants–Challenge and Future. Proceedings 2014 AWWA/AMTA Membrane Technology Conference, Las Vegas, NV, March 10-14.

Jelic, A.; Petrovic, M.; & Barcelo, D. 2012. Emerging Organic Contaminants and Human Health. Springer-Verlag Berlin Heidelberg, Published Online pp. 47–70. http://www.springer.com/us/book/9783642281310

Jeong, B.H.; Hoek, E.M.V.; Yan, Y.; Subramani, A.; Huang, X.; Hurwitz, G.; Ghosh, A.K.; & Jawor, A. 2007. Interfacial Polymerization of Thin Film Nanocomposites: A New Concept for Reverse Osmosis Membranes. *J. Mem. Sci.*, 294:1–7.

Joshi, R.K.; Carbone, P.; Wang, F.C.; Kravets, Su, Y.; Grigorieva, I.V.; Wu, H.A.; Geim, A.K.; & Nair, R.R. 2014. Precise and Ultrafast Molecular Sieving Through Graphene Oxide Membranes. *Science*, 343:752–754.

Kalra, A.; Garde, S.; & Hummer, G. 2003. Osmotic Water Transport Through Carbon Nanotube Membranes. *Proc. Natl. Acad. Sci. U.S.A.*, 100:10175–10180.

Kaufman, Y.; Berman, A.; & Freger, V. 2010. Supported Lipid Bilayer Membranes for Water Purification by Reverse Osmosis. *Langmuir*, 26:7388–7395.

Kim, K.; Ahmed, Z.; Ahn, K.; & Paeng, K. 2009. Biodegradation of Two Model Estrogenic Compounds in a Pre-Anoxic/Anaerobic Nutrient Removing Membrane Bioreactor. *Desalination*, 243:265–272.

Kim, S.G.; Hyeon, D.H.; Chun, J.H.; Chun, B.H.; & Kim, S.H. 2013. Novel Thin Nanocomposite RO Membranes for Chlorine Resistance. *Desalination and Water Treatment*, 51:6338–6345.

Kim, Y.D.; Thu, K.; Ghaffour, N.; & Ng, K.C. 2013. Performance Investigation of a Solar-Assisted Direct Contact Membrane Distillation System. *J. Membr. Sci.*, 427:345–364.

Klaysom, C.; Cath, T.Y.; Depuydt, T.; & Vankelecom, I.F.J. 2013. Forward and Pressure Retarded Osmosis: Potential Solutions for Global Challenges in Energy and Water Supply. *Chem. Soc. Rev.*, 42:6959–6989.

Kumar, M.; Grzelakowski, M.; Zilles, J.; Clark, M.; & Meier, W. 2007. Highly Permeable Polymeric Membranes Based on the Incorporation of the Functional Water Channel Protein Aquaporin Z. *Proc. Nat. Acad. Sci.*, 104:20719-20724.

Kurth, C.J.; Burk, R.L.; Green, J.; & Portillo, L. 2011. Utilizing Nanotechnology to Enhance RO Membrane Performance for Seawater Desalination. American Membrane Technology Association.

Kurth, C.J.; Koehler, J.A.; Zhou, M.; Holmberg, B.A.; & Burk, R.L. 2014. Hybrid nanoparticle TFC membranes, Patent US 8603340 B2. http://www.google.com/patents/US8603340.

Le-Minh, N.; Coleman, H.; Khan, S.; Van Luer, Y.; Trang, T.; Watkins, G.; & Stuetz, R. 2010. The Application of Membrane Bioreactors as Decentralized Systems for Removal of Endocrine Disrupting Chemicals and Pharmaceuticals. *Water Sci. Technol.*, 61:5:1081–1088.

Liberman, B. & Liberman, I. 2005. RO Membrane Cleaning–Replacing Membrane CIP by Direct Osmosis Cleaning. *Desalination & Water Reuse*, 15:28-32.

Liu, L.; Meng W.; Duo W.; & Congjie G. 2010. Current Patents of Forward Osmosis Membrane Process. 2.1:76-82.

Majumder, M.; Chopra, N.; Andrews, R.; & Hinds, B.J. 2005. Nanoscale Hydrodynamics: Enhanced Flow in Carbon Nanotubes. *Nature*, 438:44.

Mathioulakis, E.; Belessiotis, V.; & Delyannis, E. 2007. Desalination by Using Alternative Energy: Review and State-of-the-Art. *Desalination*, 203:46–365.

McCutcheon, J.; McGinnis, R.L.; & Elimelech, M. 2005. A Novel Ammonia-Carbon Dioxide Forward (Direct) Osmosis Desalination Process. *Desalination*, 174:1–11.

———. 2006. Desalination by Ammonia-Carbon Dioxide Forward Osmosis: Influence of Draw and Deed Solution Concentrations on Process Performance. *J. Membrane Sci.*, 278:114–123.

McGovern, R.K. & Lienhard, J.H. 2014. On the Potential of Forward Osmosis to Energetically Outperform Reverse Osmosis Desalination, *J. Membrane Sci.*, 469:245–250.

Mi, B. 2014. Graphene Oxide Membranes for Ionic and Molecular Sieving. *Science*, 343:740–742.

Miller, J.E. & Evans, L.R. 2006. Forward Osmosis: A New Approach to Water Purification and Desalination. Albuquerque, N.M.: Sandia National Laboratories.

MWH. 2014. Emerging Energy Reducing Technologies for Desalination Applications. Alexandria, VA: WateReuse Research Foundation.

Nair, R.R.; Wu, H.A.; Jayaram, P.N.; Grigorieva, I.V.; & Geim. A.K. 2012. Unimpeded Permeation of Water Through Helium-Leak-Tight Graphene-based Membranes. *Science*, 335:442–444.

NanoH2O, Inc. 2012. Changing the Economics of Desalination. NanoH2O, Inc. https://www.greentechmedia.com/articles/read/nanoh2o#gs._XHC4fI

———. 2011. Qfx SW 400 ES SWRO Element Data Sheet.

National Water Research Institute. 2010. Source, Fate and Trasmport of Endocrine Disruptors, Pharmaceuticals, and Personal Care Products in Drinking Water Sources in California. Fountain Valley, CA: NWRI.

New York Times (NYTimes). 2014. New Solar Pond Distillation System Designed. http://www.sciencedaily.com/releases/2010/01/100105170942.htm.

Ngheim, L.; Tadkaew, N.; & Sivakumar, M. 2009. Removal of Trace Organic Contaminants by Submerged Membrane Bioreactors. *Desalination*, 236:127–134.

Nicolai, A.; Sumpter, B.G.; & Meunier, V. 2014. Tunable Water Desalination Across Graphene Oxide Framework Membranes. *Physical Chemistry Chemical Physics*, 16:8646–8654.

O'Hern, S.C.; Boutilier, M.S.H.; Idrobo, J.C.; Song, Y.; Kong, J.; Laoui, T.; Atieh, M.; & Karnik, R. 2014. Selective Ionic Transport Through Tunable Subnanometer Pores in Single-Layer Graphene Membranes. *NanoLetters*, 14:1234–1241.

Pendergast, M.M. & Hoek, E.M.V. 2011. A Review of Water Treatment Membrane Nanotechnologies. *Energy Environ. Sci.*, 4:1946–1971.

Pendergast, M.M.; Ghosh, A.K.; & Hoek, E.M.V. 2013. Separation Performance and Interfacial Properties of Nanocomposite Reverse Osmosis Membranes. *Desalination*, 308:180–185.

Phuntsho, S.; Ho, K. S.; Seungkwan, H.; Sangyoup, L.; & Vigneswaran, S. 2011. A Novel Low Energy Fertilizer Driven Forward Osmosis Desalination for Direct Fertigation: Evaluating the Performance of Fertilizer Draw Solutions. *J. Membrane Sci.*, 375:172–181.

Qin, J.; Oo, M.; Kkkre,K.; & Liberman, B. 2010. Development of Novel Backwash Cleaning Technique for Reverse Osmosis in Reclamation of Secondary Effluent. *J. Membr. Sci.*, 346:8–14.

Sahar, E.; David, I.; Gelman, Y.; Chikurel, H.; Aharoni, A.; Messalem, R.; & Brenner, A. 2011. The Use of RO to Remove Emerging Micropollutants Following CAS/UF or MBR Treatment of Municipal Wastewater. *Desalination*, 273:142–147.

Salvetat, J.P.; Briggs, A.D.; Bonard, J.M.; Bacsa, R.R.; Kulik, A.J.; Stockli, T.; Burnham, N.A.; & Forro, L. 1999. Elastic and Shear Moduli for Single-Walled Carbon Nanotube Ropes. *Phys. Rev. Lett.*, 82:944–947.

Salvetat, J.P.; Kulik, A.J.; Bonard, J.M.; Briggs, G.A.D.; Stockli, T.; Metenier, K.; Bonnamy, S.; Beguin, F.; Burnham, N.A.; & Forro, L. 1999. Elastic Modulus of Ordered and Disordered Mutiwalled Carbon Nanotubes. *Adv. Mater.*, 11:161–165.

Semiat, R. 2008. Energy Issues in Desalination Processes. *Environ. Sci. Technol.*, 42:8193–8201.

Shaffer, D.L.; Yip, N.Y.; Gilron, J.; & Elimelech, M. 2012. Perspective: Seawater Desalination for Agriculture by Integrated Forward and Reverse Osmosis: Improved Product Water Quality for Potentially Less Energy. *J. Membrane Sci.*, 415–416:1–8.

Sholl, D.S. & Johnson, J.K. 2006. Making High-Flux Membranes With Carbon Nanotubes. *Science*, 312 95776 0:1003-1004.

Sipma, J.; Osuna, B.; Collado, N.; Monclus, H.; Ferrero, G.; Comas, J.; & Rodriquez-Roda, I. 2010. Comparison of Removal of Pharmaceuticals in MBR and Activated Sludge Systems. *Desalination*, 250:653–659.

Song, L.; Hu, J.Y.; One, S.L.; Ng, W.J.; Elimelech, M.; & Wilf, M. 2003. Emergence of Thermodynamic Restriction and Its Implications for Full-Scale Reverse Osmosis Processes. *Desalination*, 144:213–228.

Sui, H.; Han, B.G.; Lee, J.K.; Walian, P.; & Jap, B.K. 2001. Structural Basis of Water Specific Transport Through the AQP1 Water Channel. *Nature*, 414:872–878.

Tambosi, J.; Felix de Sena, R.; Favier, M.; Gebhardt, W.; Jose, H.; Schroder, H.; & Moreira, R. 2010. Removal of Pharmaceutical Compounds in Membrane Bioreactors (MBR) Applying Submerged Membranes. *Desalination*, 261:148–156.

Tang, C.Y.; Zhao, Y.; Wang, R.; Helix-Nielsen, C.; & Fane, A.G. 2013. Desalination by Biomimetic Aquaporin Membranes: Review of Status and Prospects. *Desalination*, 308: 34–40.

Tomaszewska, M. 2000. Membrane Distillation–Examples of Application in Technology and Environmental Protection. *Pol. J. Environ. Stud.*, 9:27–36.

US Bureau of Reclamation. 2009. Secondary/Emerging Constituents Report, Southern California Regional Brine-Concentrate Management Study–Phase 1, Lower Colorado Region. Washington DC: USDOI-BOR.

USEPA (US Environmental Protection Agency). 2010. Treating Contaminants of Emerging Concern: A Literature Review Database. Washington, DC: USEPA.

University of Colorado at Boulder. (2010. 2013. Gender-Bending Fish Problem in Colorado Creek Mitigated by Treatment Plant Upgrade.

Verliefde, A.; Hejiman, S.G.J.; Cornelissen, E.; Amy, G.; Van der Bruggen, B.; & Van Dijk, J.C. 2007. Rejection of Trace Organic Pollutants with High Pressure Membranes (NF/RO). *Environmental Pollution*, 146:281–289.

Walton, J.; Lu, H.; Turner, C.; Solis, S.; & Hein, H. 2004. Solar and Waste Heat Desalination by Membrane Distillation. Desalination and Water Purification Research and Development (DRIP) Program, Report No. 81. Denver, CO: USDOI-BOR

Water Desalination Report. 2014. Montana–a Ceramic Peg in a Polymeric Hole. August 11.

Westerling, K. 2014. 9 Best Bets in Membrane Technology." *Water Online*, March 27.

World Health Organization. 2012. Pharmaceuticals in Drinking Water. Geneva, Switzerland: WHO Press.

WRRF (WateReuse Research Foundation). 2006. *Rejection of Wastewater-Derived Micropollutants in High-Pressure Membrane Applications Leading to Indirect Potable Reuse: Effects of Membrane and Micropollutant Properties*. Alexandria, VA: WRRF.

———. 2013. *Predicting Future Contaminants of Concern for Indirect Potable Water Reuse*. Alexandria, VA: WRRF.

Xue, M.; Qiu, H.; & Gui, W. 2013. Exceptionally Fast Water Desalination at Complete Salt Rejection by Pristine Graphyne Monolayers. *Nanotechnology*, 24:505720.

Yangali-Quintanilla, V.; Sadmani, A.; McConville, M.; Kennedy, M.; & Amy, G. 2009. Rejection of Pharmaceutically Active Compounds and Endocrine Disrupting Compounds by Clean and Fouled Nanofiltration Membranes. *Water Res.*, 43:2349–2362.

———. 2010a. A QSAR Model for Predicting Rejection of Emerging Contaminants (Pharmaceuticals, Endocrine Disruptors) by Nanofiltration Membranes. *Water Res.* 44:373–384.

Yangali-Quintanilla, V.; Maeng, S.; Fujioka, M.; Kennedy, M.; & Amy, G. 2010b. Proposing Nanofiltration as Acceptable Barrier for Organic Contaminants in Water Reuse. *J. Membr. Sci.;* 362:334–345.

Yangali-Quintanilla, V.; Zhenyu L.; Valladeres, R.; Li, Q.; & Amy, G. 2011. Indirect Desalination of Red Sea Water with Forward Osmosis and Low Pressure Reverse Osmosis for Water Reuse. *Desalination*, 280:160–166.

Yen, S.K.; Mehnas, H.N.F.; Su, M.; Wang, K.Y.; & Chung, T.S. 2010. Study of Draw Solutes Using 2-Methylimidazole-based Compounds in Forward Osmosis. *J. Membr. Sci.*, 364:242–252.

Yoshikawa, N.; Asari, T.; Kishi, N.; Hayashi, S.; Sugai, T.; & Shinohara, H. 2008. An Efficient Fabrication of Vertically Aligned Carbon Nanotubes on Flexible Aluminum Foils by Catalyst-Supported Chemical Vapor Deposition. *Nanotechnol.*, 19:245607.

Zazouli, M.A.; Susanto, H.; Nasseri, S.; & Ulbricht, M. 2009. Influences of Solution Chemistry and Polymeric Natural Organic Matter on the Removal of Aquatic Pharmaceutical Residuals by Nanofiltration. *Water Res.*, 43:3270–3280.

Zhang, H.; Zhong, Z.; & Xing, W. 2013. Application of Ceramic Membranes in the Treatment of Oilfield-Produced Water: Effect of Polyacrylamide and Inorganic Salts. *Desalination*, 309:84–90.

Zhao, S.; Zou, L.; Tang, C.Y.; & Mulcahy, D. 2012. Recent Developments in Forward Osmosis: Opportunities and Challenges. *J. Membr. Sci.*, 396:1–21.

Zhu, F.Q.; Tajkhorshid, E.; & Schulten, K. 2004. Theory and Simulation of Water Permeation in Aquaporin-1. *Biophys. J.*, 86:50–57.

Zraick, F.; Janex-Habibi, M.; Glucina, K.; & Simon, J. 2013. Prediction Models for the Removal of Emerging Contaminants by Nanofiltration and Reverse Osmosis Membranes. Proceedings AWWA/AMTA Membrane Technology Conference & Exposition; San Antonio, Texas, February 25–February 28. Denver, CO: AWWA.

Appendix A

The Groundwater Replenishment System Design Criteria and Water Quality Tables

Table A-1 MF system key design parameters

MF	System Manufacturer	Product	Instantaneous Flux, gfd (LMH)	Nominal Recovery	Backwash Interval	CIP Interval (days)	# of Trains	Total # of Membrane Modules
	Siemens	Memcor CS	20.4 (34.7)	89%	22 minutes	21	4	17,784 (684 per train)

Table A-2 RO system key design parameters

RO	Membrane Element Manufacturer	Flux, gfd (LMH)	Recovery	Train Array	Train Size (mgd)	# of Trains	CIP Interval (months)	Total # of Elements
	Hydranautics ESPA2	12 (20.4)	85	3-stage, 78:48:24, (7 elements per vessel)	5	15	6–8	15,750

Table A-3 GWRS typical water quality

Parameter	Units	Q1	MFF	MFE	ROF	ROP	UVP	FPW	Permit Limit
Electric conductivity	umhos/cm	1,502.83	1,530.53[1]	1,530.00	1,607.867[1]	37.03[1]	51.47	79.89[1]	900
Total dissolved solids	mg/L	955.19	na	na	964.17	21.96	22.53	43.24	500[2]
Suspended solids	mg/L	7.32	8.46	<1	na	na	na	na	N/A
Turbidity	ntu	2.27	3.242[1]	0.105[1]	<118[1]	0.03[1]	na	0.070[1]	≤0.2
Ultraviolet percent transmittance @ 254 nm	%	na	na	69.38	na	98.4[1]	na	na	≤0.5
pH	units	7.45	7.26[1]	7.51	6.70[1]	5.48[1]	5.71	7.64[1]	6–9
Total hardness (as $CaCO_3$)	mg/L	312.75	na	na	305.67	<1	<1	21.42	240[2]
Calcium	mg/L	78.69	na	na	77.57	0.02	0.03	8.73	N/A
Magnesium	mg/L	28.23	na	na	27.18	<0.1	<0.1	<0.1	N/A
Sodium	mg/L	203.33	na	na	188.75	6.53	6.72	6.50	45
Potassium	mg/L	16.80	na	na	16.63	0.41	0.42	0.38	N/A
Bromide	mg/L	na	na	na	na	na	na	<0.1	N/A
Chloride	mg/L	230.33	na	na	215.75	4.56	5.23	5.23	55
Sulfate	mg/L	213.00	na	na	238.08	0.27	0.10	0.33	100
Hydrogen peroxide	mg/L	na	na	na	na	na	2.40	2.29	N/A
Bicarbonate (as $CaCO_3$)	mg/L	na	na	na	136.50	7.88	5.77	26.25	N/A
Nitrate nitrogen	mg/L	9.21	na	na	9.28	0.95	1.07	0.99	3[2]
Nitrite nitrogen	mg/L	0.52	na	na	0.15	<0.002	0.001	0.03	1[2]
Ammonia nitrogen	mg/L	2.36	na	na	2.84	0.45	na	0.33	N/A
Organic nitrogen	mg/L	0.73	na	na	0.40	0.01	na	0.02	N/A
Total nitrogen	mg/L	12.35	na	na	12.65	na	na	1.36	5
Phosphate phosphorus	mg/L	0.32	na	na	na	na	na	<0.01	N/A
Iron	µg/L	276.00	na	na	91.38	0.83	0.70	0.65	300
Manganese	µg/L	34.22	na	na	32.19	0.33	0.70	<1	50
Aluminum	µg/L	11.87	na	na	6.99	1.06	1.44	2.60	200[2]
Arsenic	µg/L	0.18	na	na	0.31	<1	<1	<1	10
Barium	µg/L	29.66	na	na	27.08	<1	<1	<1	1,000
Boron	µg/L	0.37	na	na	0.37	0.23	0.23	0.22	N/A
Cadmium	µg/L	<1	na	na	<1	<1	<1	<1	5
Chromium	µg/L	0.34	na	na	<1	<1	<1	<1	50
Copper	µg/L	4.14	na	na	15.37	0.65	0.55	<1	1,000[2]
Cyanide	µg/L	0.91	na	na	<5	<5	na	<5	150
Fluoride	mg/L	0.93	na	na	na		na	<0.1	2

(Continued)

Table A-3 GWRS typical water quality *(Continued)*

Parameter	Units	Q1	MFF	MFE	ROF	ROP	UVP	FPW	Permit Limit
Lead	µg/L	<1	na	na	8.71		<1	<1	15
Mercury	µg/L	0.16	na	na	0.17		<0.1	<0.1	2
Nickel	µg/L	5.00	na	na	4.65		0.18	<1	100
Perchlorate	µg/L	na	na	na	na	na	na	<2.5	6
Selenium	µg/L	1.60	na	na	1.58	<1	<1	<1	50
Silica	µg/L	21.21	na	na	23.08	<1	0.74	0.40	N/A
Silver	µg/L	<1	na	na	<1	<1	<1	<1	100
Zinc	µg/L	23.10	na	na	60.20	0.27	0.19	<1	5,000
N-nitrosodiumethylamine	ng/L	11.95	na	na	na	na	0.24	0.66	N/A
1,4-Dioxane	µg/L	3.47	na	na	na	na	<1	<1	N/A
Total trihalomethanes	µg/L	na	na	na	na	na	na	0.70	80
Dibromoacetic acid	µg/L	na	na	na	na	na	na	<1	60, total HAA5
Dichloroacetic acid	µg/L	na	na	na	na	na	na	<1	60, total HAA5
Monobromoacetic acid	µg/L	na	na	na	na	na	na	<1	60, total HAA5
Monochloroacetic acid	µg/L	na	na	na	na	na	na	<1	60, total HAA5
Trichloroacetic acid	µg/L	na	na	na	na	na	na	<1	60, total HAA5
Apparent color (unfiltered)	units	na	na	na	25.92	0.53	na	<3	15
Total organic carbon (unfiltered)	mg/L	9.96		na	8.17	0.15	0.20	0.16	0.5
Surfactants (MBAS)	mg/L	0.23	na	na	0.25	<0.02	na	<0.02	0.5
Total coliform (mult. tube fermentation)	mpn/100 mL	335,657	2,936	<2	na	<2	<2	<2	2.2
Fecal coliform (mult. tube fermentation)	mpn/100 mL	108,771	391	<2	na	<2	<2<2	<2	N/A

[1] On-line average

Abbreviations: Q1, Secondary Effluent (AWPF influent); MFF, microfiltration feed; MFE, microfiltration effluent; na, not analyzed; ROP, reverse osmosis product; UVP, ultraviolet UV/AOP product; FPW, finished product water; N/A, not applicable.

This page intentionally blank.

Appendix B

Broad Run Water Reclamation Facility Design Criteria and Water Quality Tables

Table B-1 Design summary—Broad Run WRF Membrane Bioreactor System

Parameter	Criteria
Design Flows	
Average daily flow	10.0 mgd
Maximum 30-day flow	11.0 mgd
Peak hydraulic flow (equalized)	18.8 mgd
Design Criteria – Biological Treatment	
Influent wastewater	Fine-screened primary effluent
Temperature range	12 to 25°C
Process configurations	5-stage or 5-stage plus de-aeration
No. of biological reactor basins (BRBs)	3 (2 + 1 spare)
Hydraulic retention time (without membrane tanks)	9.1 hours (average); 4.8 hours (peak)
Solids retention time (SRT)	20 to 23 days
Anaerobic/anoxic volume	Approximately 35% of total volume
Aerobic volume	Approximately 65% of total volume
Mixed liquor concentration in membrane tanks	7,000 to 11,000 mg/L
Provisions for chemical addition	Alum (for phosphorus removal)
	Methanol (for nitrogen removal)
	Sodium hydroxide (pH and alkalinity adjustment)
	Sodium hypochlorite (biofoam control)

(Continued)

Table B-1 Design summary—Broad Run WRF Membrane Bioreactor System *(Continued)*

Parameter	Criteria
Design Criteria – Membrane Filtration	
Membrane flux	15 to 20 gallons per square foot/day (gfd) @ average and maximum 30-day flow < 30 gfd @ peak hydraulic flow
Membrane type	Immersed
Backpulse and maintenance cleaning – frequency and duration	As recommended by the manufacturer
Estimated recovery cleaning interval	Every 4 to 6 months
Provisions for chemical addition	Sodium hypochlorite and citric acid
Design Summary – Membrane Filtration	
Number of membrane trains	12 (10 trains in service and 2 out of service for maintenance or cleaning)
Number of cassettes/train	4
Number of modules/cassette	48
Membrane area/module (sf)	340
Installed membrane area/train (sf)	65,280
Membrane tanks	Rectangular, concrete
Membrane tank dimensions (each)	12 feet (wide); 45 feet (length); 11–12 feet (depth)

Table B-2 Broad Run Water Reclamation Facility—influent and effluent characteristics

	Influent (Daily Average)		Discharge (Monthly Average)	
Parameter	Design	Actual	Requirements	Performance
TSS, mg/L	200	310	1	< DL
COD, mg/L	500	680	10.0	< 5.0 (DL)
TKN, mg/L	36	48	1.0	0.5
TN, mg/L	--------	--------	4.0	3.5
TP, mg/L	7	6.3	0.1	0.03
Turbidity (NTU)	---------	--------	0.5	0.1

NOTES: DL = Detection Limit.
Summarized data are typical of monthly averages for 2011–2012

Appendix C

Big Spring Raw Water Production Facility Design Criteria and Water Quality Tables

Table C-1 Membrane system design criteria

	Microfiltration	Reverse Osmosis
Manufacturer	Pall Corporation	Pall Corporation
Model	Aria AP8	
Number of units	2	2
Modules/unit	78	36
Membranes used	Microza UNA-620A	Toray TML20-400
Design flux	32.9 gfd	10.5 gfd
Design recovery	95%	75%
	Cleaning Frequencies	
Feed flush/air scrub	20 minutes	N/A
Enhanced flux maintenance	48 hours	N/A
Clean in place	30 days	60 days
Feed chemicals	none	Sulfuric acid Anti-scalant Chloramine
Cleaning chemicals	Sodium hypochlorite Caustic soda Citric acid	Caustic soda Citric acid

Abbreviation: N/A, not applicable.

Table C-2 Feed water quality (July 2008–March 2013)

	Total Dissolved Solids (mg/L)	Chloride (mg/L)	Sulfate (mg/L)	Total Hardness (mg/L)	Total Alkalinity (mg/L)
Mean	2,822	1,050	667	911	113
Minimum	1,540	571	352	493	67
Maximum	4,250	1,820	992	1,320	207

Appendix D

Edward C. Little Water Recycling Facility Design Criteria and Water Quality Tables

Table D-1 Feed water quality

Constituent	Permit Limit	Average	Maximum	Minimum
Turbidity (NTU)	None	8.1	10.7	5.5
pH	None	6.8	6.9	6.6
Total organic carbon (mg/L)	None	11.8	14.0	8.0
Total suspended solids (mg/L)	30	22.0	28.0	13.0
cBOD (mg/L)	25	11.6	19.0	7.0
NH_3-N (mg/L)	None	38	53	23
TDS (mg/L)	None	854	1020	684

Table D-2 Ozone system design criteria

Description	Units	Value
Design flow	mgd	30
Ozone dose range	mg/L	3 - 16
Ozone dose average	mg/L	12
Number of ozone generators	No.	2
Design ozone gas concentration	% by wt.	10
Capacity at 10% by weight	lbs/day	2,000

Table D-3 Microfiltration design criteria

Description	Phase II	Phase III	Phase IV	Phase V
Process area	Barrier	Industrial boiler feed	Barrier	Barrier/Industrial
Design capacity (mgd)	2.6	4.6	11.5	7.6
Manufacturer	Siemens	Siemens	Siemens	Pall
Membrane material	PP	PP	PP	PVDF
Nominal pore size (um)	0.2	0.2	0.2	0.1
Module type	CMF	CMF	CMF-S	Microza
Number of units	5	10	6	6
Normally operating	5	9	5	5
Stand-by	0	1	1	1
No. of MF modules per unit	90	90	416	140
Membrane area per module (sq ft)	360	360	335	538
Filtrate flow per unit (gpm)	368	357	1,596	1,048
Instantaneous design flux (gfd)	19.0	18.5	18.7	27.2
Recovery (%)	90	90	90	90
Ferric chloride pre-treatment	No	No	No	up to 10 mg/L
Enhanced flux maintenance	None	None	None	NaOCl once per day, citric acid every 10 days
Design CIP Interval	3 weeks	3 weeks	3 weeks	3 weeks
Typical CIP Chemicals	NaOH, Citric acid	NaOH, Citric acid	NaOH, Citric acid	NaOCl/NaOH, Citric acid

Table D-4 Reverse osmosis design criteria

Description	Trains 1, 2	Train 3	Trains 4, 5	Trains 6, 7, 8	Train 9	Trains 10, 11
Process area	Barrier	Barrier	First pass Industrial Boiler feed	Second pass Industrial Boiler feed	Barrier	Barrier
Design capacity per train (mgd)	3.0	3.0	2.5	1.3 (on standby)	3.5	2.5
Membrane material	Polyamide	Polyamide	Polyamide	Polyamide	Polyamide	Polyamide
Design feed pressure range (psi)	125–300	125–300	125–300	125–300	125–300	125–300
Train flux rate (gfd)	10	10	11.4	19	11.6	11.9
Recovery (%)	85	85	85	90	85	85
No. of stages per train	3	3	2	2	2	3
Array	72:36:18	60:36:12	48:24	18:6	72:36	43:20:12
Pressure vessel per train	126	108	72	24	108	75
Elements per pressure vessel	6	7	7	7	7	7
No. of elements per train	756	756	504	168	756	525
Membrane area per element (sq ft)	400	400	440	400	400	400

Table D-5 Advanced oxidation process

Description	Value
System design rlow (mgd)	17.5
Hydrogen peroxide dose (mg/L)	3
Type of UV system	Low pressure, high intensity
No. of trains	4
No. of reactors per train	4
No. of lamps per reactor	72
No. of lamps per train	288
Nominal input power (watts per lamp)	257
Design flow per train, all trains in service (mgd)	4.4
Max flow per train, one train out of service (mgd)	5.8
Minimum UV dose delivery at max. flow rate (mJ/cm^2)	50
Design UV dose delivery at max flow rate (mJ/cm^2)	>115
Reduction of NDMA, all trains in service	1.3 log
Reduction of NDMA, one train out of service	1.0 log

Table D-6 Barrier recycled water constituents (2012)

Constituent	Units	Limit	Average	Max	Min
Turbidity (continuous)	NTU	0.2 avg/ 0.5 max	0.04	0.06	0.01
Daily flow	mgd	-	5.84	6.70	4.73
pH (continuous)	pH units	6.5–8.5	7.8	8.0	7.7
Conductivity (continuous)	μS/cm	300.0	72.4	90.4	50.1
Minimum			60.0	78.3	38.9
Maximum			83.3	100.1	57.5
Miscellaneous (Daily/Biweekly/Weekly Grab)					
Total coliform (daily grab)	MPN	2.2/7 days	ND	ND	ND
TOC (weekly grab)	mg/L	0.5 divided by max. average recycled water contribution	0.23	0.28	0.19
Temperature (weekly grab)	deg F	100	78.75	85.00	73.00
Oil and grease (weekly grab)	mg/L	-	ND	ND	ND
cBOD (weekly composite)	mg/L	-	ND	ND	ND
Nitrate (twice weekly grab)	mg/L N	-	0.62	0.88	0.37
Nitrite (twice weekly grab)	mg/L N	-	0.15	0.25	0.12
TKN (twice weekly grab)	mg/L N	-	2.6	3.3	1.8
Ammonia (twice weekly grab)	mg/L N	-	2.3	2.8	1.7
Total organic nitrogen	mg/L N	-	0.2	1.3	0.2
Total nitrogen	mg/L N	5 divided by max. average RWC	1.2	2.20	0.30

(Continued)

Table D-6 Barrier recycled water constituents (2012) *(Continued)*

Constituent	Units	Limit	Average	Max	Min
Trace Metals/Inorganic Chemicals (Quarterly Grab)					
Aluminum	mg/L	1	0.0014	0.0022	0.0015
Antimony	mg/L	0.006	ND	ND	ND
Arsenic	mg/L	0.01	ND	ND	ND
Barium	mg/L	1	0.00035	0.00058	0.00037
Beryllium	mg/L	0.004	ND	ND	ND
Cadmium	mg/L	0.005	ND	ND	ND
Chromium	mg/L	0.05	0.00153	0.00240	0.00180
Copper	mg/L	1.3	0.0021	0.0041	0.0006
Iron	mg/L	0.3	ND	ND	ND
Lead	mg/L	0.015	0.00010	0.00024	0.00018
Mercury	μg/L	2	ND	ND	ND
Manganese	mg/L	0.05	0.00012	0.00028	0.00022
Nickel	mg/L	0.1	0.00084	0.00260	0.00032
Selenium	mg/L	0.05	ND	ND	ND
Silver	mg/L	0.10	ND	ND	ND
Thallium	mg/L	0.002	ND	ND	ND
Vanadium	μg/L	50	0.389	0.610	0.425
Zinc	mg/L	5	0.0012	0.0019	0.0005
Chemicals of Concern (Annual Grab)					
Chlorate	μg/L	800	99	160.00	97.00
n-Nitrosodimethylamine (NDMA)	ng/L	10	5.3	16.00	1.10
n-Nitrosodiethylamine (NDEA)	ng/L	10	0.9	4.70	ND
Formaldehyde	μg/L	100	55	110.00	19.00
General Physical (Quarterly Grab)					
Color	-	15	ND	ND	ND
Corrosivity	-	non-corrosive	-0.6	-0.24	-0.98
Foaming agents (MBAS)	mg/L	1	ND	ND	ND
Odor	odor units	3	1		
General Mineral (Quarterly Grab)					
Asbestos	MFL	7.0	ND	ND	ND
Total dissolved solids	mg/L	800	82	110.00	31.00
Calcium	mg/L	-	16	20.00	5.60
Magnesium	mg/L	-	0.017	0.03	0.02
Sodium	mg/L	-	14	16.00	12.00
Potassium	mg/L	-	0.9	1.20	0.54
Total hardness (as $CaCO_3$)	mg/L	-	41	51.00	14.00
Total alkalinity (as $CaCO_3$)	mg/L	-	60	78.00	25.00
Bicarbonate (as $CaCO_3$)	mg/L	-	60	77.00	25.00
Carbonate (as $CaCO_3$)	mg/L	-	0.4	0.80	0.10
Chloride	mg/L	-	14	21.00	9.00
Cyanide	mg/L	0.15	ND		
Sulfate	mg/L	500	ND	ND	ND
Fluoride (@ 64 to 71°F)	mg/L	2	0.56		

Abbreviation: ND, not detected.

Table D-7 Single pass low pressure boiler feed water

Parameter	Units	DL	Average	Maximum	Minimum
TDS	mg/L	5	19	38	10
Conductivity	µS/cm	1	37	50	26
Calculated TDS	mg/L	0.6	22.0	30.0	15.6
pH	unit	1.0	7.3	7.7	6.9
Alkalinity (as $CaCO_3$)	mg/L	2	11.3	25.0	8.0
Bicarbonate (as $CaCO_3$)	mg/L	2	11.3	24.9	8.0
Carbonate (as $CaCO_3$)	mg/L	0.1	ND	ND	ND
Hydroxide (as $CaCO_3$)	mg/L	0.01	0.01	0.03	0.01
Calcium	mg/L	0.005	0.019	0.035	0.006
Magnesium	mg/L	0.005	0.008	0.017	0.005
Hardness (as $CaCO_3$)	mg/L	0.03	0.06	0.14	0.03
Sodium	mg/L	0.01	4.45	6.04	2.99
Potassium	mg/L	0.01	0.45	0.61	0.30
Silica	mg/L	0.04	0.29	0.53	0.15

Abbreviations: DL, detection limit; ND, not detected.

Table D-8 Double pass high pressure boiler feed water

Parameter	Units	DL	Average	Maximum	Minimum
TDS	mg/L	5	14	22	10
Conductivity	µS/cm	1	5	8	3
Calculated TDS	mg/L	0.6	2.9	4.7	1.8
pH	Units	1.0	6.5	7.1	6.1
Alkalinity (as $CaCO_3$)	mg/L	2	ND	ND	ND
Bicarbonate (as $CaCO_3$)	mg/L	2	ND	ND	ND
Carbonate (as $CaCO_3$)	mg/L	0.1	ND	ND	ND
Hydroxide (as $CaCO_3$)	mg/L	0.01	ND	ND	ND
Calcium	mg/L	0.005	0.005	0.005	0.005
Magnesium	mg/L	0.005	0.008	0.010	0.006
Hardness (as $CaCO_3$)	mg/L	0.03	0.04	0.04	0.04
Sodium	mg/L	0.01	0.36	0.70	0.20
Potassium	mg/L	0.01	0.023	0.057	0.010
Silica	mg/L	0.04	0.11	0.23	0.05

Abbreviations: DL, detection limit; ND, not detected.

This page intentionally blank.

Appendix E

Singapore NEWater Facilities Process Design Criteria and Water Quality Tables

Table E-1 Design characteristics of treatment processes used at NEWater Demonstration Plant

Treatment Processes		
Microfiltration (MF)	Pore size	0.2 μm
	Water recovery	> 90 %
	No. of trains	5
	Capacity of each train	115 m^3/h
	Backwash interval	30 minutes
	Cleaning frequency	21 days, minimum
Reverse osmosis (RO)	Water recovery	85 %
	No. of trains	2
	Capacity of each train	5,000 m^3/d
	Cleaning frequency	1 month, minimum
Ultra-violet (UV) disinfection	Type	Medium pressure
	No. of units	2 duty, 1 standby
	Dosage (minimum)	80 mJ/cm^2

Table E-2 Feedwater characteristics of Bedok NEWater Demonstration Plant

Parameter	Value, mg/L
BOD_5	20
TSS	30
NH_3-N	6
TDS	400–1,600
TOC	12

Table E-3 Design criteria for plant processes and equipment at Bedok NEWater factory

Plant Capacity		Phase 1 – 32,000 m^3/day (installed 2003)	Phase 2 – 50,000 m^3/day (installed 2008)
MF/UF system	Manufacturer name & model no.	Zenon Zeeweed 500C (UF)	Asahi Kasei Microza UNA620A (MF)
	No. of trains; modules or cassettes per train	5 trains; 14 cassette/train; 36 modules/cassette	6 trains; 2 skids/train; 102 modules/skid
	Flux & recovery, backwash frequency & duration	30.38 (operating) & 37.98 (max) LMH (when 1 train offline); 90% recovery; backwash 30 s every 12 minutes	59 LMH/rack (max) & 25 LMH/rack (min); 90.2% recovery; backwash 90 s every 278 m^3 permeate produced
	Clean-in-place (CIP) type and frequency	Maintenance clean every week for 50 minutes, recirculation using citric acid; Recovery clean every 1,000 hours for 5 hours, soaking using sodium hypochlorite	Maintenance clean every 72 air scrubbing reverse filtration cycles (or 20016 m^3 of filtrate produced) for 43 minutes, recirculation using sodium hypochlorite Recovery clean every 6 months for 6 hours, recirculation with citric acid (low pH) and sodium hypochlorite & sodium hydroxide (high pH)
RO system	Element manufacturer name & model no.	Hydranautics ESPA2	Hydranautics ESPA2 Saehan CSM RE-BLR (16-inch)
	No. of trains; vessels per train; elements per vessel	4 trains; 50:23 vessels; 7 elements/vessel	7 trains; 12:6 vessels; 4 elements/vessel
	Flux & recovery	17.6 LMH; 75% recovery	40.3 LMH; 75% recovery
	Clean-in-place (CIP) type and frequency	Stage 1: 1–3 months Stage 2: 1–2 months Using Citric acid, caustic + SLS + EDTA + biocide	Stage 1: 1–3 months Stage 2: 1–2 months Using citric acid, caustic + SLS + EDTA + biocide
UV system	Manufacturer name & model no.	Hanovia Photon V2	Wedeco BX 1000 UV-C
	No. of trains	3 trains (2 duty, 1 standby)	6 trains (4 duty, 2 standby)
	Type	Medium pressure	Low pressure
	No. of lamps per unit	6	12
	Flow	16,000 m^3/day (max per train)	14,208 m^3/day

Table E-4 Design criteria for plant processes and equipment at Kranji NEWater factory

Plant Capacity		Phase 1 - 40,000 m^3/day (installed 2003)	Phase 2 - 42,000 m^3/day (installed 2006/07)
MF System	Manufacturer name & model no.	MEMCOR-CMF-S 480S10T	MEMCOR-CMF-S 480S10T
	No. of trains; modules or cassettes per train	6 trains; 14 racks/train; 32 modules/rack	6 trains; 15 racks/train; 32 modules/rack
	Flux & recovery, backwash frequency & duration	35.8 LMH (Max); 90% recovery; Backwash 150s every 20–30 minutes	35.8 LMH (Max); 90% recovery; backwash 150s every 20–30 minutes
	Clean-in-place (CIP) type and frequency	Recovery clean every operation > 21 days, using NaOH and citric acid	Recovery clean every 500 running hours, using NaOH and citric acid
RO System	Element manufacturer name & model no.	Hydranautics ESPA2	Hydranautics ESPA2 & Woongjin RE8040
	No. of trains; vessels per train; elements per vessel	5 trains; 49:24 vessels; 7 elements/vessel	ESPA2 - 2 trains; 49:24 vessels; 7 elements/vessel RE8040 - 2 trains; 78:40 vessels; 7 elements/vessel
	Flux & recovery	17.5 LMH; 75% recovery	17.5 LMH; 75% recovery
	Clean-In-Place (CIP) Type and Frequency	Stage 1: 1–3 months Stage 2: 1–2 months Using Citric acid, caustic + SLS + EDTA + biocide	Stage 1: 1–3 months Stage 2: 1–2 months Using citric acid, caustic + SLS + EDTA + biocide
UV system	Manufacturer name & model no.	Hanovia Photon V2	Wedeco BX650
	No. of trains	3 trains (2 duty, 1 standby)	6 trains (5 duty, 1 standby)
	Type	Medium pressure	Low pressure
	No. of lamps per unit	6	8
	Flow	20,000 m^3 per day (per unit)	9,000 m^3 per day (per unit)

Table E-5 Details of plant processes and equipment at Ulu Pandan & Changi NEWater factories

		Ulu Pandan	Changi
Plant capacity, contract period & concession company		227,300 m^3/day; from 2007 to 2027 Keppel-Seghers	145,500 m^3/day; from 2010 to 2035 Sembcorp Utilities
MF/UF system	Manufacturer name & model no.	Asahi Kasei MICROZA UNA-620A (MF)	Siemens Memcor CP 960 L20V (UF)
	No. of trains; modules or cassettes per train	10 trains; 4 racks/train; 80 modules/rack	7 trains; 8 blocks/train; 120 modules/block
	Flux & recovery, backwash frequency & duration	66 LMH; 90% recovery; backwash 90 s every 30 min	46.4 LMH; 93% recovery; backwash 240 s every 26 min
	Clean-in-place (CIP) type and frequency	Maintenance clean every 2 days or 48 cycles of backwash with NaOCl	Maintenance clean every 150 hrs with NaOCl and sulphuric acid
		Recovery clean every 30 days or 16 cycles of maintenance clean with NaOCl, NaOH, citric acid	Recovery clean every 630 hrs with citric acid, NaOCl and sulphuric acid
RO system	Element manufacturer name & model no.	Hydranautics ESPA 2+	Toray TML 20-430
	No. of trains; vessels per train; elements per vessel	13 trains; 64:36 vessels; 7 elements/vessel	10 trains; 135:63 vessels; 7 elements/vessel
	Flux & recovery	18.8 LMH; 80% recovery	17.3 LMH; 80% recovery
	Clean-in-place (CIP) type and frequency	Every 2–6 months	Every 6 months
UV system	Manufacturer name & model no.	Wedeco	Trojan D30
	No. of trains	5 trains (4 duty, 1 standby)	6 trains (5 duty, 1 standby)
	Type	Low pressure high intensity	Low pressure
	No. of lamps per unit	32	30
	Flow	38,200 m^3 per day	45,600 m^3/day

Table E-6 Treated water quality criteria for NEWater treatment processes

Process	Design Criteria
MF/UF	Turbidity: < 0.1 NTU 95% of the time < 0.2 NTU 100% of the time Silt density index (SDI): <3.0
RO	To achieve a minimum (%) rejection: Reactive ammonia (NH3-N) 90 Nitrate (NO3-N) 80 Sodium chloride 95 Total dissolved solids 97 To achieve the following targets: Aluminium (as Al) < 0.05 mg/L Barium (as Ba) < 0.02 mg/L Boron (as B) < 0.1 mg/L Calcium (as Ca) < 1 mg/L Copper (as Cu) < 0.05 mg/L Iron (as Fe) < 0.05 mg/L Manganese (as Mn) < 0.05 mg/L Sodium (as Na) < 10 mg/L Strontium (as Sr) < 0.1 mg/L Zinc (as Zn) < 0.1 mg/L Chloride (as Cl) < 10 mg/L Fluoride (as F) < 0.5 mg/L Nitrate (as NO_3) < 10 mg/L Sulphate (as SO_4) < 0.5 mg/L Ammonical nitrogen (as N) < 0.5 mg/L Silica (as SIO_2) < 0.5 mg/L Total dissolved solids < 50 mg/L Turbidity < 0.2 NTU Total organic carbon < 0.1 mg/L Total trihalomethanes < 0.02 mg/L Formaldehyde < 0.1 mg/L All constituents and parameters meet 2004 WHO guidelines values
UV	Achieved a heterotrophic plate count (HPC) of less than 5 colony forming units per milliliter (cfu/mL)

Table E-7 Typical NEWater quality

Parameters	Units	Typical NEWater Quality
Aluminium	mg/L	<0.05
Ammoniacal nitrogen (as N)	mg/L	<0.5
Barium	mg/L	<0.02
Boron	mg/L	<0.2
Calcium	mg/L	<1
Chloride	mg/L	<10 (if feedwater chloride ≤350) <20 (if 350 < feedwater chloride ≤550)
Color	Hazen units	<5
Conductivity	µS/cm	<100 (if feedwater conductivity ≤1580) <150 (if 1580 < feedwater conductivity ≤2100)
Copper	mg/L	<0.05
Enterovirus	PFU/mL	Not detectable
Fluoride	mg/L	<0.5
Formaldehyde	mg/L	<0.1
Heterotrophic plate count (35°C, 48 hrs)	CFU/mL	<100
Iron	mg/L	<0.01
Manganese	mg/L	<0.05
Nitrate (as NO_3)	mg/L	<10
pH*	-	7.0–8.5
Silica* (as SiO_2)	mg/L	<0.5
Sodium*	mg/L	<20
Strontium	mg/L	<0.1
Sulphate (as SO_4)	mg/L	<1.0
Total alkalinity* (as $CaCO_3$)	mg/L	<50
Total coliform	Counts/100mL	Not detectable
Total dissolved solids*	mg/L	<60 (if feedwater TDS ≤1000) <90 (if 1000 < feedwater TDS ≤1330)
Total hardness (as $CaCO_3$)	mg/L	<5
Total organic carbon (TOC)**	mg/L	<0.1
Total residual chlorine (as monochloramine)	mg/L	<2
Total trihalomethane	mg/L	<0.02
Turbidity*	NTU	<0.2
Zinc	mg/L	<0.1

* These values shown are readings after the caustic dosing (after UV stage) where the pH of NEWater produced is adjusted to a range of 7–8.5 for transmission and distribution of NEWater to customers.

** Typical on-line TOC reading for RO permeate is 70–85 ppb.

Appendix F

Western Corridor Water Recycling Scheme Design Criteria and Water Quality Tables

Table F-1 Raw water quality for Bundamba Advanced Water Reclamation Facility

		WWTPs							
		1	2	3	4	1	2	3	4
		90th percentile				50th percentile			
Flow percentage		33%	21%	5%	41%	33%	21%	5%	41%
Parameter	Units								
Al	mg/L	0.60	1.70	0.03	0.11	0.05	0.04	0.03	0.05
NH_3-N	mg/L	3.07	5.90	0.97	5.00	0.35	0.40	0.14	0.50
Ba	mg/L	0.02	0.02	0.01	0.03	0.01	0.01	0.01	0.02
BOD	mg/L	3.0	4.0	3.5	9.0	2.0	2.0	2.5	5.0
Ca	mg/L	35	36	31	31	31	31	29	29
Cl	mg/L	174	164	150	170	138	123	150	164
Conductance at 25°C	uS/cm	967	871	890	1156	827	809	815	1064
F	mg/L	0.58	1.05			0.33	0.18		
Fe	mg/L	0.11	0.14	0.032	0.29	0.070	0.048	0.025	0.16
Mg	mg/L	15	29	16	19	14	21	15	16
Mn	mg/L	0.060	0.060	0.030	0.15	0.035	0.053	0.021	0.070
NO_3-N	mg/L	3.80	1.10	2.50	2.50	2.3	0.50	1.25	1.25
NO_2-N	mg/L	0.10	0.10	0.50	0.50	0.10	0.10	0.25	0.25
Organic N	mg/L	1.95	1.74	3.00	3.00	1.40	1.30	1.50	1.50
O-PO_4	mg/L	2.65	7.84	7.00	7.00	0.50	7.3	3.50	3.50
pH									
K	mg/L	19.0	17.0	19.2	24.7	16.0	14.0	17.8	21.3
SiO_2	mg/L								
Na	mg/L	126	108	126	159	110	98	114	143
Sr	mg/L	0.23	0.25			0.19	0.205		

(Continued)

Table F-1 Raw water quality for Bundamba Advanced Water Reclamation Facility *(Continued)*

		WWTPs							
		1	2	3	4	1	2	3	4
		90th percentile				50th percentile			
Flow percentage		33%	21%	5%	41%	33%	21%	5%	41%
Parameter	Units								
S as SO_4	mg/L	111	72	70	90	75	57	70	90
TSS	mg/L	7	13	6	18	4	5	3	9
TDS	mg/L	933	690			497	484		
TOC	mg/L	15	16			10	10		
Total n	mg/L	6.7	4.4	8.0	5.0	4.4	3.0	5.0	3.0
Total p	mg/L	4.1	9.5	10	10	1.3	7.0	7.0	7.0
Turbidity	ntu	3.5	6.9	4.8	9.4	2.1	3.0	3.1	7.3

Table F-2 Treated water quality requirements for Bundamba Advanced Water Reclamation Facility (selected parameters, in addition to Australian Drinking Water Guidelines)

Parameter	Units	Water Quality Standard (Values to be less than or equal to values listed, in 90% of samples, unless otherwise noted)		
Nutrients				
Total nitrogen during minimum output test	mg/L	<	0.8	Annual, 50th percentile
Total phosphorus	mg/L	<	0.13	Annual, 50th percentile
Clostridium perfringens				
Median	cfu/100 ml	<	1	
95th percentile	cfu/100 ml	<	10	
Minerals				
Hardness, as $CaCO_3$	mg/L	>	50	21 day average
Alkalinity	mg/L	>	40	21 day average
Corrosiveness				
Calcium carbonate precipitation potential (CCPP)	-		0 to −5	
Total organic carbon (TOC)	mg/L	<	1	
Nitrosodimethylamine (NDMA)	ng/L	<	10	
Endocrine disrupter compounds				
EDC Estrone (E1)	ng/L	<	3	
EDC 17-beta estradiol (E2)	ng/L	<	1	
EDC Ethinylestradiol (EE2)	ng/L	<	0.1	
EDC total PNEC	ng/L	<	1	
EDC Nonlyphenol (NP)	ng/L	<	7	

Table F-3 Concentrate water quality requirements for Bundamba AWTF 1A

Parameter	Water quality standard (90th percentile values)
Nutrients (mg/L)	
Nitrate (as N)	1
Total phosphorus (as P)	4
Total suspended solids (mg/L)	10

Table F-4 MF design criteria and performance at Bundamba AWTF

Parameter	Design Value	Actual Value
Backwash interval	25 min	25 min
CIP interval	30 days	30 to 60 days
Recovery	92%	95%
Filtrate turbidity	<0.1% 95th percentile <0.3 100th percentile	Based on lab analyses, turbidity was consistently below 0.1 ntu
Filtered water SDI	<3.1 95th percentile <4.0 100th percentile	<0.95 95th percentile <1.42 100th percentile

Table F-5 Gibson Island AWTP feed water design basis

Parameter	Average	Maximum
BOD (mg/L)	3.5	10
TOC (mg/L)	10	15
TSS (mg/L)	5	18
TDS (mg/L)	1,020	1,440
Total nitrogen (mg/L)	5.4	--
Total phosphorus (mg/L)	4.9	10
Total alkalinity (mg/L as $CaCO_3$)	210	300
Sodium (mg/L)	304	424
Chloride (mg/L)	370	570
Sulphate (mg/L)	75	110
Calcium (mg/L)	36	42
Magnesium (mg/L)	18	23
Silica (mg/L)	9	16
Aluminium (mg/L)	0.03	0.07
Iron (mg/L)	0.14	0.47
Manganese (mg/L)	0.06	0.19

Table F-6 Gibson Island AWTP treated water quality requirements

Parameter	Limit for Treated Water
pH	7 to 8
TDS (mg/L)	500
Total nitrogen (mg/L)	0.8
Total phosphorus (mg/L)	0.13
Sodium (mg/L)	180
Chloride (mg/L)	250
Sulphate (mg/L)	250
Total hardness (mg/L as $CaCO_3$)	> 50
Total alkalinity (mg/L as $CaCO_3$)	> 40
Calcium carbonate precipitation potential	0 to −5
Clostridium perfringens (cfu/100 mL)	1
Total organic carbon (mg/L)	1
Nitrosodimethylamine (NDMA) (ng/L)	10
Estrone (ng/L)	3
17-beta estradiol (ng/L)	1
Aluminium (mg/L)	0.2
Iron (mg/L)	0.3
Manganese (mg/L)	0.1

Table F-7 Gibson Island AWTP UF system design criteria

Membrane module	LV20
Number of trains	2 x 61 mL/d
Number of units per train	5 duty plus 1 standby
Membrane modules per unit	348
Filtration area per membrane module (m^2)	38.1
Flux during filtration (LMH)	47.3
Average flux (LMH)	38.9
Recovery	91%
Filtration interval (minutes)	22
Maintenance cleaning chemicals	Sodium hypochlorite Sulphuric and citric acid
Maintenance cleaning frequency	Alternating chlorine and acid every 48 hours of filtration
CIP chemicals	Sodium hypochlorite Sulphuric and citric acid
CIP frequency	30 days
Integrity testing	Daily pressure decay test

Table F-8 Gibson Island AWTP RO system design criteria

Membrane module	Hydranautics ESPA 2
Membrane module diameter (inch)	8
Number of units	7 (6 duty, 1 standby)
Permeate flow per unit (ML/d)	17
Recovery	85%
Stages	3
Membrane elements/pressure vessel	7
Pressure vessels/unit	154 (88:44:22)
Average flux (LMH)	17.3
Fouling factor	0.7
Salt passage allowance (% per year)	15
Flux balancing	Inter-stage boost pumps
CIP chemicals	Sodium hydroxide Citric acid
CIP frequency	6 months

Table F-9 Gibson Island AWTP treated water quality requirements (ADWG and owner) and typical values

Parameter	Limit for Treated Water	Typical Treated Water Value
pH	7 to 8	7.6
TDS (mg/L)	500	150
Total nitrogen (mg/L)	0.8	0.3
Total phosphorus (mg/L)	0.13	0.01
Sodium (mg/L)	180	19
Chloride (mg/L)	250	19
Sulphate (mg/L)	250	15
Total hardness (mg/L as $CaCO_3$)	> 50	72
Total alkalinity (mg/L as $CaCO_3$)	> 40	74
Calcium carbonate precipitation potential	0 to −5	−3
Clostridium perfringens (cfu/100 mL)	1	ND
Total organic carbon (mg/L)	1	<0.5
Nitrosodimethylamine (NDMA) (ng/L)	10	<5
Estrone (ng/L)	3	<1
17-beta estradiol (ng/L)	1	<1
Aluminium (mg/L)	0.2	0.03
Iron (mg/L)	0.3	0.03
Manganese (mg/L)	0.1	0.01

Table F-10 Gibson Island AWTP performance vs. reclaimed water requirements

Parameter	Units	Standard	Typical Treated Water Value
	Microbial		
Campylobacter		Nil	< 1
Clostridium perfringens	cfu/100 mL	Nil	< 1
Cryptosporidium sp		Nil	< 1
E-Coli	cfu/100 mL	Nil	< 1
F-RNA phages	pfu/100 mL	Nil	< 1
Giardia sp		Nil	< 1
Salmonella		Nil	< 1
Shigella	in 1 L	Nil	< 1
Somatic coli phages	pfu/100 mL	Nil	< 1
Vibrio cholerae		Nil	< 1
	Hormones and Industrial Organics		
17-alpha-ethynylestradiol	ng/L	1.5	< 1
17-beta-estradiol	ng/L	175	< 1
Androsterone	ng/L	14,000	< 1
Estriol	ng/L	50	< 1
Estrone	ng/L	30	< 1
Predicted estradiol equivalent	ng/L	1	< 1
Testosterone	ng/L	7,000	< 1
4-t-Octylphenol	ng/L	50,000	< 10
Bisphenol A	ng/L	200,000	25
Cholesterol	ng/L	7,000	11.5
EDTA	μg/L	250	< 50
Nonylphenol	ng/L	500,000	< 100
	Nitrosamines		
NDEA	ng/L	10	<10
NDMA	ng/L	10	< 5
	Pesticides		
2,4-D dichlorophenoxy AAC	μg/L	30	<0.01
2,4-DP dichlorophenoxypropionic	μg/L	100	<0.01
3,4 DiCl Aniline	μg/L	0.35	<0.01
Ametryn	μg/L	50	<0.01
Atrazine	μg/L	40	<0.01
Bromacil	μg/L	300	<0.01
Carbaryl – ug/L	μg/L	30	<0.01
Desethyl atrazine	μg/L	40	<0.01
Desisopropyl atrazine	μg/L	40	<0.02
Diazinon	μg/L	3	<0.02
Dicamba	μg/L	100	<0.01
Diuron	μg/L	30	<0.01
Fluometuron	μg/L	50	<0.02
Glyphosate – ug/L	μg/L	1,000	<0.5
Haloxyfop	μg/L	1.05	<0.01
Haloxyfop methyl	μg/L	0.175	<0.01
Hexazinone	μg/L	300	<0.01
MCPA	μg/L	2	<0.01
Mecoprop	μg/L	10	<0.01
Metolachlor	μg/L	300	<0.03
Picloram	μg/L	300	<0.02

Table F-10 Gibson Island AWTP performance vs. reclaimed water requirements *(Continued)*

Parameter	Units	Standard	Typical Treated Water Value
Prometryn	µg/L	105	<0.01
Propoxur	µg/L	70	<0.01
Simazine	µg/L	20	<0.01
Terbutryn	µg/L	300	<0.01
Triclopyr	µg/L	10	<0.01
	Pharmaceuticals		
Acetylsalicylic acid	µg/L	29	<0.01
Atenolol	µg/L	25	<0.01
Atorvastatin	µg/L	5	<0.01
Caffeine	µg/L	0.35	<0.02
Carbamazepine	µg/L	100	<0.01
Cephalaxin	µg/L	35	<0.01
Chloramphenicol	µg/L	175	<0.1
Chlorotetracycline	µg/L	105	<0.2
Ciprofloxacin	µg/L	250	<0.15
Citalopram	µg/L	4	<0.01
Codeine	µg/L	50	<0.1
Dapsone	µg/L	0.5	<0.01
DEET	µg/L	2,500	<0.01
Diazepam (Valium)	µg/L	2.5	<0.01
Diclofenac	µg/L	1.8	<0.01
Enrofloxacin	µg/L	22	<0.02
Erythromycin	µg/L	17.5	<0.01
Fluoxetine	µg/L	10	<0.01
Fluvastatin	µg/L	0.1	<0.01
Gemfibrozol	µg/L	600	<0.01
Hydrochlorthiazide	µg/L	6	<0.07
Ibuprofen	µg/L	400	<0.01
Indomethacin	µg/L	25	<0.01
Iopromide	µg/L	750	<0.2
Lincomycin	µg/L	3,500	<0.01
Metoprolol	µg/L	25	<0.01
Naproxen	µg/L	220	<0.1
Norfloxacin	µg/L	400	<0.05
Oxazepam	µg/L	15	<0.01
Oxycodone	µg/L	10	<0.01
Oxytetracycline (Terramycin)	µg/L	105	<0.4
Paracetamol	µg/L	175	<0.02
Praziquantel	µg/L	70	<0.01
Propranolol	µg/L	40	<0.01
Ranitidine	µg/L	26	<0.01
Roxithromycin	µg/L	150	<0.02
Salicylic Acid	µg/L	105	0.1
Sulfsalazine	µg/L	500	<0.01
Sulphadiazine	µg/L	35	<0.01
Sulphamethoxazole	µg/L	35	<0.01
Sulphathiazole	µg/L	35	<0.01
Temazepam	µg/L	5	<0.01
Tetracycline	µg/L	105	<0.1

Table F-10 Gibson Island AWTP performance vs. reclaimed water requirements *(Continued)*

Parameter	Units	Standard	Typical Treated Water Value
Triclosan	μg/L	0.35	<0.01
Trimethoprim	μg/L	70	<0.01
Tylosin	μg/L	1,050	<0.05
Venlafaxine	μg/L	75	<0.01
Warfarin	μg/L	1.5	<0.01
Phenolic Compounds			
2,4,5-Trichlorophenol	μg/L	350	<0.25
2,4,6-Trichlorophenol	μg/L	20	<0.25
2,4-Dichlorophenol	μg/L	200	<0.25
2,6-Dichlorophenol	μg/L	10	<0.25
2-Chlorphenol	μg/L	300	<0.25
4-Methylphenol	μg/L	600	<0.25
4-Nitrophenol	μg/L	30	<0.5
Pentachlorophenol	μg/L	10	<0.5
Phenol	μg/L	150	<0.25
Volatile Organic Compounds			
1,1-Dichloroethene	μg/L	30	<2.0
1,2,3-Trichlorobenzene	μg/L	30	<2.0
1,2-Dibromoethane (EDB)	μg/L	1	<1.0
1,2-Dichlorobenzene	μg/L	1,500	<2.0
1,2-Dichloroethane	μg/L	3	<2.0
1,2-Dichloroethene (cis)	μg/L	60	<2.0
1,4-Dichlorobenzene	μg/L	40	<2.0
Benzene	μg/L	1	<1.0
Carbon tetrachloride	μg/L	3	<2.0
Chlorobenzene	μg/L	300	<2.0
Ethylbenzene	μg/L	300	<2.0
Hexachlorobutadiene	μg/L	0.7	<0.7
Methylene chloride	μg/L	4	<4.0
Ortho-xylene	μg/L	600	<2.0
Styrene (Vinylbenzene)	μg/L	30	<2.0
Tetrachloroethene	μg/L	50	<2.0
Toluene	μg/L	800	<2.0

Table F-11 Luggage Point AWTP influent (feed water) quality design basis (selected parameters)

Parameter, mg/L[1]	50th Percentile	90th Percentile	Maximum
Aluminium	0.03	0.06	0.2
Ammonia	0.15	0.85	1.0
Barium	0.017	0.022	0.7
Bicarbonate	198	224	NE[2]
Calcium	42	46	NE
Chloride	665	793	1000
Fluoride	0.9	1.4	1.5
Iron	0.13	0.17	1.0
Manganese	0.16	0.68	0.68
Nitrate	11	22	50
Nitrite	NE	NE	3
Potassium	27	32	NE
Sodium	333	422	1,000
Silica	10	11	NE
Sulphate	100	120	200
Temperature, °C	25	20 (minimum)	30
TDS	NA	1,558	3,000
Total nitrogen	5.0	7.0	NE
Total phosphorous	1.0	2.0	14
pH, units	6.6	7.0	NE
Turbidity, NTU	0.1	0.3	10
Clostridium perfingens, cfu/100 mL	1,300	NE	1,900
Total organic carbon	13	20	25
Total suspended solids	10	20	50
NDMA, ng/L	10	15	100
Estrone, ng/L	NA	NA	3
17-beta estradiol, ng/L	NA	NA	1
Ethinylestradiol, ng/L	NA	NA	0.1
Total PNEC, ng/L	NA	NA	1
Nonlyphenol (NP), ng/L	NA	NA	7

[1] Unless otherwise indicated.

[2] NE = Not established.

Table F-12 Luggage Point AWTP treated water quality requirements (supplement to ADWG)

Parameter	Water Quality Standard (90th percentile)
Total nitrogen	<1.2 mg/L
Total phosphorous	<0.13 mg/L
Total organic carbon	<1 mg/L
Alkalinity, as $CaCO_3$	>40 mg/L (21 day average)
Hardness, as $CaCO_3$	>50 mg/L (21 day average)
CCPP[1], mg/L as $CaCO_3$	>0 and <5
Clostridium perfringens	Median – 1 cfu/100 mL; 95th percentile – 10 cfu/100mL
Nitrosodimethylamine (NDMA)	10 ng/L
Endocrine Disrupter Compounds	
Estrone	3 ng/L
17 beta estradiol (E2)	1 ng/L
Ethinylestradiol (EE2)	0.1 ng/L
Total PNEC	1 ng/L

[1] Calcium carbonate precipitation potential

Table F-13 Luggage Point AWTP MF system design criteria

System	Pall Corporation Microza
Membrane module	Asahi UNA-620A
Number of trains	11 (10 duty/1 standby)
Flow per train (maximum)	111 L/sec
Membrane modules per train	122
Filtration area per membrane module (m^2)	50
Instantaneous flux (LMH/GFD)	65
Net flux (LMH)	57
Recovery (%)	95
Filtration interval (minutes)	21
Enhanced flux maintenance chemicals	Sodium hypochlorite Sulphuric and citric acid
Maintenance cleaning frequency	Acid, acid, chlorine (3, 3, 3 day intervals, respectively)
CIP chemicals	Sodium hypochlorite and caustic Sulphuric and citric acid
CIP frequency	>60 days
Integrity testing	Daily pressure decay test

Table F-14 Luggage Point AWTP RO system design criteria

Membrane module	Toray TML20-430
Membrane module diameter (inch)	8
Number of units	4 (3 duty, 1 standby)
Permeate flow per unit (ML/d)	17
Recovery	85%
Stages	3
Membrane elements/pressure vessel	7
Pressure vessels/unit	210 (120:60:30)
Average flux (LMH)	17.8
Flux balancing	Stage 1: permeate backpressure Stage 2: feed pressure boost via hydraulic turbocharger
CIP chemicals	Sodium hydroxide Citric acid EDTA[1] SLS[1]
CIP frequency	6 months

[1] EDTA, ethylene diamine tetraacetic acid; SLS, sodium lauryl sulfate.

Table F-15 Luggage Point AWTP treated water quality: requirements vs. plant performance test

Parameter	Requirement	Performance Test Value
Alkalinity	> 40 mg/L (21-day average)	59 mg/L (average over entire test)
Clostridium perfringens	Median: 1 cfu/100mL; 95th percentile: 10 cfu/100mL	<1 cfu/100mL in 100% of samples
Hardness as $CaCO_3$	>50 mg/L (21-day average)	57 mg/L (average over entire test)
N-Nitrosodimethylamine (NDMA)	≤10 ng/L in 90% of samples	≤10 ng/L in 100% of samples
Estrone	≤3 ng/L in 90% of samples	≤1 ng/L in 100% of samples
17-Beta estradiol	≤1 ng/L in 90% of samples	≤1 ng/L in 100% of samples
Ethinylestradiol	≤0.1 ng/L in 90% of samples	≤1 ng/L in 100% of samples
Total PNEC[1]	≤1 ng/L in 90% of samples	≤1 ng/L in 100% of samples
Nonylphenol (NP)	≤7 ng/L in 90% of samples	≤1 ng/L in 100% of samples when raw water boundary influent limit was met
Calcium carbonate precipitation potential (CCPP)	Between 0 to −5 in 90% of samples	Between 0 and −5 in 27 of 28samples (96.4%)
Total organic carbon	≤1 mg/L in 90% of samples	≤1 mg/L in 100% of samples
Total nitrogen	0.8 mg/L (50th pecentile)	0.65 mg/L (50th percentile)
Total phosphorus	0.13 mg/L (50th percentile)	0.016 mg/L (50th percentile)

[1] Predicted No Effect Concentration for estrogenic compounds

This page intentionally blank.

Appendix G

CAPCO Industrial Wastewater Reclamation Facility Design Criteria and Water Quality Tables

Table G-1 Wastewater effluent characteristics and discharge requirements

	Cooling Water B/D	OWWTS Effluent	Effluent Discharge Standards
pH	7.32	8.45	6–9
TSS (mg/L)	10	24	30
Total hardness (mg/L) as $CaCO_3$	980	39	--
TDS (mg/L)	3,000	3,544	--
Cond (μS/cm)	3,690	4,260	--
SiO_2 (mg/L)	71.6	3.5	--
COD (mg/L)	51	74	100
Co (mg/L)	--	2.2	--
Mn (mg/L)	--	1.3	10

Table G-2 Effluent water quality

Average Product Water Quality	
Conductivity, µS/cm (>94.8 percent removal)	<240
Hardness, mg/L as $CaCO_3$	<5
SiO_2, mg/L	<2
pH, su	6.9
CO_2, mg/L	5
Mn, mg/L	Trace
Co, mg/l	Trace
COD, mg/L	Trace
Average Recovery Rates	
UF	92 percent
RO	80 percent
UF/RO	73.6 percent

Index

Note: *f.* indicates figure; *t.* indicates table

S

T

U

V

W

AWWA Manuals

M1, *Principles of Water Rates, Fees, and Charges*, #30001
M2, *Instrumentation and Control*, #30002
M3, *Safety Management for Water Utilities*, #30003
M4, *Water Fluoridation Principles and Practices*, #30004
M5, *Water Utility Management*, #30005
M6, *Water Meters—Selection, Installation, Testing, and Maintenance*, #30006
M7, *Problem Organisms in Water: Identification and Treatment*, #30007
M9, *Concrete Pressure Pipe*, #30009
M11, *Steel Pipe—A Guide for Design and Installation*, #30011
M12, *Simplified Procedures for Water Examination*, #30012
M14, *Backflow Prevention and Cross-Connection Control: Recommended Practices*, #30014
M17, *Fire Hydrants: Installation, Field Testing, and Maintenance*, #30017
M19, *Emergency Planning for Water Utilities*, #30019
M20, *Water Chlorination/Chloramination Practices and Principles*, #30020
M21, *Groundwater*, #30021
M22, *Sizing Water Service Lines and Meters*, #30022
M23, *PVC Pipe—Design and Installation*, #30023
M24, *Planning for the Distribution of Reclaimed Water*, #30024
M25, *Flexible-Membrane Covers and Linings for Potable-Water Reservoirs*, #30025
M27, *External Corrosion Control for Infrastructure Sustainability*, #30027
M28, *Rehabilitation of Water Mains*, #30028
M29, *Water Utility Capital Financing*, #30029
M30, *Precoat Filtration*, #30030
M31, *Distribution System Requirements for Fire Protection*, #30031
M32, *Computer Modeling of Water Distribution Systems*, #30032
M33, *Flowmeters in Water Supply*, #30033
M36, *Water Audits and Loss Control Programs*, #30036
M37, *Operational Control of Coagulation and Filtration Processes*, #30037
M38, *Electrodialysis and Electrodialysis Reversal*, #30038
M41, *Ductile-Iron Pipe and Fittings*, #30041
M42, *Steel Water-Storage Tanks*, #30042
M44, *Distribution Valves: Selection, Installation, Field Testing, and Maintenance*, #30044
M45, *Fiberglass Pipe Design*, #30045
M46, *Reverse Osmosis and Nanofiltration*, #30046
M47, *Capital Project Delivery*, #30047
M48, *Waterborne Pathogens*, #30048
M49, *Butterfly Valves: Torque, Head Loss, and Cavitation Analysis*, #30049
M50, *Water Resources Planning*, #30050
M51, *Air Valves: Air-Release, Air/Vacuum and Combination*, #30051
M52, *Water Conservation Programs—A Planning Manual*, #30052
M53, *Microfiltration and Ultrafiltration Membranes for Drinking Water*, #30053
M54, *Developing Rates for Small Systems*, #30054
M55, *PE Pipe—Design and Installation*, #30055
M56, *Nitrification Prevention and Control in Drinking Water*, #30056
M57, *Algae: Source to Treatment*, #30057
M58, *Internal Corrosion Control in Water Distribution Systems*, #30058
M60, *Drought Preparedness and Response*, #30060
M61, *Desalination of Seawater*, #30061
M62, *Membrane Applications for Water Reuse* #30062
M63, *Aquifer Storage and Recovery*, #30063
M65, *On-Site Generation of Hypochlorite*, #30065
M66, *Cylinder and Vane Actuators and Controls—Design and Installation*, #30066
M68, *Water Quality in Distribution Systems*, #30068

www.ingramcontent.com/pod-product-compliance
Lightning Source LLC
LaVergne TN
LVHW080639011026
849160LV00014B/1263